多物理场相场裂缝数值模拟
——建模、离散及求解

Multiphysics Phase-Field Fracture
Modeling, Adaptive Discretizations, and Solvers

〔德〕托马斯·威克（Thomas Wick） 著

侯 冰 戴一凡 译

科学出版社

北京

图字：01-2023-2354 号

内 容 简 介

相场法(phase field method)近年来在裂缝扩展问题的研究中崭露头角，为工程和科学领域带来了一项重大技术进步。与传统模拟方法相比，该方法引入了一个相场参数来表征模型中空间点的物理状态，可利用该参数作为指示函数搭建耦合系统。同时该方法采用基于能量最小化原理判断裂缝扩展，因此在处理复杂地质结构、模拟裂缝分叉、转向等问题时不需要添加额外的判别准则，计算过程简单，结果准确率高，计算量相对较小。本书针对多物理场相场裂缝扩展模型的基本概念、模型建立、时间/空间离散、求解及自适应方法进行了讨论。旨在帮助地质学家、工程师和数值计算人员理解并掌握这一强大的数值模拟工具，以解决裂缝扩展问题和相关多物理场耦合问题。

本书可供高等院校石油类研究生阅读，也可供油田现场数值模拟等工程技术人员参考使用。

Wick, Thomas. Multiphysics Phase-Field Fracture: Modeling, Adaptive Discretizations, and Solvers, 2020 © Walter de Gruyter GmbH Berlin Boston. All rights reserved.
This work may not be translated or copied in whole or part without the written permission of the publisher (Walter De Gruyter GmbH, Genthiner Straße 13, 10785 Berlin, Germany).

图书在版编目(CIP)数据

多物理场相场裂缝数值模拟：建模、离散及求解/(德) 托马斯•威克 (Thomas Wick) 著；侯冰，戴一凡译. —北京：科学出版社，2024.3
书名原文：Multiphysics Phase-Field Fracture: Modeling, Adaptive Discretizations, and Solvers
ISBN 978-7-03-077763-8

Ⅰ. ①多… Ⅱ. ①托… ②侯… ③戴… Ⅲ. ①断裂力学-数值模拟 Ⅳ. ①O346.1

中国国家版本馆 CIP 数据核字(2024) 第 020678 号

责任编辑：刘信力 杨 探／责任校对：杨聪敏
责任印制：赵 博／封面设计：无极书装

科学出版社 出版
北京东黄城根北街 16 号
邮政编码：100717
http://www.sciencep.com

北京中科印刷有限公司印刷
科学出版社发行 各地新华书店经销
*
2024 年 3 月第 一 版 开本：720×1000 1/16
2025 年 1 月第二次印刷 印张：17 1/2
字数：348 000
定价：188.00 元
(如有印装质量问题，我社负责调换)

中 文 版 序

本书原作于 2020 年由德古意特出版社 (de Gruyter) 出版。很高兴相场裂缝模型在中国学术界与工业界受到广泛关注。感谢侯冰教授，以及戴一凡、常智、赵智强、王财宝、郑振权、窦浩原等同学为本书翻译所做出的工作，感谢科学出版社的大力支持。

托马斯·威克

2023 年 1 月

汉诺威

Preamble

This book is the translated version from the original English edition published from de Gruyter into Chinese. I am happy to see interest in multiphysics phase-field fracture by Chinese academic friends and colleagues as well as in the Chinese University and Industry research landscape. I am thankful to Prof. Dr. Bing Hou, Mr. Yifan Dai, Dr. Zhi Chang, Mr. Zhiqiang Zhao, Mr. Caibao Wang, Mr. Zhenquan Zheng and Mr. Haoyuan Dou for the paramount work of the translation and the friendly cooperation of Science Press.

Thomas Wick
January 2023
Hannover

译 者 序

Multiphysics Phase-Field Fracture: Modeling, Adaptive Discretizations, and Solvers 一书主要介绍了相场裂缝模型的数学建模、算法设计及数值模拟等相关内容。原著作者 Thomas Wick 是德国汉诺威大学应用数学专业教授。主要研究领域为自适应数学建模、设计与分析、非平稳、非线性、耦合偏微分的数值方法和高性能求解方法。研究重点包括流固耦合模型、相场裂缝模型、多孔介质模型等。

相场法是一项基于能量最小化原理提出的新型数值模拟方法，多应用于马氏体相变、沉淀相变及凝固等方面。Thomas Wick 教授根据其在博士后期间以及后续的相关研究，成功地将该方法应用于裂缝扩展模型中。相场模拟并无确定的商业化软件，需要研究者逐行编写代码，对研究者的材料学、数学基础和计算机程序设计基础有着较高要求，这也在很大程度上阻碍了此方法的应用活跃度。

水力压裂已成为石油工程领域开发致密储层的主要手段之一。致密储层常处于高温高压的复杂地质环境下，影响裂缝扩展的因素较多。现有的多场耦合模型在处理裂缝交叉、分裂等复杂情况时常需要用到多种判别方法，模型复杂，工程应用难度较大。相场裂缝模型通过裂缝表面能最小化的原则模拟裂缝扩展形态，在求解多孔介质多场耦合的工程问题时有独到的优势。本书重点介绍了基于相场法的裂缝扩展数模问题，包括模型建立、离散、求解等关键步骤。为读者详细介绍了相场模型中的误差分析、自适应网格及界面处理方法。随着压裂技术的不断发展，酸化压裂、超临界二氧化碳压裂等新技术的不断出现，石油工程领域对多场耦合裂缝扩展数值模型的需求也在不断增大。基于该方法建立的数值模型求解速度快，判别准则简单且有效，在工程应用方面有着广阔的前景。

本书的出版获得汉诺威大学应用数学研究所、中国石油大学 (北京) 克拉玛依校区以及科学出版社的鼎力支持。感谢科学出版社刘信力编辑在此期间的辛勤工作，感谢常智、赵智强、王财宝、郑振权、窦浩原等各位同学为本书提供的帮助。本书翻译过程中为便于读者理解，在征得原书作者同意的情况下，对部分章节内容及参考文献进行了修订。如有疑问之处，还请各位读者批评指正。

侯 冰、戴一凡

2024 年 2 月

新疆克拉玛依

相关机构与课程

本书中部分内容受到特别班 (Spezialvorlesung) 和暑期学校所开展的报告及讨论的协助。在此感谢 Johannes Kepler University，Sorin Iliu Pop (Hasselt University)、Ulrich Langer (JKU Linz) 以及 Marc-André Keip(University of Stuttgart)。

1. 连续介质力学的数值方法 (2020 年夏学期，汉诺威大学)

    http://www.thomaswick.org/links/topics_NumContMech_SoSe_2020.txt

2. 面向目标的误差分析与自适应有限元方法 (2019/2020 冬学期，汉诺威大学)

    http://www.thomaswick.org/links/topics_online_dwr_ws_19_20.txt

3. 适用于多孔介质的相场法 (2019 年 6 月 26 日，哈塞尔特大学 (比利时)，暑期讲座)

    https://www.uhasselt.be/summer-school-phase-field-modeling-2019

4. 接触问题的数值方法：相场裂缝扩展变分模型中的应用 (2018/2019 冬学期，汉诺威大学)

    http://www.thomaswick.org/links/ankuendigung_vpff_wick_mang_noii.pdf

5. 变分相场断裂问题的数值方法 (2018 年 3 月，Johannes Kepler University，集体讲座)

    https://www.dk-compmath.jku.at/Courses/2018s/phase-field-fractureproblems

6. 多物理场相场裂缝的数值模拟方法及其适应性 (2016 年秋，斯图加特大学，暑期讲座)

    http://www.thomaswick.org/commas_stuttgart_fall_2016.html

课题来源与相关工作

在此对过去 8 年所获得的项目支持表达感谢：

1. German Priority Program 1748(DFG SPP 1748)

https://www.uni-due.de/spp1748/

压力驱动的三维裂缝相场模型自适应方法

https://www.uni-due.de/spp1748/adaptive_enriched_galerkin_methods

2. 奥地利科学基金项目 (FWF)P-29181 面向目标的相场断裂多物理场问题目标分析

http://www.numa.uni-linz.ac.at/Research/Projects/P29181/index.shtml

3. German Priority Program 1962 (DFG SPP 1962)

https://spp1962.wias-berlin.de/

基于相场法优化裂缝扩展模拟

https://spp1962.wias-berlin.de/project.php?projectID=15

4. The Cluster of Excellence PhoenixD(EXC 2122)

https://www.phoenixd.uni-hannover.de/en/

5. 印度-德国高等教育合作伙伴 2019~2023 年，汉诺威大学-印度工业大学

6. Alexander von Humboldt 基金会 Feodor Lynen Research 奖学金 (2013~2014)

https://www.oden.utexas.edu/about/news/228/

7. ICES 博士后研究基金 (2013 年)

https://www.oden.utexas.edu/media/uploaded-images/351.jpg

本书中部分内容涉及作者与其团队相关刊出论文，包括期刊论文与会议论文等，另外还包括一些未经评审的工作 (GAMM Leitartikel 或课堂讲稿)。作者在相场裂缝多物理场扩展方面的成果大致可分为以下几类：

(1) 相场裂缝数值模拟中的基本概念。

(a) 裂缝不可逆性约束 [318]、增广拉格朗日法 [424]、原始-对偶方法 [217,296]；

(b) 高阶有限元研究 [296,245]；

(c) 刚性介质裂缝扩展方程 [296,42]；

(d) 位移/相场全耦合模型 [438,439]、拟全耦合模型 [217,438] 和分部耦合模型 [86,152,243]；

(e) 非线性求解器 [438,439];

(f) 多尺度网格并行预处理器 [218,246,245]。

(2) 自适应建模、误差控制与局部网格细化。

(a) 预测-校正网格自适应方法 [217](2D), [442,281](3D);

(b) 基于残差的误差分析 [293,42];

(c) 面向目标的误差分析 [436];

(d) 多目标函数 [151];

(e) 非侵入式全局-局部自适应方法 [336,9];

(f) 加载步长自适应方法 [437,157]。

(3) 相场裂缝模拟方法优化。

(a) 最优控制方法 [331,332];

(b) 返回值估计方法 [255]。

(4) 工程应用。

(a) 螺钉断裂与损伤模拟 [427];

(b) 不可压缩固体中断裂损伤模拟 [294,296];

(c) 多孔介质中单向耦合模拟 [444];

(d) 现场裂缝初始化方法 [283];

(e) 基于相场法的 I&II 型裂缝起裂与扩展模拟 [158]。

(5) 多物理场建模及应用。

(a) 多孔介质中加压裂缝数值模拟 [319,318,442,281,9];

(b) 多孔介质中流固耦合裂缝数值模拟 [316,317,281,282,14];

(c) 多孔介质非线性相场裂缝模型 [419];

(d) 多孔介质相场裂缝中的流体流动模拟 [282];

(e) 多孔介质相场裂缝中的两相流动模拟 [276];

(f) 多孔介质中加压裂缝温度场模拟 [337];

(g) 流固耦合相互作用模拟 [435,437];

(h) IPACs: 基于相场法的多物理场裂缝扩展模型框架 [425];

(i) 模拟软件开发 [218,425,203,180,443,218]。

(6) 课堂讲稿。

(a) 相场裂缝课堂讲义 [295,440];

(b) 第 9、11~14 章部分内容 [441]。

目　　录

中文版序
译者序
相关机构与课程
课题来源与相关工作
第 1 章　简介 ··· 1
　1.1　多物理场 ··· 1
　1.2　相场裂缝变分模型 ·· 2
　1.3　界面定义与处理方法 ··· 3
　　1.3.1　固定界面 ··· 4
　　1.3.2　移动界面 ··· 4
　　1.3.3　扩展界面 ··· 5
　　1.3.4　界面处理方法 ·· 5
　1.4　工程问题：螺钉断裂和损坏 ······································· 7
　1.5　相关文献 ··· 7
　1.6　章节介绍 ··· 8
第 2 章　数学符号、模型条件和原始模型问题 ······························ 11
　2.1　数学符号 ·· 11
　　2.1.1　空间域、边界条件、裂缝 ·································· 11
　　2.1.2　加载时间/加载步 ·· 12
　　2.1.3　模型、材料以及离散化参数 ································ 12
　　2.1.4　解变量 ··· 12
　　2.1.5　梯度、散度、迹、拉普拉斯算子 ·························· 13
　　2.1.6　函数空间 ·· 14
　　2.1.7　内积、双线性型、半线性型的符号表示 ·················· 17
　　2.1.8　参数符号表 ·· 18
　2.2　微分与积分的前提条件 ··· 19
　　2.2.1　高斯-格林定理/散度定理 ··································· 19
　　2.2.2　分部积分与格林公式 ······································· 19
　　2.2.3　Lebesgue 积分 ·· 20

2.2.4　Banach 空间中的微分问题 ···················· 20
　　　2.2.5　链式法则 ······································· 22
　2.3　原始相场模型公式 ·· 23
　2.4　原始相场模型问题 ·· 26
　　　2.4.1　简单障碍问题 ··································· 26
　　　2.4.2　能量方程 ······································· 27
　2.5　动态相场裂缝模型 ·· 27

第 3 章　偏微分方程及变分不等式的分类 ························ 29
　3.1　线性/非线性偏微分方程 ···································· 29
　　　3.1.1　微分方程的一般定义 ······························ 29
　　　3.1.2　偏微分方程分类 ·································· 30
　　　3.1.3　示例 ·· 30
　3.2　变分方程与变分不等式 ···································· 31
　　　3.2.1　变分不等式 ······································ 31
　　　3.2.2　函数极小化问题 ·································· 32
　3.3　耦合问题与多物理场偏微分方程 ······························ 33
　　　3.3.1　域耦合 ·· 35
　　　3.3.2　界面耦合 ·· 36
　　　3.3.3　耦合方法 ·· 37
　3.4　弹性力学 Biot 方程耦合问题 ································ 41
　　　3.4.1　符号与方程 ······································ 41
　　　3.4.2　控制方程 ·· 42
　　　3.4.3　典型示例 ·· 43
　　　3.4.4　变分公式 ·· 43
　　　3.4.5　离散，建模，结果分析 ····························· 44
　3.5　原型准静态模型问题 ······································ 45
　3.6　原型动态模型问题 ·· 46

第 4 章　相场裂缝模型 ·· 47
　4.1　连续介质力学介绍与本构关系 ································ 47
　4.2　简单脆性断裂建模 ·· 49
　4.3　Griffith 模型 ·· 49
　4.4　Francfort-Marigo 脆性断裂变分模型 ·························· 50
　　　4.4.1　表面能与体积势能 ································· 50
　　　4.4.2　模型准则 ·· 51
　　　4.4.3　系统总能量的具体表现形式 ·························· 52

  4.4.4 准静态脆性相场裂缝问题弱形式···58
 4.5 热力学扩展···58
  4.5.1 裂缝拉伸或压缩的能量问题···58
  4.5.2 基于应力分解的相场裂缝模型··61
  4.5.3 利用应变场替换不可逆约束的相场裂缝模型··61
 4.6 变分能量公式与 Euler-Lagrange 系统总结···64

第 5 章 数值建模 I：正则化与离散化···66
 5.1 裂缝不可逆性约束··66
  5.1.1 应变场函数··66
  5.1.2 简单补偿法与增广 Lagrangian 补偿法···66
  5.1.3 原始-对偶活动集法··68
 5.2 时间离散化···70
  5.2.1 单次 θ 分解法与分步 θ 分解法···70
  5.2.2 例题··72
  5.2.3 非线性时间导数···72
  5.2.4 准静态相场裂缝···73
  5.2.5 非稳态多物理场模拟过程中的时间步设置···73
 5.3 空间离散化···75
  5.3.1 有限元空间··75
  5.3.2 准静态脆性相场裂缝模型空间离散弱形式···76
 5.4 模型离散参数介绍··77
  5.4.1 ε 与 h 的关系···78
  5.4.2 简化原型数值模型分析···79
  5.4.3 正则化参数取值建议··85
 5.5 动态相场裂缝模型··87

第 6 章 数值模拟 I：相场裂缝域··90
 6.1 研究对象与特征··90
 6.2 建模方案··90
 6.3 边界条件··91
 6.4 初始条件··91
 6.5 模型参数··91
 6.6 解变量与相关参数··92
 6.7 本章小结··92

第 7 章 数值建模 II：线性/非线性求解器及线性解·······························98
 7.1 分部耦合和全耦合··98

7.2	线性化方案	99
7.3	不动点迭代法	99
7.4	函数迭代法	101
7.5	PDE 系统分部耦合模型	102
7.6	相场裂缝分部耦合模型	102
7.7	相场裂缝全耦合模型	104
	7.7.1 模型挑战	104
	7.7.2 准全耦合方案	105
	7.7.3 外推迭代算法	106
7.8	牛顿迭代法概述	107
	7.8.1 单调性测试	108
	7.8.2 基于残差的牛顿法基本算法	108
	7.8.3 全耦合方案概述及其数值解-图片公式化	109
7.9	牛顿迭代法在相场裂缝问题中的运用	111
	7.9.1 面向残差的牛顿迭代法算法	111
	7.9.2 面向误差的牛顿迭代法	114
	7.9.3 修正 Jacobian 矩阵的牛顿迭代法	115
	7.9.4 基于修正 Jacobian 矩阵的面向误差牛顿迭代法	118
	7.9.5 基于原始-对偶活动集的牛顿迭代法	119
7.10	基于牛顿迭代法的全耦合相场裂缝问题线性求解	121
	7.10.1 预处理矩阵 P^{-1}	121
	7.10.2 外推格式的块-对角预处理器	122
	7.10.3 无矩阵几何多重网格求解器	123
	7.10.4 自适应网格并行求解	123
7.11	增广 Lagrangian 补偿法与原始-对偶活动集法	124
第 8 章	**数值模拟 II：单边剪切实验**	**126**
8.1	模型框架	126
8.2	测试案例	126
8.3	建模方案	126
8.4	边界条件	127
8.5	初始条件	128
8.6	模型参数	128
8.7	本章小结	128
	8.7.1 不同耦合方案下模拟结果研究	128
	8.7.2 不同加载步条件下模拟结果研究	131

第 9 章　数值建模 Ⅲ：自适应方法 ·· 133
9.1　自适应方法简介 ·· 133
9.1.1　时间/空间自适应相场裂缝模型简介 ························· 134
9.1.2　相关研究进展 ··· 134
9.1.3　有效性，可靠性以及基本自适应算法 ······················· 134
9.1.4　误差估计量 η ·· 135
9.2　面向目标的稳态相场裂缝后验误差分析 ···························· 136
9.2.1　问题陈述与设置 ··· 136
9.2.2　非线性问题中的双重加权残差法 ······························ 137
9.2.3　局部误差估计 ··· 138
9.2.4　基于伴随误差的有限元近似问题 ······························ 139
9.2.5　PU-DWR 准静态相场裂缝误差分析 ························· 140
9.2.6　网格细化与标记方法 ··· 141
9.3　静态二维 Navier-Stokes 模型误差分析 ······························· 142
9.3.1　基本方程 ··· 142
9.3.2　目标函数 ··· 144
9.3.3　基于对偶的后验误差估计 ·· 144
9.3.4　二维几何模型 ··· 145
9.3.5　模型参数 ··· 145
9.3.6　算例 1 ·· 145
9.3.7　算例 2 ·· 146
9.3.8　模拟云图 ··· 146
9.4　面向多目标的后验误差估计 ··· 148
9.4.1　多目标函数介绍 ··· 148
9.4.2　多目标函数组合方法 ··· 148
9.4.3　面向多目标函数的稳态相场模型后验误差估计 ······· 150
9.4.4　面向多目标函数的自适应算法 ·································· 151
9.5　预测-矫正-自适应方法 ·· 153
9.5.1　扩展裂缝周边局部网格细化方案 ······························ 153
9.5.2　非侵入性式自适应方法 ··· 154
9.6　自适应时间步长控制算法 ··· 155
第 10 章　数值模拟 Ⅲ：面向目标的相场狭缝误差控制和螺杆时间控制 ····· 158
10.1　基于双加权残差法的相场裂缝模拟 ································· 158
10.1.1　目标函数 ··· 158
10.1.2　面向目标的后验误差分析 ·· 158

10.2 非相场法多目标误差分析 ································· 161
10.2.1 模型设置 ·· 161
10.2.2 研究对象 ·· 161
10.2.3 单目标函数误差分析 ·· 162
10.2.4 多目标函数误差分析 ·· 163
10.3 螺钉断裂实验中的误差控制 ····································· 163
10.3.1 模型配置 ·· 164
10.3.2 边界条件与初始条件 ·· 164
10.3.3 模型参数 ·· 164
10.3.4 模拟结果 ·· 165

第 11 章 多物理场相场裂缝模型 ·· 167
11.1 界面定义与界面条件 ··· 167
11.2 水平集方法 ··· 167
11.3 缝宽与裂缝总体积计算 ·· 169
11.4 相场裂缝模型建模 ··· 172
11.4.1 弹性固体压裂模型 ·· 172
11.4.2 非等温裂缝模型 ··· 177
11.4.3 面向目标的误差分析 ·· 181
11.5 多孔介质中的压力传播方程 ····································· 182
11.6 连续介质力学简介 ··· 183
11.6.1 Euler-Lagrange 坐标系 ···································· 185
11.6.2 Nanson 公式和 Piola 变换 ······························· 189
11.6.3 Reynolds 传输定理及动量守恒 ························· 190
11.6.4 本构关系 ·· 192
11.7 固体变形中的裂缝问题 ·· 192
11.7.1 动态裂缝的 Lagrangian 描述 ··························· 192
11.7.2 动态压裂裂缝的 Lagrangian 描述 ····················· 193
11.7.3 固定网格下的欧拉框架 ····································· 194
11.8 流固耦合相场裂缝问题 ·· 195
11.8.1 Lagrangian-Eulerian 耦合框架 ························· 195
11.8.2 全 Eulerian 固定网格模型 ································· 202
11.8.3 Navier-Stokes 流固耦合相场裂缝模型 ··············· 204
11.8.4 模拟结果 ·· 206
11.9 多物理相场裂缝模型难点 ··· 208

第 12 章　数值模拟 IV ·· 209
12.1　准静态裂缝模型 ··· 209
12.1.1　Sneddon 模型 ······································ 209
12.1.2　混合边界条件 ······································ 211
12.2　非等温相场裂缝模型 ····································· 213
12.2.1　模型配置 ·· 213
12.2.2　例 1：二维 Sneddon 模型 ··························· 213
12.2.3　例 2：二维 Sneddon 定压恒温模型 ··················· 213
12.2.4　例 3：二维 Sneddon 非等温模型 ····················· 214
12.3　多孔介质中的裂缝扩展模型 ································ 214
12.3.1　主要概念 ·· 214
12.3.2　模拟结果 ·· 215
12.4　Navier-Stokes 流固耦合相场裂缝模型 ······················ 216
12.4.1　例 1：定流量模型 ·································· 216
12.4.2　例 2：窦性流 ······································ 216
第 13 章　数值模拟研究软件 ····································· 218
13.1　团队建设 ·· 219
第 14 章　结论及问题 ·· 220
第 15 章　结束语 ·· 222
参考文献 ··· 225
索引 ·· 255

第 1 章 简 介

本书主要研究多物理场相场裂缝数值模型。在弹性材料领域，在 20 年前就已经有了应用变分法解决裂缝问题的相关研究 [171,70,71]。在过去的 10 年中，世界各地的研究团队均对基于相场法的多场耦合裂缝扩展问题展开了讨论。本书讨论了关于相场裂缝扩展模型的基本概念、离散、求解以及自适应方法。讨论的重点包括多场耦合问题的分类、数值模型以及对应的模拟软件开发进展。

1.1 多 物 理 场

在多场问题中，几种物理场会相互影响，例如，固体场、流体场、温度场和电磁场。本书中介绍的控制方程基于连续介质力学，在模型中通常会出现**非平稳非线性耦合偏微分方程组**问题，其中可能包含非线性不等约束。由于我们将使用变分法，因此在本书中最终系统被称为**耦合变分不等式系统** (CVIS)。该系统具有以下特点：

(1) 至少包含两个或以上偏微分方程 (PDE) 或变分不等式 (VI)；
(2) 模型基于拉格朗日或欧拉坐标系；
(3) 存在物理学守恒定律，如质量守恒、动量守恒、角动量守恒、能量守恒等；
(4) 遵循热力学第二定律：采用熵不等式描述能量传递方向，用以区分不可逆或可逆过程；
(5) 在不同的偏微分方程/耦合变分不等式系统之间通过以下方法交换参数：
(a) 系数；
(b) 线性或非线性的耦合项；
(c) 右手项；
(d) 接口 (内部边界)。

备注 1 (多物理场耦合问题)　相场裂缝模型作为本书中讨论的多物理问题的关键部分，由两个相互作用的子系统组成。然而，相场裂缝不是一个普通的多物理场系统，因为系统中采用一个偏微分方程描述位移状态，同时采用变分不等式描述另外一个物理状态 (相态)。该变分不等式只起到一个指示功能，表示一个损坏或断裂的不连续的状态。因此在某种程度上，相场裂缝只是一个拟耦合问题。

1.2 相场裂缝变分模型

1. 核心思想

相场法的核心思想是用一个相场参数 φ 来近似描述低维现象 (例如, 裂缝或材料不连续状态)。该参数通过求解一个定义在整个系统中的函数得到。同时利用正则化参数来获取裂缝宽度 $(\varepsilon > 0)$。如图 1.1 所示, 相场参数 φ 的取值范围从 0 (断裂状态) 到 1 (非断裂状态)。

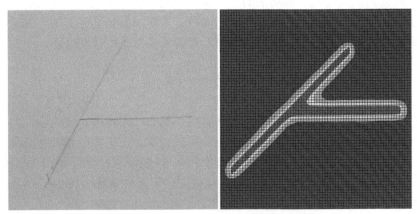

图 1.1　裂缝 (左) 和与其对应的固定网格相场模型 (右)
过渡区 (绿色) 表示裂缝的近似宽度

2. 优点

本书介绍的相场裂缝模型是一种正则化的界面分析方法, 它有以下优点。

(1) 相场模型是一个变分模型 [19,25,26,27,53,54,70,171];

(2) 是一个热力学一致问题 [312,308];

(3) 是一种不需要重新划分网格或更新基础函数的固定网格方法 (欧拉法);

(4) 相场模型能够完成复杂裂缝的计算和预测 (见图 1.2 和图 1.3), 包括裂缝交汇与分叉现象, 判别准则较简单;

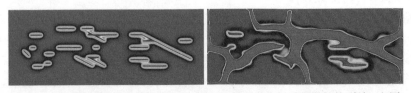

图 1.2　使用相场模型的裂缝网络 (左图为初始几何) 和不断增长的裂缝 (右图)
红色表示裂缝区域 $(\varphi = 0)$, 蓝色为未破碎区域 $(\varphi = 1)$。过渡区用黄色/绿色表示。此计算使用文献 [318] 中提出的模型完成

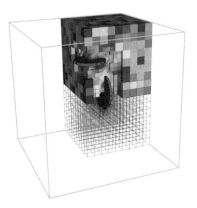 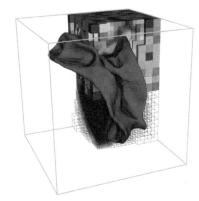

图 1.3　三维裂缝扩展模拟：在非均匀介质中，两个层状裂缝扩展并连接，随后分支
模拟结果表明，相场模型可以处理复杂的三维裂缝。模型计算过程采用高性能并行计算求解方法 (7.10.4 节)，采用自适应网格预测并矫正局部网格 (9.5.1 节)。该模拟结果请参考文献 [442]

(5) 基于水平集方法可准确判断缝宽等相关参数 (第 4 章)；

(6) 可以进一步利用相场参数的指示特性来搭建基于相场参数的多物理场函数 (第 11 章)。

3. 缺点与挑战

(1) 在处理多物理场时，多组方程和离散化参数之间存在相互作用，因此需要提前设定求解顺序。

(2) 在研究界面区域时，需要在界面区域进行高精度离散化处理，界面条件设定复杂，在精确计算裂缝开度、尖端应力以及裂缝扩展速度等参数方面存在挑战。同时在裂缝起裂条件等方面有待进一步优化。

(3) 模型是非凸的，只能保证顺梯度方向是局部最优解，这对数值算法的理论和设计提出了挑战。

(4) 求解尖锐物理表面时需要数学上的一些特殊函数。在稳定性和收敛性分析方面具有挑战性。

研究人员在相场裂缝模型领域进行了大量研究 (见 1.5 节)，为解决多物理场相场裂缝模拟提供了许多新思路，而这些研究成果往往会为该领域带来新的问题以及新的研究方向 [31,33,34,36,48]。

1.3　界面定义与处理方法

大部分多场耦合模型的一个特征是将整个系统分成至少两个子物理场系统。裂缝扩展也属于这一类问题之一。然而，模型内部不同介质的界面可能以不同的方式出现，因此有必要对它们进行分类。

定义 1 (界面类型) 界面在本书中用 Γ (或 Γ_i、C) 表示，它们是将至少两个介质分开的低维界面。随着复杂程度的增加，它们可以分为三种类型：

(1) 在空间、时间和面积 (长度) 上都固定的固定界面。例如，在上覆和下覆岩层地下建模中进行区域分割时遇到的相关问题。

(2) 在空间、时间上变化且不 (或仅轻微) 改变其区域的移动界面。例如，两相流和部分流固耦合模型中的相关问题。

(3) 在空间、时间上变化并显著扩展其面积 (长度) 的传播界面。例如，变形、损坏和裂缝扩展中的相关问题。

备注 2 三个符号 (Γ、Γ_i 或 C) 均为不同场景下的界面符号。

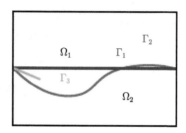

Ω_1：介质 1；Ω_2：介质 2；Γ_i, $i=1,2,3$ 三种界面

1.3.1 固定界面

固定界面的代表性例子是地质模型中地层之间的界面，其中往往涉及流固耦合问题。这一类界面位置往往是固定的，不随时间变动，见图 1.4。

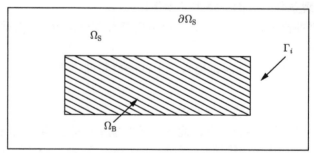

图 1.4 以 Ω_i 为代表的两种介质具有孔弹性，两者之间的界面是固定的

1.3.2 移动界面

移动界面的一个著名例子是流固相互作用模型 [91,167,184,90,46,62,371,175]，其中 Navier-Stokes 方程与 (非线性) 弹性力学相耦合，如图 1.5 所示。在这里，界面具体位置随时间移动且需要进行跟踪计算 (参见 1.3.4 节)。

1.3 界面定义与处理方法

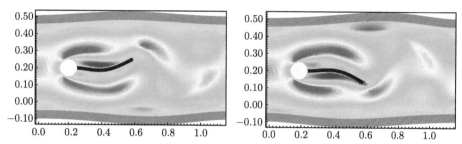

图 1.5 考虑流固相互作用的不可压缩流体 Navier-Stokes 方程与非线性弹性耦合
在这里，界面随着空间和时间的变化而变化。本书在图 1.7 基于该模型进行了进一步讨论 [430]

1.3.3 扩展界面

扩展界面的代表性例子为裂缝扩展问题。与第二类移动界面类似，界面位置会不断变化。但同时这类界面往往会跟随一个移动间断点 (即裂缝尖端) 而延伸其长度或面积，如图 1.6 所示。裂缝尖端区域的计算精度要求更高，界面位置预测难度更大，这要求我们随时更新界面位置。这给数学分析、数值方法的设计以及软件开发的可靠性带来了挑战。我们将在本书中讨论各种相关算例。

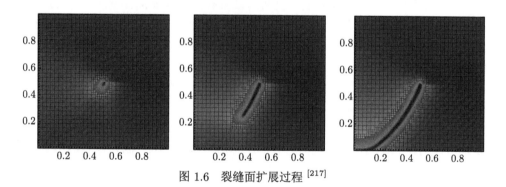

图 1.6 裂缝面扩展过程 [217]

1.3.4 界面处理方法

对于断裂力学中的界面问题，有两种基本的处理方法 (如图 1.7 所示)：
(1) 界面跟踪法；
(2) 界面捕获法。

定义 2 (界面跟踪法) 在界面跟踪法中，界面是显式描述的。在这里，我们需要能够允许我们跟踪界面变化的网格。然后，随着界面的进一步变化，网格需要随之更新。其中一个缺点是网格单元可能会变形太多，如果更新后网格没有妥善处理，会导致模型求解失败。

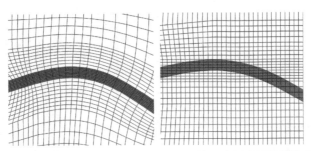

图 1.7 基于拉格朗日法的界面跟踪法 (左) 和基于欧拉法的界面捕获法 (右)
在界面跟踪法中，界面位于网格边缘。在界面捕获法中，界面通过网格元素切割，
并通过附加的标记功能实现捕获

定义 3 (界面捕获法) 在界面捕获法中，模型的求解是在一个固定的空间域上进行的，界面是通过附加函数隐式定义的。界面函数捕获界面并标记其位置。网格更新不再是难点，但界面标记函数的稳定性至关重要。

界面跟踪法同时还包括时间/空间离散化 [405,406] 的拉格朗日-欧拉耦合法 [131,230,239,132,166]。界面捕获法的代表性方法则是水平集方法 [340,396] 和相场法，相关研究可追溯到文献 [96, 12]。同时部分研究 [38,387,104,143,162,262,457,39,389,44,5,407,430,433] 中将两种技术进行了结合。此外，还存在拟合和非拟合方法。拟合法的代表性方法是局部修正有限元法 [176]、界面拟合子空间投影法 [43]。非拟合法有扩展/广义有限元 [323,181,305,32]、单元切割法 [94]、子空间法 [200]、单元侵入法 [345] 或有限单元法 [343,353]。

决定是使用界面跟踪法还是界面捕获法，需要在算法、方程实现和计算量、所需界面近似的准确性及它们的数学证明之中进行权衡。虽然一般来说，界面捕获法可以更容易地处理移动和变化的界面，但该方法计算精度与网格尺寸有关 (如图 1.8 所示)。在处理扩展界面时，界面跟踪法难度会更大，因为模型网格必须能够适配扩展界面，这一点在三维模型中尤为重要。

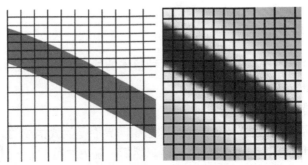

图 1.8 低界面分辨率和高界面分辨率的对比

1.4 工程问题：螺钉断裂和损坏

本节我们简要介绍相场裂缝模型的一项工程应用：螺钉拉伸的断裂与损伤过程模拟 [427]。部分模拟结果如图 1.9 所示。

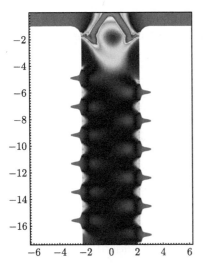
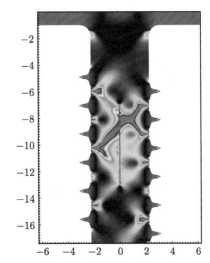

图 1.9 左图：螺钉的拉伸实验。在螺钉头部施加拉力，同时固定底部，在头部附件区域出现高应力，最后螺钉沿红色区域发生断裂。右图：内部裂纹导致螺钉损坏，螺钉沿红色区域破坏 [427,426]

螺钉损伤现象一般指螺钉侧翼螺纹、接缝或空心处出现闭合的裂纹，这通常由加工错误造成。在生产过程中，螺钉的螺纹是通过轧制一个未加工的毛坯所产生的。我们开发的弹塑性材料的相场模型能够有效预测螺钉的断裂区域。这些数值模拟有助于更好地帮助加工人员理解螺钉的损坏机理。

1.5 相关文献

本节提供了与裂缝变分方法和相场裂缝方法有关的文献调研情况。由于世界各地的研究团队正在同步推进各项研究，因此调研情况难免有所遗漏，还请见谅。

1) 文章概览

在文献 [72, 453, 425] 中，作者总结了目前相场裂缝研究的进展。在文献 [72] 中，作者从理论导向方面进行了开创性的总结。在文献 [453] 中，作者提供了相场裂缝研究的综述。在文献 [425] 中，作者构建了一个多孔介质中多物理场裂缝扩展模型的并行自适应框架，该数模框架包括基于各物理场的离散化和耦合模块。

2) 专题概览

在这个列表中，存在某一出版物同时属于多个主题的现象。

(1) 数学建模与分析 [171,71,120,119,172,170,169,194,329]。

(2) 热力学解释 [312,308,452]。

(3) 相场裂缝与损伤力学的关系 [125,452,136,297,187]。

(4) 裂缝扩展的分析与计算 [106,421,266,427,439,17,306,438]。

(5) 其他离散化技术：IGA(等几何分析) [65,64,221,222,18]，特殊基函数 [263]，非连续伽辽金法研究 [153,321,320]，有限元研究 [7,8]。

(6) 基于有限元的多场裂缝起裂与扩展耦合研究 [196,186]。

(7) 裂缝正则化 (Ambrosio-Tortorelli) 函数的不同表达形式 [21,74,346,159]。

(8) 退化函数 [381,187,29,452]。

(9) 最优控制，参数估计，拓扑/形状优化 [331,332,255,254,451,11,378]。

(10) 空间网格适应性 [30,217,436,35,281,92,93,456,344]。

(11) 时间步长自适应和多速率耦合研究 [462,437,14,243]。

(12) 线性求解器研究 [159,218,246,462]。

(13) 非线性求解器 [189,438,439,261]。

(14) 耦合迭代，最小化迭代研究 [70,68,306,92,308,86,155]。

(15) 动态断裂过程研究 [65,272,73,273,231,383,460,187]。

(16) 热应力影响下塑性断裂过程研究，黏弹性固体，各向异性固体中裂缝扩展研究 [74,311,400,307,18,16,416,265,10,388,130,403,136,6,8]。

(17) 相场裂缝宽度计算 [333,282,110,454]。

(18) 压力驱动和流体驱动的下裂缝扩展研究 [424,313,69,318,444,337,442,459,316,317,310,309,298,214,334,380,215,281,278,277,98,283,9,110,460,447,461]。

(19) 流固耦合研究 [435,437]。

(20) 多尺度相场研究方法 [191,336,9,186,344]。

除此之外，其他参考文献请参考文献目录。

1.6 章节介绍

第 2 章 在这一章中，介绍了这本书中使用的主要符号、相关公式以及一些数学上的先决条件。另外，介绍了相场裂缝的典型问题。这些有助于从多个角度介绍和讨论多物理场相场裂缝问题。

第 3 章 在当今应用数学和工程中，多场耦合问题占据着重要地位。我们试图基于耦合策略提供一个针对耦合问题的分类，并提供几个代表性例子。

1.6 章节介绍

第 4 章　本章回顾了 Francfort 和 Marigo 提出的一个经典的脆性材料准静态变分断裂模型。重点介绍模型中所采用的 Ambrosio-Tortorelli 泛函逼近方法，包括针对这些问题的一些历史注释。同时推导出相应的欧拉-拉格朗日系统。最后，简要讨论了一些热力学扩展问题。

第 5 章　本书主要分为三个算法章节和一个多物理场建模章节，并且中间穿插了四个数值模拟章节。第 5 章为第一个算法章节，重点介绍了变分的正则化、不等式约束，然后是时间和空间离散化。虽然准静态过程研究中时间离散化不太重要，但由于多物理场裂缝扩展过程通常与时间有关，因此仍然引入了目前最先进的时间步迭代方法。在空间离散化过程中，我们回顾了 Galerkin 有限元法的基本概念。然后对简化模型中相场正则化参数与空间离散化参数之间的关系进行了数值分析。

第 6 章　在本数值模拟章节中介绍了一个稳态狭缝模型的数值算例，并分析了空间离散化过程中模型收敛性问题。

第 7 章　在这一章中重点讨论了非线性解问题。着重讨论了子模块构建和整合技术。给出了不动点迭代、泛函迭代和牛顿迭代法的几种非线性求解方法。详细讨论了几种不同的牛顿迭代法。最后对线性解作了简要讨论。在这里，我们省略了一些技术细节，建议读者参考相关文献。

第 8 章　在第二个数值模拟章节中，基于单边缺口剪切实验，对几种非线性求解方法进行了比较讨论，同时给出了关于应力加载步骤的改进方案。

第 9 章　本章将考虑适应性问题。首先，我们讨论了针对稳态裂缝目标导向的误差分析和空间网格自适应。这些内容对多物理场中多目标导向误差分析的概念进行了补充。然后讨论了自适应网格预测-校正方法。它们用于探索解决网格自适应和局部精细度问题。最后，探索性地提出了一种面向目标的时间误差控制方法。

第 10 章　本章给出了三个数值算例。首先，我们基于第 6 章的狭缝模型提出了面向目标的误差分析方法。然后，我们进行了无相场狭缝的多目标误差分析。相关结果可以与 (未来的) 相场模型的解决方案相比较。最后展示了时间/加载步优化在螺杆仿真中的相关研究结果。

第 11 章　一方面，在以前的章节中，我们只关注耦合问题，还没有实现真正的多物理场应用。另一方面，大多数提出的算法都会扩展应用到多物理场问题上。本书作者自从博士论文[428]以来，在研究流固耦合过程中，做了大量多物理场问题的相关工作，因此，我们已经在之前的章节讨论了多重物理场研究所需要面临的相关问题。在本章中，我们现在考虑一些具体的部分。

首先，我们强调了相场法和水平集方法的相似性。水平集概念是本章的关键，也是贯穿本部分的主线。具体地说，我们定义了指示函数，允许我们在固定网格域中区分不同的物理状态。

然后，我们研究了界面相关的物理量，如裂缝宽度的计算。接下来，我们将相场裂缝和多物理场问题相结合。首先考虑相场裂缝模型在多孔介质的应用，然后是流体-固体的相互作用。大多数研究结果已经发表在同行评审的期刊上，但在本书中也增加了新的补充材料。这尤其适用于将所有方程都定义在一个欧拉系统固定网格上的、完全欧拉描述的相场裂缝问题。

这一欧拉相场固体描述法最终将被扩展为完全欧拉描述的流固耦合相场裂缝的初始概念。该方法充分利用了基于固定网格法的相场法的优点。我们使用了水平集函数来区分流体和固体的子域，展示了考虑多物理问题时相场指示函数的优势与重要性。

第 12 章 在本章中，我们进行了一些特征模拟来佐证第 11 章的内容，并且展示了我们过去几年在多物理场相场裂缝应用方面的努力与成果。

第 13 章 本书作者工作的一个重要部分是相关研究软件的开发。本章对作者的代表性成果进行了总结。还提供了开源代码的链接。

第 14 章 我们在这一章中总结了这本的主要内容。最后对今后的研究方向进行了简要的讨论。

第 15 章 发现问题，总结规律，并在书面作品中将其展示出来是一件快乐的事情，是一段需要专注力、创造力、语言能力、思维能力的过程，同时也是一段令人失望和消极的过程。恰恰是最初的动机和希望支撑我们去实现这些目标。作者在本章展示了自己进行这方面研究的期望与动力。

第 2 章 数学符号、模型条件和原始模型问题

2.1 数 学 符 号

本章将介绍本书的常见符号。后续如有添加，会补充对应的说明。

2.1.1 空间域、边界条件、裂缝

本书中，d 表示空间维度。令 $B \subset \mathbb{R}^d, d=2,3$ 为全空间域，其中 $C := C(t) \subset \mathbb{R}^{d-1}$ 表示一条或多条 (扩展的) 裂缝。$\Omega := \Omega(t) \subset \mathbb{R}^d$ 是全空间点集 (图 2.1)。此时模型中任意一点可表示为

$$x = (x_1, x_2, x_3, \cdots, x_d)$$

其中 $\Omega_F := \Omega_F(t) \subset \mathbb{R}^d$ 代表高维裂缝。由于缝宽过渡区 $\varepsilon > 0$，因此裂缝增加了一个维度。此时系统中引入了一个内部边界区域 $\partial C := \partial C(t) := \bar{\Omega}_F \cap \bar{\Omega}$，该区域表示基质与裂缝之间的过渡区。本书在 11.2 节中详细介绍了如何在实际模型中搭建并处理 Ω_F 区域。

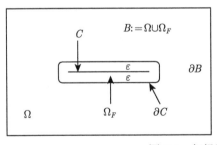 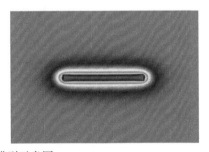

图 2.1 相场裂缝模型示意图

未断裂区域 Ω；裂缝 C；后者可在定义域 Ω 中近似表示为 Ω_F，其厚度的一半，即半缝宽过渡区可用 ε 表示。裂缝边界为 ∂C，而整个系统的外部边界是 ∂B。右图给出了相应的相场实现方法，裂缝区域用相场变量 $\varphi = 0$ 表示，过渡区 $0 < \varphi < 1$，其宽度可用 ε 表示。因此，整个 Ω_F 区域可以用相场参数 $\varphi = 0$ 的区域来表示

边界条件可归纳为 $\partial B = \partial B_{HD} \cup \partial B_{ND} \cup \partial B_{HN} \cup \partial B_{NN}$，其中 ∂B_{HD} 代表齐次狄利克雷 (Dirichlet) 边界条件，∂B_{ND} 代表非齐次狄利克雷边界条件，∂B_{HN} 代表齐次诺伊曼 (Neumann) 边界条件，∂B_{NN} 代表非齐次诺伊曼边界条件。

2.1.2 加载时间/加载步

加载时间用 $I := (0, T)$ 表示,其中 $T > 0$ 代表结束时间。本书主要研究准静态断裂问题,这意味着在每个时间步中,模型均处于热力学平衡状态。

备注 3 符号 I 也可用于表示单位矩阵/张量,本书使用时会通过上下文对其进行区分。

2.1.3 模型、材料以及离散化参数

1) 正则化参数和补偿量

在相场模型中设定正则化参数 ε 与 κ ($\varepsilon > 0, \kappa > 0$)。在某些模型中,我们使用 γ 表示补偿量 ($\gamma > 0$)。

2) 材料参数

关于拉梅 (Lamé) 常量,我们用 μ 与 λ 表示。$\mu > 0$,为剪切模量;$\lambda = -\dfrac{2}{3}\mu$;单位为 $\text{Pa} = \text{N/m}^2 = \text{kg}/(\text{m} \cdot \text{s}^2)$。我们采用 σ_S (通常简化为 σ) 来表示固体中的应力张量,其单位为 Pa/m^2。临界能量释放率用 G_c 表示,其单位为 $\text{J/m}^2 = \text{N/m}$。泊松比为 ν_s,取值范围为 $-1 < \nu_s < 0.5$。杨氏模量为 E_Y,取值为正,单位为 $\text{Pa} = \text{N/m}^2 = \text{kg}/(\text{m} \cdot \text{s}^2)$。后续相关参数会在对应章节中补充介绍。如 3.4 节中的多孔弹性介质问题,第 11 章中的流体问题等。

3) 空间离散化和时间离散化

时间 (加载) 步长我们用 $k_n = t_n - t_{n-1}$ 表示,本书中提到的时间步长一般为最大步长,即 $k = \max(k_n)$。其中时间 (加载) 步与加载时间的关系为

$$0 := t_0 < t_1 < t_2 < \cdots < t_n := T$$

时间点 t_n 之间的间隔也可以不均匀。空间离散化参数,即单元格尺寸参数,通常用 h 表示。更多关于 h 的说明可见 5.4 节。

2.1.4 解变量

相场裂缝模型中解变量有:

(1) 以向量值表示的位移变量 $u := u(x, t) : B \times I \to \mathbb{R}^d$;
(2) 以标量值表示的相场参数 $\varphi := \varphi(x, t) : B \times I \to [0, 1]$。

相场参数 φ 通过数值变化的方式描述裂缝的起裂与否:

(1) $\varphi = 0$ 表示裂缝区域;
(2) $\varphi = 1$ 表示未破碎的材料 (通常为基质);
(3) $0 < \varphi < 1$ 表示根据正则化参数 ε 构成的平滑过渡区。在工程中,ε 通常是长度尺度参数。该区域表示材料的断裂或者劣化区域,代表介于未破碎与完全损坏状态之间的物理过渡区 (其中的部分思路可参考 5.5 节与 3.6 节)。

备注 4 在材料学等相关领域，ε 就是所谓的相互作用长度。在两相流中，ε 称为界面厚度。在地质研究中，ε 被称为地层过渡层。

在第 11 章中，我们进一步研究了其他解变量，如：

(1) 以矢量值表示的速度 $v := v(x,t) : B \times I \to \mathbb{R}^d$;

(2) 以标量值表示的压力 $p := p(x,t) : B \times I \to \mathbb{R}$;

(3) 以标量值表示的温度 $\theta := \theta(x,t) : B \times I \to \mathbb{R}$。

由于一个模型中通常需要同时研究多个解变量，因此常使用 U 作为解变量集。例如，$U = (u, \varphi)$ 或 $U = (v, p, u)$ 或者 $U = (v, u, p, \varphi)$。同时我们注意到解变量与正则化参数 ε 相关，即

$$U := U_\varepsilon$$

当然为了简化表示，常省略下标 ε，使用 U 代表解变量集。

2.1.5 梯度、散度、迹、拉普拉斯算子

使用倒三角算子[85]来定义空间导数。单值函数 $v : \mathbb{R}^d \to \mathbb{R}$ 的梯度表示为

$$\nabla v = \begin{pmatrix} \partial_1 v \\ \vdots \\ \partial_d v \end{pmatrix}$$

其中 $\partial_i v := \dfrac{\partial v}{\partial x_i}$，$x_i$ 代表第 i 个方向的方向分量。

而向量值函数 $v : \mathbb{R}^d \to \mathbb{R}^m$ 的梯度，即雅可比 (Jacobian) 矩阵，表示为

$$\nabla v = \begin{pmatrix} \partial_1 v_1 & \cdots & \partial_1 v_1 \\ \vdots & & \vdots \\ \partial_d v_m & \cdots & \partial_d v_m \end{pmatrix}$$

向量值 $v \in \mathbb{R}^d$ 的散度为

$$\mathrm{div}\, v := \nabla \cdot v := \nabla \cdot \begin{pmatrix} v_1 \\ \vdots \\ v_d \end{pmatrix} = \sum_{j=1}^d \partial_j v_j$$

张量 $\sigma \in \mathbb{R}^{d \times d}$ 的散度为

$$\nabla \cdot \sigma := \left(\sum_{j=1}^d \frac{\partial \sigma_{ij}}{\partial x_j} \right)_{1 \leqslant i \leqslant d}$$

矩阵 A 的迹为

$$\mathrm{tr}(A) = \sum_{j=1}^{d} a_{jj}$$

定义 4 (拉普拉斯算子) 二阶连续可微标量 $u : \mathbb{R}^d \to \mathbb{R}$ 的拉普拉斯算子可表示为

$$\nabla \cdot (\nabla u) = \Delta u = \sum_{j=1}^{d} \partial_{jj} u$$

定义 5 向量值函数 $u : \mathbb{R}^d \to \mathbb{R}^m$ 的拉普拉斯算子可表示为

$$\Delta u = \Delta \begin{pmatrix} u_1 \\ u_2 \\ \vdots \\ u_m \end{pmatrix} = \begin{pmatrix} \sum_{j=1}^{d} \partial_{jj} u_1 \\ \sum_{j=1}^{d} \partial_{jj} u_2 \\ \vdots \\ \sum_{j=1}^{d} \partial_{jj} u_m \end{pmatrix}$$

2.1.6 函数空间

2.1.6.1 通用符号

我们在 Lebesgue 和 Sobolev 空间中一般使用标准表示法[449,2]，如 $D := B$ 或 $D := \Omega$ 或 $D := \Omega_F$。通过定义 $L^p(D), 1 \leqslant p \leqslant \infty$，我们可以表示包含了函数 u 的标准 Lebesgue 空间，其中 u 的 p 次方是 Lebesgue 可积的。此时集合 $L^p(D)$ 与 $\|u\|_{L^p(D)}$ 构成了一个 Banach 空间：

$$\|u\|_{L^p(D)} := \left(\int_D |u(x)|^p \mathrm{d}x \right)^{\frac{1}{p}}, \quad 1 \leqslant p \leqslant \infty$$

$$\|u\|_{L^\infty(D)} := \mathrm{ess\ sup} |u(x)|$$

当 $p = 2$ 时，可得 Hilbert 空间 $L^2(D)$，此时

$$(u, v)_{L^2(D)} := \int_D u(x) v(x) \mathrm{d}x$$

Sobolev 空间 $W^{m,p}(D), m \in \mathbb{N}, 1 \leqslant p \leqslant \infty$ 为 $L^p(D)$ 中具有 m 阶分布导数的函数空间，该空间符合以下条件：

$$\|u\|_{W^{m,p}(D)} := \left(\sum_{|\alpha| \leqslant m} \|D^\alpha u\|_{L^p(D)}^p \right)^{\frac{1}{p}}, \quad 1 \leqslant p < \infty$$

$$\|u\|_{W^{m,\infty}(D)} := \max_{|\alpha| \leqslant m} \|D^\alpha u\|_{L^\infty(D)}$$

定义 6 符号 $\alpha = (\alpha_1, \cdots, \alpha_d) \in \mathbb{N}^d$ 所代表的值具有以下属性：

$$|\alpha| := \sum_{j=1}^{d} \alpha_j, \quad D^\alpha := \frac{\partial^{|\alpha|}}{\partial x_1^{\alpha_1} \cdots \partial x_d^{\alpha_d}}$$

若 $k \in \mathbb{N}_0$，我们定义所有 k 阶偏导数的集合：

$$D^k u(x) := \{D^\alpha u(x) : |\alpha| = k\}$$

$p = 2$ 时，$H^m(D) := W^{m,2}(D)$ 是一个 Hilbert 空间，其内积为

$$(u, v)_{H^m(D)} := \sum_{|\alpha| \leqslant m} (D^\alpha u, D^\alpha v)_{L^2(D)}$$

范数为 $\|\cdot\|_{H^m(D)}$，半范数为

$$\|u\|_{W^{m,p}(D)} := \left(\sum_{|\alpha| = m} \|D^\alpha u\|_{L^p(D)}^p \right)^{\frac{1}{p}}, \quad 1 \leqslant p < \infty$$

$$\|u\|_{W^{m,\infty}(D)} := \max_{|\alpha| = m} \|D^\alpha u\|_{L^\infty(D)}$$

最后我们用 $W_0^{m,p}(D)$ 表示在 ∂X 上迹为 0 的 $W^{m,p}(D)$ 子空间，$H_0^1(D) = \{u \in H^1(D) : u = 0|\Gamma_D \subset \partial X\}$。

2.1.6.2 时空函数空间

分析时间相关的弱形式时，采用 Sobolev 函数空间，即所谓的 Bochner 空间 [122,449]。L^2 在 $I = (0, T)$ 中的积分可以在时间连续时近似为 I。

定义 7 令 $\Omega \subset \mathbb{R}^d$ 是一个 Lebesgue 可测的开放域，可测函数几乎在任何位置均有定义 (即除了某一组 Lebesgue 测度以外)，简言之，在 Ω 中：

$$f(x) = g(x)$$

当 $Z \subset \Omega$ 时，Z 的 Lebesgue 测度是 0，

$$f(x) = g(x), \quad x \in \Omega \backslash Z$$

定义 8 在时间连续 $I = (0, T)$ 模型中，函数可在几乎所有 $t \in I$ 时近似。

备注 5 显然，在时间离散化之后，可将实际问题离散为 $n = 1, 2, 3, \cdots, N$ 后进行分析。

2.1.6.3 相场裂缝问题相关符号

在本书中，我们经常使用简化符号：

$$V_B := H^1(B), \quad V_B^0 := H_0^1(B)$$

对于向量值函数空间 (例如，位移向量 $u : B \to \mathbb{R}^d$)，我们使用相同的符号进行表示：

$$V_B := H^1(B), \quad V_B^0 := H_0^1(B)^d$$

结合上下文，可以知晓同样的符号代表标量值还是向量值。相场裂缝问题的基本空间定义如下。

定义 9 位移场的空间定义为

$$V_u^D := V_u^D(B) := \{u \in u_D, \text{在 } \partial \Omega_D \backslash \partial B \text{ 上}\}$$

位移场测试空间是经典的 $H_0^1(B)$ 空间，定义为

$$V_u^0 := V_u^0(B) := \{u \in V_u \mid u = 0, \text{在 } \partial B \text{ 上}\}$$

相场变量的测试空间为

$$V_\varphi := V_\varphi(B) = H^1(B)$$

这些空间将作为原型模型空间。在实际模拟中，位移场使用齐次 Neumann 条件 ∂B_N，后续章节会进行具体说明。

备注 6 在一般算法研究中，其边界条件通常不会引起特定的变化，我们仅简单研究位移场与相场。

在相场裂缝问题中，我们主要研究位移量

$$u \in V_u^D$$

以及相场变量

$$\varphi \in V_\varphi$$

2.1 数学符号

在此基础上，我们引入了凸集:

$$K := K\left(\varphi^{n-1}\right) := \{\varphi \in V_\varphi : 0 \leqslant \varphi^n \leqslant \varphi^{n-1} \leqslant 1, \quad \text{a.e.} B\}$$

这是考虑时间的变分不等式约束，即所谓的不可逆性约束。

定义 10 解变量乘积空间如下:

$$X = V_u^D \times V_\varphi$$

由此可得 $U \in X$，即 $(u,\varphi) \in V_u^D \times V_\varphi$，同时有 $X^D = X = V_u^D \times V_\varphi$, $X^0 = V_u^0 \times V_\varphi$。

2.1.7 内积、双线性型、半线性型的符号表示

我们简要地介绍变分弱形式的重要符号:

(1) 令 $f, g \in V^d$，则内积表示为

$$(f, g) = \int_B f \cdot g \mathrm{d}x, \quad f, g \in V^d$$

(2) 令 $F, G \in V^{d \times d}$，则矩阵的内积为

$$(F, G) = \int_B F \cdot G \mathrm{d}x, \quad F, G \in V^{d \times d}$$

(3) 通常令 $u \in V$:

$$a(u, \varphi) = l(\varphi), \quad \forall \varphi \in V$$

在这里 $a(u, \varphi) : V \times V \to \mathbb{R}$ 是一个双线性形式，$l(\varphi) \in V^*$ 是线性形式 (线性泛函)，V^* 表示对偶空间 $V^* = L(V, \mathbb{R})$。

(4) 在非线性问题中，解变量 $u \in V$ 是非线性的，但函数仍然是线性的，此时令 $u \in V$:

$$a(u)(\varphi) = l(\varphi), \quad \forall \varphi \in V$$

这里的 $a(u)(\varphi)$ 是半线性形式，其中第一个参数 $u \in V$ 是非线性的，第二个参数 $\varphi \in V$ 是线性的。

(5) 在线性偏微分系统中，定义空间 X，如 $X1 = V_u^D \times V_\varphi$，令 $U = (u, \varphi) \in X$，有

$$A(U, \psi) = F(\psi), \quad \forall \psi := (w, \varphi) \in X^0 := V_u^0 \times V_\varphi$$

(6) 在非线性偏微分系统中，令 $U \in X$，有

$$A(U)(\psi) = F(\psi), \quad \forall \psi \in X^0$$

备注 7 本书中，我们并未使用粗体字母、箭头或类似的符号来区分标量值、向量值和张量值变量。所有变量都是非粗体字体。我们希望读者结合上下文来区分它们。

2.1.8 参数符号表

本书中常用参数符号与单位如表 2.1 所示。

表 2.1 相场裂缝模型常用参数表

符号	含义	单位
d	空间维度	
x	位置	m
t	时间；准静态 (增量) 问题加载时间	s
B	总域：$B = \overline{\Omega} \cup \overline{C}$，即 $\overline{\Omega} \cup \overline{\Omega}_F$	m^3/m^2
C	裂缝 (低维度)	m^2/m
I	$I = (0, T)$：时间/负载间隔	s
T	结束时间	s
h	空间离散性参数 (网格尺寸)	m
k	$k_n = t_n - t_{n-1}$：时间步长	s
U	解变量集，如 $U = (u, \varphi)$	
V_u^D	位移变量函数空间	
V_φ	相场变量函数空间	
κ	相场正则化参数	无量纲
ε	相场正则化参数	m
γ	补偿参数	
u	位移	m
φ	相场变量	无量纲
∇u	位移梯度	无量纲
Δu	$\Delta u = \nabla \cdot \nabla u$	1/m
$\partial_t u$	$\partial_t u = v$：速度 (时间衍生的位移)	m/s
$\partial_t v$	加速度 ($\partial_t v = \partial_{tt}^2 u$)	m/s^2
$E(\cdot)$	能量函数	$J = kg \cdot m^2/s^2$
v	速度	m/s
∇v	速度梯度/流体的内摩擦力	1/s
p	压力	$Pa/m = N/m^3 = kg/(m^2 \cdot s^2)$
∇p	压力梯度	$Pa = N/m^2 = kg/(m \cdot s^2)$
$\hat{F}$	变形梯度	无量纲
n	法向量	m
σ_f	流体应力张量	$Pa = N/m^2 = kg/(m \cdot s^2)$
$\sigma \cdot n$	曳引力	$Pa \cdot m = N/m$
ρ	密度	kg/m^3
v	运动流体黏度	m^2/s
f	力 (如重力)	$N/kg = m/s^2$
σ	应力张量	Pa/m^2

续表

符号	含义	单位
σ^+	裂缝驱动应力	Pa/m^2
σ^-	非裂缝驱动应力	Pa/m^2
G_c	临界能量释放率	J/m^2 = N/m
μ	拉梅参数	Pa = N/m^2 = kg/(m·s^2)
λ	拉梅参数	Pa = N/m^2 = kg/(m·s^2)
ν_s	泊松比	无量纲
E_Y	杨氏模量	Pa = N/m^2 = kg/(m·s^2)
η	误差估计值	
TOL	求解器容差	

2.2 微分与积分的前提条件

本节介绍了微分与积分的前提条件与基础假设。

2.2.1 高斯-格林定理/散度定理

高斯-格林 (Gauss-Green) 定理，通常称为散度定理。令 $\Omega \subset \mathbb{R}^d$ 是一个有界的、开放的域，其边界条件为 $\partial\Omega$ 属于 C^1。

定理 1 (散度定理) 假设 $u := u(x) \in C^1(\bar{\Omega})$，其中点坐标为 $x = (x_1, \cdots, x_d)$，然后有

$$\int_\Omega u_{x_i} \mathrm{d}x = \int_{\partial\Omega} u n_i \mathrm{d}s, \quad i = 1, \cdots, d$$

此处 n_i 是法向量 n 的第 i 个分量，有

$$\int_\Omega \mathrm{div} u \mathrm{d}x = \int_{\partial\Omega} u \cdot n \mathrm{d}s$$

对于每一个向量 $u \in C^1(\bar{\Omega}; \mathbb{R}^d)$ 均成立。

证明请参考文献 [260]。

2.2.2 分部积分与格林公式

根据散度定理，可知:

命题 1 (分部积分) 令 $u, v \in C^1(\bar{\Omega})$，则有

$$\int_\Omega u_{x_i} v \mathrm{d}x = -\int_\Omega u v_{x_i} \mathrm{d}x + \int_{\partial\Omega} u v n_i \mathrm{d}s, \quad i = 1, \cdots, d$$

可简化表达为

$$\int_\Omega \nabla u v \mathrm{d}x = -\int_\Omega u\nabla v \mathrm{d}x + \int_{\partial\Omega} uvn \mathrm{d}s$$

证明可根据散度定理推导。

我们可以继续从分部积分中推导出部分结论，这些结论会在后续研究中发挥作用。

命题 2 (格林公式) 令 $u,v \in C^1(\bar{\Omega})$，则有

$$\int_\Omega \Delta u \mathrm{d}x = \int_{\partial x} \partial_n u \mathrm{d}s$$

$$\int_\Omega \nabla u \cdot \nabla v \mathrm{d}x = -\int_\Omega \Delta u v \mathrm{d}x + \int_{\partial\Omega} v\partial_n u \mathrm{d}s$$

证明可根据分部积分推导。

2.2.3 Lebesgue 积分

定理 2 (换元积分) 令 $\hat{B} \subset \mathbb{R}^d$ 为开，令 $\hat{T}: \hat{B} \to \mathbb{R}^d$ 为 $\mathbb{R}^d$ 中的一个微分同胚映射。函数 $f: \hat{T}(\hat{B}) \to \mathbb{R}$ 在 $\hat{T}(\hat{B})$ 上 Lebesgue 可积，当且仅当函数 $\hat{x} \in \hat{B} \to f(\hat{T}(\hat{x}))|\det(\hat{\nabla}\hat{T}(\hat{x}))| \in \mathbb{R}$ 在 $\hat{B}$ 上 Lebesgue 可积。然后我们有

$$\int_{\hat{T}(\hat{B})} f(x)\mathrm{d}x = \int_{\hat{B}} f(\hat{T}(\hat{x}))|\det(\hat{\nabla}\hat{T}(\hat{x}))|\mathrm{d}\hat{x}$$

定理 3 令 $B \subset \mathbb{R}^m$ 可测，$A \subset \mathbb{R}^d$ 为开。函数 $f: A \times B \to \mathbb{R}$ 在 B 上对于任意一点 $x \in A$ 都 Lebesgue 可积，此外令 f 对于任意 $x \in A$ 以及几乎全部 $y \in B$ 均连续可微。假设边界 $g: B \to \mathbb{R}$ 上，对任意 $x \in A$ 及几乎全部 $y \in B$ 均有 $\|\nabla_x f(x,y)\| \leqslant g(y)$，则我们有：

(1) $\nabla_x f(x,y)$ 在 B 上对于任意 $x \in A$ 均 Lebesgue 可积；

(2) $F(x) = \int_B f(x,y)\mathrm{d}y$ 连续可微，且

$$\nabla F(x) = \int_B \nabla_x f(x,y)\mathrm{d}y$$

2.2.4 Banach 空间中的微分问题

在某些领域我们需要使用到 Banach 空间中的微分。根据参考文献 [413, 114, 423]，我们总结了部分已知成果。

2.2 微分与积分的前提条件

定义 11 (Banach 空间中的方向导数) 令 V 与 W 是赋范向量空间,$U \subset V$ 为非空集。令 $f: U \to W$ 是一个给定的映射,若存在

$$f'(u)(\phi) := \lim_{h \to 0} \frac{f(u+h\phi) - f(u)}{h}, \quad u \in U, \quad \phi \in V$$

那么 $f'(u)(\phi)$ 被称为映射 f 在 u 点 ϕ 方向的方向导数,若对所有 $\phi \in V$ 均存在方向导数,则称 f 在 u 点方向可微。

备注 8 为了突出方向与变量 u 相关,我们会将 ϕ 用 δu 表示。该方法常在处理多个解变量与多个方向导数并存的问题中使用。

备注 9 Banach 空间中方向导数的定义与空间 $\mathbb{R}$ 中导数的定义是一致的 [259]:

$$f'(x) = \lim_{h \to 0} \frac{f(x+h) - f(x)}{h}$$

在 $\mathbb{R}^d$ 中有 [260]

$$f'(x)(\phi) := \lim_{h \to 0} \frac{f(x+h\phi) - f(x)}{h}$$

当一个函数对于点 $x \in \mathbb{R}^d$ 的所有方向导数均存在时,该函数称为可微函数,该方向的导数为 $e_i, i = 1, \cdots, d$。在标准基坐标中这就是偏导数。

定义 12 (Gâteaux 导数) 方向可微映射的定义可参考定义 11。对于 $u \in U$,如果存在线性连续映射 $A: U \to W$,则

$$f'(x)(\phi) = A(\phi)$$

对于任意 $\phi \in U$ 均成立,则称其为 Gâteaux 可微。A 是 f 在 u 的 Gâteaux 导数,写作 $A = f'(u)$。

备注 10 Gâteaux 导数需借助方向导数计算,$f'(u) \in L(U, W)$,如 $W = \mathbb{R}$,则 $f'(u) \in U^*$。

定义 13 (Fréchet 导数) 若存在算子 $A \in L(U, W)$,映射 $r(u, \cdot): U \to W$。对于任意 $\phi \in U, u + \phi \in U$,有

$$f(u + \phi) = f(u) + A(\phi) + r(u, \phi)$$

与

$$\frac{\|r(u, \phi)\|_W}{\|\phi\|_U} \to 0, \quad \|\phi\|_U \to 0$$

则映射 $f: U \to W$ 在 $u \in U$ 是 Fréchet 可微的。

算子 $A(\cdot)$ 是 f 在 u 的 Fréchet 导数,写作 $A = f'(u)$。

定义 14 (方向导数求导公式) 在上述表达中，我们使用了多种方法来计算方向导数：

$$f'(u)(\phi) = \lim_{h \to 0} \frac{f(u+h\phi) - f(u)}{h}$$
$$= \frac{\mathrm{d}}{\mathrm{d}h} f(u+h\phi) \Big|_{h=0}$$
$$= f(u+\phi) - f(u) - r(u,\phi)$$

例 1 双线性形式 $a(u,\varphi) = (\nabla u, \nabla \varphi)$ 在参数 u 上 Fréchet 可微 (当然在参数 φ 也是，但是 u 是研究参数)

$$a(u+\phi, \varphi) = (\nabla(u+\phi), \nabla\varphi) = (\nabla u, \nabla\varphi) + (\nabla\phi, \nabla\varphi)$$

其中
$$(\nabla u, \nabla\varphi) = a(u,\varphi)$$
$$(\nabla\phi, \nabla\varphi) = a'_u(u,\varphi)(\phi)$$

这里余项为 0，即 $r(u,\phi) = 0$，因为双线性形式在 u 上是线性的，因此 $a(u,\varphi) = (\nabla u, \nabla\varphi)$ 的 Fréchet 导数为 $a'_u(u,\varphi)(\phi) = (\nabla\phi, \nabla\varphi)$。

2.2.5 链式法则

在域 $\mathbb{R}^{d+1}$ 中的链式法则内容如下。

定义 15 令函数 $g: (a,b) \to \mathbb{R}^{d+1}$ 与 $f: \mathbb{R}^{d+1} \to \mathbb{R}$，给定 $h = f \circ g \in \mathbb{R}$，令 $g(t,x) := (t,x) := (t, x_1, x_2, \cdots, x_d)$:

$$\frac{\mathrm{d}}{\mathrm{d}h} h(t,x) = \frac{\mathrm{d}}{\mathrm{d}t} f(g(t,x)) = \frac{\mathrm{d}}{\mathrm{d}t} f(t,x)$$
$$= \sum_{j=0}^{d} \partial_j f(g(x)) \cdot \partial_t g_j$$
$$= \sum_{j=0}^{d} \partial_j f(t, x_1, x_2, \cdots, x_d) \cdot \partial_t x_j, \quad 其中 \ x_0 := t$$
$$= \partial_t f \cdot \partial_t t + \sum_{j=1}^{d} \partial_j f(t, x_1, x_2, \cdots, x_d) \cdot \partial_t x_k$$
$$= \partial_t f + \nabla f \cdot (\partial_t x_1, \cdots, \partial_t x_d)^{\mathrm{T}}$$
$$= \partial_t f + \nabla f \cdot v$$

2.3 原始相场模型公式

当研究包含时间 t 的三维问题时，需要进行四维设置，即 (t,x,y,z)。

备注 11 链式法则的定义请参考文献 [260]，Banach 空间中的链式法则可参考文献 [413, 114, 423]。

定义 16 令 U,V,W 为 Banach 空间。有映射 $F:U\to V$ 与 $G:V\to W$。F 在 $u\in U$ 可微，G 在 $F(u)\in V$ 可微。$H=G\circ F$，$H(u)=G(F(u))$，$H:U\to W$，在 $u\in U$，有

$$H'(u)=G'(F(u))F'(u)$$

具体来说，在方向 $\delta u\in U$ 上，有

$$H'(u)(\delta u)=G'(F(u))F'(u)(\delta u)$$

例 2 给定算子 $T:U\to W$，基于链式法则

$$T(u)=u^3$$

然后有

$$T(u+\phi)=(u+\phi)^3=u^3+3u^2\phi+3u\phi^2+\phi^3$$

由于 $A(\phi)$ 是 ϕ 的线性算子，因此有 $A(\phi)=3u^2\phi$。余项可表示为 $r(u,\phi)=3u\phi^2+\phi^3$。

为了保证 $T(u)=u^3$ 是 Fréchet 可导的，当 $\|\phi\|_U\to 0$ 时，$\dfrac{\|r(u,\phi)\|_W}{\|\phi\|_U}\to 0$。

这里我们需要注意，u 是固定的，我们仅变换 ϕ。尽管我们有 ϕ^2 与 ϕ^3，但 ϕ 为基础，我们只需要确定 $T(u)=u^3$ 是 Fréchet 可导的。

例 3 给定解变量 $v\in V_1$ 与 $u\in V_2$，忽略 v 的控制方程，令 u 由半线性形式确定：

$$A(v,u)(\varphi):=(v\nabla u,\nabla\varphi)+(u^3,\varphi)$$

方向 $(\delta v,\delta u)\in V_1\times V_2$ 上的方向导数为

$$A'(v,u)((\delta v,\delta u),\varphi)=(\delta v\nabla u+v\nabla\delta u,\nabla\varphi)+(3u^2\delta u,\varphi)$$

根据链式法则，当 $v\in V_1$ 已知时，有

$$A'(v,u)((\delta v,\delta u),\varphi)=(v\nabla\delta u,\nabla\varphi)+(3u^2\delta u,\varphi)$$

2.3 原始相场模型公式

本节将初步介绍原始相场模型。该模型可用于模拟准脆性相场裂缝问题。为了求解两个解变量 u 和 φ，我们需要两个平衡方程，并根据弹性力学方程计算位

移参数。为使方程表达尽可能简单,采用拉普拉斯算子表示。令 ∂B 是域 B 的边界,并假设光滑[77]。令 $u := B \to \mathbb{R}$,在边界 ∂B 上有

$$-\nabla \cdot (a\nabla u) = f \tag{2.1}$$

其中系数 $a > 0$,与材料本身性质有关,在相场裂缝模型中,参数 a 取决于位置参数 $x \in B$ 以及相场参数 φ。

相场参数 φ 可用来描述裂缝扩展路径 ($\varphi = 1$ 表示未破碎状态,$\varphi = 0$ 表示破碎状态)。我们拟采用 Ambrosio-Tortorelli 椭圆泛函来处理[70]。其中包括一个反应项,参数 $\varepsilon > 0$,$\varphi : B \to \mathbb{R}$,在边界 ∂B 上有

$$-\varepsilon \Delta \varphi - \frac{1}{\varepsilon}(1 - \varphi) = g \tag{2.2}$$

这里 $g : B \to \mathbb{R}$ 为右手项。公式 (2.2) 在 H^1 范数中有界。

命题 3 已知方程 (2.2) 并根据定理 5 假设其为 H^1 范数,$0 \leqslant \varphi \leqslant 1$,有:

(1) 参数 φ 实际上是一个取值范围在 0 到 1 之间的指示函数;

(2) 梯度 $\nabla \varphi$ 在 $\varphi = 1$ 与 $\varphi = 0$ 之间平滑过渡。当正则化参数 ε 较大时,过渡区域也更平滑 (更大);当 ε 较小时,过渡区域也更小,梯度更大。$\varphi = 1$ 与 $\varphi = 0$ 均为取值极限,此时我们控制 $\varepsilon \to 0$。

证明参考定理 5,图 2.2 为描述性插图。

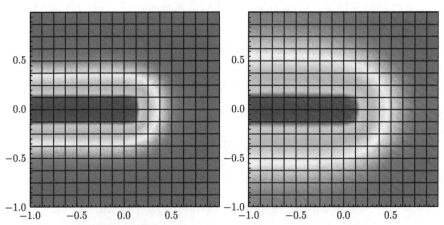

图 2.2 相场裂缝近似模型。左侧正则化参数 $\varepsilon = h$,右侧正则化参数 $\varepsilon = 2h$

此外,根据裂缝扩展的不可逆性 (裂缝无法愈合),我们还需添加一条约束:

$$\partial_t \varphi \leqslant 0 \tag{2.3}$$

2.3 原始相场模型公式

该问题也因此非常类似于障碍模型[140,256,257,412,411]。然而,相场裂缝模型中两个主要的 PDEs 不包括时间导数,仅在约束中含有时间导数,与标准障碍问题存在一定差异。

借助于这些约束,我们可以得到简单相场裂缝模型的公式,这有助于我们理解整个相场裂缝模型的特征与特点。为此,我们首先研究:

公式 1 域 B 边界条件为 $\partial B := \partial\Omega_{\text{HD}} \cup \partial\Omega_{\text{ND}} \cup \partial\Omega_{\text{HN}}$。位移函数为 $u: B \times I \to \mathbb{R}$ 以及相场参数为 $\varphi: B \times I \to [0,1]$。此时有

$$-\nabla \cdot (\varphi^2 \nabla u) = f \quad (\text{位移 } u \text{ 方程,域 } B \times I) \tag{2.4}$$

$$\varphi|\nabla u|^2 - \varepsilon\Delta\varphi - \frac{1}{\varepsilon}(1-\varphi) \leqslant 0 \quad (\text{相场参数 } \varphi \text{ 方程,域 } B \times I) \tag{2.5}$$

$$\partial_t \varphi \leqslant 0 \quad (\text{裂缝不可逆约束,域 } B \times I) \tag{2.6}$$

$$\left[\varphi|\nabla u|^2 - \varepsilon\Delta\varphi - \frac{1}{\varepsilon}(1-\varphi)\right] \cdot \partial_t \varphi = 0 \quad (\text{互补条件,域 } B \times I) \tag{2.7}$$

此外需要补充以下边界条件与初始条件:

$$u(x,t) = u_D(x,t), \quad \text{边界 } \partial\Omega_{\text{ND}} \times I$$

$$u(x,t) = 0, \quad \text{边界 } \partial\Omega_{\text{HD}} \times I$$

$$\varphi^2 \nabla u \cdot n = 0, \quad \text{边界 } \partial\Omega_{\text{HN}} \times I$$

$$\varepsilon \partial_n \varphi = 0, \quad \text{边界 } \partial B \times I$$

$$\varphi(x,0) = \varphi_0, \quad \text{域 } B \times \{0\}$$

φ_0 为初始裂缝,$\varepsilon > 0$ 为相场正则化参数。

备注 12 (耦合项) 根据式 (2.1) 与 (2.2),令 $a = \varphi^2, g = -\varphi|\nabla u|^2$,可得公式 1。

备注 13 PDE 方程 (2.4) 和 (2.5) 与时间无关,但约束条件 (2.6) 与时间相关,该约束条件可以通过后向差商离散为

$$\partial_t \varphi \approx \frac{\varphi^n - \varphi^{n-1}}{k} \leqslant 0$$

其中 $\varphi^n := \varphi(t_n)$,时间步长 $k = t_n - t_{n-1}$,这就要求 $\varphi^n \leqslant \varphi^{n-1}$。关于约束条件的相关研究见第 5 章。

2.4 原始相场模型问题

2.4.1 简单障碍问题

障碍问题 [140,411,256,257,205,412] 与我们之前论述的相场裂缝问题之间有几点相似之处。因此，本节将简要介绍相关问题的模型与结论。

经典泊松问题有以下条件，当 $u: B \to \mathbb{R}$ 时

$$-\Delta u = f, \quad 域 \ B \ 中$$

$$u = 0, \quad 边界 \ \partial B \ 上$$

在障碍问题中 (如图 2.3 所示的晾衣绳问题) 有

$$-\Delta u \geqslant f, \quad 域 \ B \ 中 \ (\text{PDE})$$

$$u \geqslant g, \quad 域 \ B \ 中 \ (不等式约束)$$

$$(f + \Delta u)(u - g) = 0, \quad 域 \ B \ 中 \ (互补条件)$$

$$u = 0, \quad 边界 \ \partial B \ 上 \ (边界条件)$$

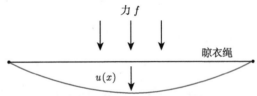

图 2.3 晾衣绳问题 (一维泊松问题)：以力 f 按压晾衣绳，导致产生位移 $u(x)$

障碍问题是一个自由边界问题 (图 2.4)。我们可以将域 B 分割为两部分：N (非活动集)，A (活动集)。

(1) 非活动集 N：若 $f + \Delta u = 0$，则求解 PDE 方程以及不等式约束，此时 $u \geqslant g$，换而言之，障碍条件未触发。

(2) 活动集 A：若 $f + \Delta u < 0$，则触发障碍条件，此时 $u = g$ 恒成立。

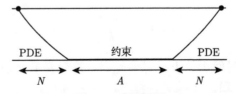

图 2.4 障碍问题

一条晾衣绳在活动集 A 中触地，触发障碍条件。在非活动集 N 中，求解 PDE 方程

2.4.2 能量方程

PDE 方程 $-\nabla \cdot (\varphi^2 \nabla u) = f$ 在齐次 Dirichlet 条件下，变分形式为：给定 $\varphi^2 \in L^\infty(B)$，当 $u \in H_0^1(B)$ 时，对于任意 $w \in H^1(B)$，有

$$(\varphi^2 \nabla u, \nabla w) = (f, w) \tag{2.8}$$

当边界条件为 Dirichlet 形式时，引入

$$E_b'(u,\varphi)(w) := (\varphi^2 \nabla u, \nabla w) - (f,w) = \int_B \varphi^2 \nabla u \nabla w \mathrm{d}x - \int_B f \cdot w \mathrm{d}x$$

然后，给定 $\varphi^2 \in L^\infty$，当 $u \in H_0^1(B)$ 时，有

$$E_b'(u,\varphi)(w) = 0$$

系统能量积分为

$$E_b(u,\varphi) = \frac{1}{2}\int_B \varphi^2 |\nabla u|^2 \mathrm{d}x - \int_B f \cdot u \mathrm{d}x \tag{2.9}$$

由此转化为最小化问题，给定 $\varphi^2 \in L^\infty(B)$，求解

$$\min_{u \in H_0^1(B)} E_b(u,\varphi)$$

考虑到式 (2.5) 中 φ 为整个系统的一个解变量，四阶表达式为

$$\min_{u \in H_0^1(B), \varphi \in K} E_b(u,\varphi)$$

2.5 动态相场裂缝模型

本节将介绍动态裂缝原型模型，在该模型的 PDE 和 VI 中引入时间导数。对位移方程进行了二阶求导，得到了双曲型波动方程。相场不等式问题推广为抛物型问题。这种抛物型相场方程的求解可参考 Allen-Cahn 模型 [12,161,457,41]。

公式 2 令 $B \subset \mathbb{R}^d$ 为开集，$I := (0,T), T > 0$ 为结束时间。当 $U: B \times I \to \mathbb{R}, \varphi: B \times I \to \mathbb{R}$ 时，有

$$\rho_s \partial_t^2 u - \nabla \cdot (\varphi^2 \nabla u) = f, \quad 域 B \times I 中$$

$$\partial_t \varphi + \varphi |\nabla u|^2 - \varepsilon \Delta \varphi - \frac{1}{\varepsilon}(1-\varphi) \leqslant 0, \quad 域 B \times I 中$$

$$\partial_t \varphi \leqslant 0, \quad \text{域 } B \times I \text{ 中}$$

$$\left[\partial_t \varphi + \varphi |\nabla u|^2 - \varepsilon \Delta \varphi - \frac{1}{\varepsilon}(1-\varphi)\right] \cdot \partial_t \varphi = 0, \quad \text{域 } B \times I \text{ 中}$$

$$u = 0, \quad \text{边界 } \partial B \times I \text{ 上}$$

$$\varepsilon \partial_n \varphi = 0, \quad \text{边界 } \partial B \times I \text{ 上}$$

$$u = u_0, \quad B \times \{0\} \text{ 时 (初始时刻)}$$

$$\partial_t u = v_0, \quad B \times \{0\} \text{ 时 (初始时刻)}$$

$$\varphi = \varphi_0, \quad B \times \{0\} \text{ 时 (初始时刻)}$$

备注 14 本书在 5.6 节对动态相场裂缝模型进行了简单介绍。在第 11 章中，主要分析了时间相关的动态大变形相场裂缝模型和流固耦合问题。

第 3 章 偏微分方程及变分不等式的分类

本章我们将对偏微分方程及变分不等式进行分类。

定义 17 微分方程分类主要依照以下特点:

(1) 自变量 $(x, y, z, t, \cdots)$;若只有一个自变量,则为 ODE (常微分方程),否则为 PDE (偏微分方程)。
(2) 时间相关;若自变量中不包括时间,则是稳态问题,否则为瞬态问题。
(3) 物质守恒定律,如质量、动量、扩散、能量等。
(4) PDE 求解以及耦合顺序 (在时间、空间或两者同时)。
(5) 单方程组与偏微分方程组。
(6) 非线性问题中:
(a) PDE 中的非线性通过状态变量;
(b) 非线性本构关系;
(c) 不等式约束。
(7) 耦合 PDE 系统与 CVIS 中:
(a) 模型或材料参数耦合;
(b) 线性或非线性耦合项;
(c) 右手项;
(d) 界面问题。

3.1 线性/非线性偏微分方程

3.1.1 微分方程的一般定义

本书微分方程多采用定义 6 中的符号表示 [156]。

定义 18 令 $\Omega \subset \mathbb{R}^n$ 为一个时空域,其中 n 为包含时间在内的总维度,即 $n = d+1$, $\Omega = B \times I$。令 $k \geqslant 1$ 为微分方程的阶数。隐式微分方程可表示为 $u : \Omega \to \mathbb{R}$

$$F\left(D^k u, D^{k-1} u, \cdots, D^2 u, Du, u, x\right) = 0, \quad x \in \Omega$$

其中

$$F : \mathbb{R}^{n^k} \times \mathbb{R}^{n^{k-1}} \times \cdots \times \mathbb{R}^{n^2} \times \mathbb{R}^n \times \mathbb{R} \times \Omega \to \mathbb{R}$$

定义 19 (PDE 系统求解) 对于 F_i 中的 PDE 系统, 若需求解未知量 $u_i(i = 1, 2, 3, \cdots, s)$, 则需要 PDE 系统中有 s 个方程。

3.1.2 偏微分方程分类

本节我们将介绍线性和非线性偏微分方程的分类。简单地说: 凡是不符合线性偏微分方程的, 都被称为非线性偏微分方程。然而, 非线性偏微分方程可以进一步细化分类。

定义 20 微分方程线性/非线性分类依据如下:

(1) 若微分方程形式符合

$$\sum_{|\alpha| \leqslant k} a_\alpha(x) D^\alpha u - f(x) = 0$$

则称其为线性的。

(2) 若微分方程形式符合

$$\sum_{|\alpha| \leqslant k} a_\alpha(x) D^\alpha u + a_0 \left(D^{k-1}u, \cdots, D^2 u, Du, u, x\right) = 0$$

则称其为半线性的。此类非线性微分方程需满足阶数 $|\alpha| < k$, 当最高阶 $|\alpha| = k$ 时, 为完全线性。

(3) 若微分方程形式符合

$$\sum_{|\alpha| \leqslant k} a_\alpha(x) \left(D^{k-1}u, \cdots, D^2 u, Du, u, x\right) + a_0 \left(D^{k-1}u, \cdots, D^2 u, Du, u, x\right) = 0$$

则称其为拟线性的。此类非线性微分方程需满足阶数 $|\alpha| < k$, 当最高阶 $|\alpha| = k$ 时, 为完全线性。

(4) 当以上条件均不满足时, 则该微分方程可称为完全非线性。

备注 15 方程的非线性程度越高, 分析也就越困难, 越难找到适用的数值方法。

3.1.3 示例

下面我们以一些流体力学中的公式为例。

(1) 欧拉方程 (Navier-Stokes 方程中黏度为 0)。此时 $n = 2$ (空间维度) + 1 (时间维度) = 3。有两个速度分量 (V_x, V_y) 和一个压力量 p。因此, 欧拉方程中的动量部分为

$$\partial_t v + v \cdot \nabla v + \nabla p = f$$

此处最高阶 $k=1$ (在时间变量和空间变量中),在空间导数前我们乘以 0 阶矩阵 v。最终解变量的一个低阶项与最高阶导数相乘,因此该方程为拟线性的。

(2) Navier-Stokes 动量方程

$$\partial_t v - \rho_f v_f \Delta v + v \cdot \nabla v + \nabla p = f$$

此处最高阶 $k=2$,但是最高项前的系数并不依赖 v,因此该方程既不是完全非线性也不是拟线性的。由于存在一阶对流项 $v \cdot \nabla v$,因此该方程原则上是半线性的。但是当 $v_f \to 0$ (高雷诺数) 时,该方程在极限条件下为拟线性的。

(3) 完全非线性方程 (仅作举例):

$$\partial_t v - \rho_f v_f (\Delta v)^2 + v \cdot \nabla v + \nabla p = f$$

3.2 变分方程与变分不等式

3.2.1 变分不等式

令 $(V, \|\cdot\|)$ 为 Hilbert 空间,K 是一个非空、封闭的凸子集。定义映射 $A: K \to V^*$,其中 V^* 是 V 的对偶空间。

公式 3 给定 $u \in K$,有

$$A(u)(\varphi - u) \geqslant 0, \quad \forall \varphi \in K$$

若 K 是 V 的线性子空间,则

$$\varphi := u \pm w \in K$$

对于任意 $w \in K$,由于半线性形式 $A(\cdot)(\cdot)$ 在第二个参数中是线性的,我们有

$$A(u)(w) \geqslant 0 \text{ 且 } A(u)(-w) \geqslant 0$$

因此

$$A(u)(w) = 0, \quad \forall w \in K$$

这表明弱形式包含在变分不等式的一般集合中。

在相关文献以及后续 5.2.3 节与 7.9.5 节中,常用 $\langle \cdot, \cdot \rangle$ 标记描述弱形式和变分不等式。这个符号更好地突出了问题所在的空间,定义如下。

公式 4 有 $u \in K$

$$A(u)(\varphi - u) \geqslant 0, \quad \forall \varphi \in K$$

这里有 $A(u) \in V^*, \varphi - u \in V$。

若 K 是 V 的线性子空间，则

$$\varphi := u \pm w \in K$$

对于任意 $w \in K$，由于半线性形式 $A(\cdot)(\cdot)$ 在第二个参数中是线性的，有

$$\langle A(u), w \rangle \geqslant 0 \text{ 且 } \langle A(u), -w \rangle \geqslant 0$$

因此

$$\langle A(u), (w) \rangle = 0, \quad \forall w \in K$$

3.2.2 函数极小化问题

我们总结了一些关于能量公式的结论[256]。令 $(V, \|\cdot\|)$ 为一个自反的 Hilbert 空间，K 是一个非空、封闭的凸子集，$F: K \to \mathbb{R}$ 是一个定义在 K 上的实函数。我们常采用 F 代表能量函数，即 $F := E$。

公式 5 (最小化问题) 当 $u \in K$ 时

$$F(u) = \inf_{v \in k} F(v)$$

换而言之：

$$F(u) \leqslant F(v), \quad \forall v \in K$$

定义 21 (凸函数) 函数 $F: K \to \mathbb{R}$，当且仅当满足

$$F(\theta u + (1-\theta)v) \leqslant \theta F(u) + (1-\theta)F(v), \quad u, v \in K, \quad 0 \leqslant \theta \leqslant 1$$

时，该函数是凸的。

当符合

$$F(\theta u + (1-\theta)v) < \theta F(u) + (1-\theta)F(v), \quad u, v \in K, \quad 0 \leqslant \theta \leqslant 1$$

时，则是严格凸的。

定义 22 函数 F 的 Gâteaux 导数为

$$\lim_{h \to 0} \frac{\mathrm{d}}{\mathrm{d}h} F(u + hv) = F'(u)(v) = \langle F'(u), v \rangle$$

定理 4 令 K 为赋范线性空间 V 的子集，函数 $F: K \to \mathbb{R}$ 是 Gâteaux 可导的。若 u 是 F 在 K 上的一个极小值，则有

(1) 若 K 是 V 上的非空闭凸子集，则

$$F'(u)(v-u) \geqslant 0, \quad \forall v \in K$$

(2) 若 K 是 V 上的非空闭凸子集且 $u \in \text{int}(K)$，则

$$F'(u)(v) \geqslant 0, \quad \forall v \in K$$

(3) 若 K 是 w 的一个线性簇 (即 K 是一个以参数 w 相对于原点移动的线性子空间)，则

$$F'(u)(v) = 0, \quad \forall v \in K$$

且 $v - w \in K$。

(4) 若 K 是 V 的线性子空间，则

$$F'(u)(v) = 0, \quad \forall v \in K$$

证明详见文献 [256]。

3.3 耦合问题与多物理场偏微分方程

耦合问题是应用数学中的一个重要研究方向。随着计算机技术、数值分析技术的发展，研究人员逐渐在这一领域取得显著进展。本节提供了耦合问题的基本定义，并为第 7 章的求解算法提供了基础。

定义 23 (耦合 PDE 系统) 耦合 PDE 系统或 CVIS 要求至少包括两个 PDE/VI 方程，并通过以下方式耦合：

(1) 模型或材料参数；
(2) 线性或非线性耦合项；
(3) 右手项；
(4) 界面接口。

在求解过程中，一个 PDE 方程的解会影响到另外一个 PDE 方程的求解。

备注 16 前面的定义同样适用于 ODE 系统，然而，在 PDE 系统中，界面耦合条件将非常复杂。

耦合项在设计数值算法的过程中至关重要。通常耦合项会带来非线性因素，增大求解难度。

定义 24 (多物理场 PDE/CVIS 系统) 多物理场问题为耦合 PDE 系统问题，其中至少有两组物理场的 PDE 方程相互作用。如果其中考虑到不等式约束条件，则称其为多物理场 CVIS 问题。

命题 4 我们对以下三种情况进行区分：

(1) 基本耦合问题；

(2) 内部物理耦合问题；

(3) 多物理场 PDE/CVIS 系统。

并有如下的实例：

(1) 相场裂缝问题 (4.4.4 节) 是一个耦合问题，包含一个不等式约束以及一个偏微分方程系统。不属于多物理场 PDE 耦合问题，而是 CVIS 系统耦合问题。

(2) 不可压缩流体 Navier-Stokes 方程 [404,360,198] 是一个内部物理耦合问题 (压力和速度相互作用)，但不是一个多物理场 PDE 方程问题。

(3) 多孔介质 Biot 方程问题 [57,58,117,408] 中，位移和压力相互作用，是一个多物理场问题，因而也是一个耦合问题。

(4) 热孔弹性 [117,418,420] 问题，是热-流-固多场耦合的多物理场问题。

(5) 流固耦合问题 [371,91,184,46] 是一个多物理问题。在这里，速度/压力与固体位移相互作用。

说明：

(1) 在相场裂缝问题中，相场变分不等式没有任何物理意义，仅标志裂缝的位置，在这一问题中唯一的物理参数是位移参数。

(2) 在不可压缩流体 Navier-Stokes 方程问题中，两个物理状态量 (压力与速度) 在方程内耦合，用来描述同一物理现象，即流体流动问题。因此，该问题不是多物理场问题，而是内部物理耦合问题。

(3) 多孔介质 Biot 方程问题中有两个物理量，固体位移与流体流动 (根据达西定律)，参考 3.4 节，是一个线性多物理场问题。

(4) 在热孔弹性问题中，Biot 方程还与温度耦合，整个系统共有三个物理参数互相影响。

(5) 在流固耦合问题中，存在两个物理现象，三个解变量 (p,v,u) 借助 Navier-Stokes 方程相互耦合。

定义 25 (单向耦合与双向耦合) 令问题中的 PDE_1 与 PDE_2 给定，当问题中的参数传输是单向时，即仅有 PDE_1 的解变量进入 PDE_2，或者仅有 PDE_2 的解变量进入 PDE_1，则称该问题为单向耦合问题。若同时存在 PDE_1 的解变量进入 PDE_2，PDE_2 的解变量进入 PDE_1，则为双向耦合问题。

例 4 单向耦合的一个典型例子如下：

$$\partial_t u - \nabla \cdot (\varphi^2 \nabla u) = f(\varphi) \quad (\text{PDE}_1)$$

$$\partial_t \varphi - \Delta \varphi = g \quad (\text{PDE}_2)$$

PDE$_2$ 的解变量进入 PDE$_1$，但反之不成立，这一典型问题可参考 2.3 节。

定义 26 (域耦合与界面耦合) 域耦合：其中一个 PDE 的解作为域变量 (定义在整个域中) 进入另一个 PDE 求解过程中。界面耦合：在界面 Γ 上交换参数消息。

由于界面是一个低维空间，因此界面耦合问题的数学规律性和数值离散性通常较差，需要比域耦合更复杂的求解工具，因此界面耦合问题具有更大的挑战性。

备注 17 (域-界面耦合问题) 这两种耦合类型可以同时出现在一个问题中，例如在任意的拉格朗日-欧拉流固耦合问题[138,428,371,175]（另见第 11 章) 或关于耦合弹性和孔隙弹性的广义 Mandel's 问题[197]（参考 3.4 节) 中。此外，在流固耦合相场裂缝问题中，两种类型的耦合也同时出现了 (参见第 11 章)。

备注 18 (体积耦合) 域耦合也称为体积耦合。

3.3.1 域耦合

前文详细讲解了多种耦合类型，本节我们详细讨论第一类域耦合。

公式 6 令 V_1 与 V_2 是描述 PDE 方程的 Hilbert 空间或 Sobole 空间，并给定两个弱形式：

$$\text{给定 } u_2, \text{ 有 } u_1 \in V_1, \text{ 令 } A_1((u_1, u_2))(\varphi_1) = F_1(\varphi_1), \quad \forall \varphi_1 \in V_1$$

$$\text{给定 } u_1, \text{ 有 } u_2 \in V_2, \text{ 令 } A_2((u_1, u_2))(\varphi_2) = F_2(\varphi_2), \quad \forall \varphi_2 \in V_2$$

这里 A_1 和 A_2 是半线性形式，F_1 和 F_2 是右手项。

定义 27 (模型参数耦合) 控制方程如下：

$$A_1((u_1, u_2))(\varphi_1) = (\mu(u_2)\nabla u_1, \nabla \varphi_1)$$

其中，在 A_2 中的解变量 u_2 作为系数 $\mu(u_2)$ 进入 A_1 中。

详见本书中相场裂缝模型问题或非牛顿流体流动问题[229]。

定义 28 (线性或非线性耦合项耦合) 具体来说

$$A_1((u_1, u_2))(\varphi_1) = (\nabla u_2 + \nabla u_1, \nabla \varphi_1)$$

或

$$A_1((u_1, u_2))(\varphi_1) = (\nabla u_2 \cdot \nabla u_1, \nabla \varphi_1)$$

该定义与定义 27 很相似，不同之处在于 u_2 不通过系数输入，而是直接输入 A_1 中。

例如，多孔介质建模的线性 Biot 方程[58,117,408]（参见 3.4 节) 或者 Stokes 方程问题[404,360,198]（参见 9.3.1 节与第 11 章)。

定义 29 (右手项耦合) 当 $u_1 \in V_1$ 时，有 $A_1((u_1, u_2))(\varphi_1) = F_1(u_2)(\varphi_1)$。这里解变量 u_2 进入右手项 F_1，因此 F_1 依赖于 u_2 的值，并不是严格意义上的右手项。我们需要将其尽可能移到左手项。

当 $u_1 \in V_1$ 时，有 $A_1((u_1, u_2))(\varphi_1) - F_1(u_2)(\varphi_1) = 0$

此时，A_2 的具体形式不重要，整个问题可以简化为之前定义过的线性或非线性域耦合问题。

3.3.2 界面耦合

接下来我们讨论定义 23 中的第 4 种耦合形式，此时系统中会有两个或以上的位于不同子域上的 PDE/CVIS。

定义 30 (界面) 令 $\Omega_1 := \Omega_1(t) \subset \mathbb{R}^d, \Omega_2 := \Omega_2(t) \subset \mathbb{R}^d$ 均为随时间变化的开放域，总域定义为 $B := \Omega_1 \cup \Omega_2$，界面可定义为

$$\Gamma(t) := \bar{\Omega}_1(t) \cap \bar{\Omega}_2(t)$$

我们常使用简化标注 $\Gamma := \Gamma(t)$。此外，我们假设两个子域的边界条件 ∂B 与界面 Γ 均光滑，由此可定义法向量 n。

定义 31 (标量情况下界面耦合) 令 $u_1 : \Omega_1 \to \mathbb{R}, u_2 : \Omega_2 \to \mathbb{R}$，在二阶空间方程中 (如二维泊松问题)，有以下两个界面条件：

$$u_1 = u_2$$

$$\sigma_1(u_1) n = \sigma_2(u_2) n$$

此处 n 为法向量，$n := n_1$ 以及 $n = -n_2$。第二个界面条件是接触力平衡条件，即动力条件，第一个为类 Dirichlet 条件，需要强加在函数空间中。第二个是 Neumann 型条件，在 PDE 方程中通过积分实现。

例 5 动态耦合条件下，有二阶泊松算子 $\nabla \cdot (\alpha \nabla u)$ 与 $\nabla \cdot (\beta \nabla u)$，此时

$$\alpha \nabla u_1 n = \beta \nabla u_2 n$$

定义 32 (向量情况下界面耦合条件) 令 $u_1 : \Omega_1 \to \mathbb{R}^d, u_2 : \Omega_2 \to \mathbb{R}^d$，对于二阶空间方程 (如二维泊松问题)，有以下界面条件：

$$u_1 = u_2$$

$$\sigma_1(u_1) n = \sigma_2(u_2) n$$

此处 $\sigma_1(u_1), \sigma_2(u_2) \in \mathbb{R}^{d \times d}$。

例 6　令 $d=3$，按照分量表示法：

$$u_1 = \left(u_{1,(1)}, u_{1,(2)}, u_{1,(3)}\right)^{\mathrm{T}} \in \mathbb{R}^3, \quad u_2 = \left(u_{2,(1)}, u_{2,(2)}, u_{2,(3)}\right)^{\mathrm{T}} \in \mathbb{R}^3$$

$$\begin{pmatrix} a_{1,(11)} & a_{1,(12)} & a_{1,(13)} \\ a_{1,(21)} & a_{1,(22)} & a_{1,(23)} \\ a_{1,(31)} & a_{1,(32)} & a_{1,(33)} \end{pmatrix} \in \mathbb{R}^{3\times 3}, \quad \begin{pmatrix} a_{2,(11)} & a_{2,(12)} & a_{2,(13)} \\ a_{2,(21)} & a_{2,(22)} & a_{2,(23)} \\ a_{2,(31)} & a_{2,(32)} & a_{2,(33)} \end{pmatrix} \in \mathbb{R}^{3\times 3}$$

$$n = \left(n_{(1)}, n_{(2)}, n_{(3)}\right)^{\mathrm{T}}$$

此处 $\sigma_1(u_1)n, \sigma_2(u_2)n \in \mathbb{R}^3$。

3.3.3　耦合方法

实现数模中的多场耦合时存在多种数值方法[61,87,88,95,115,116,119,139]，本节我们从概念上介绍两种数值方法，并在本书的第 7 章提供具体算法。算法设计时存在两种思路：

(1) 分部耦合 (partitioned approaches)：在给定的 PDE/CVI 之间迭代，同时只求解特定 PDE/CVI 中的实际解变量，并修正所有其他变量。

(2) 全耦合 (monolithic approaches)：对给定的未知数同时求解方程。

第 7 章中图 7.1 提供了两种思路的流程图。

备注 19　当同时考虑两个或以上的 PDE/CVIS 系统时，可以采用两种耦合策略相结合的方式。

接下来我们介绍几个耦合案例。

定义 33 (变分全耦合法)　在变分全耦合法中，耦合条件在弱形式 (变分形式) 中具体体现。

当 $U = (u_1, u_2) \in V_1 \times V_2$ 时，有

$$A_1(U)(\psi_1) = F(\psi_1), \quad \forall \psi_1 \in V_1 \tag{3.1}$$

$$A_2(U)(\psi_2) = F(\psi_2), \quad \forall \psi_2 \in V_2 \tag{3.2}$$

3.3.3.1　分部耦合方案

我们在处理耦合问题时常采用分部耦合法，其主要思路是在每一个时间步内通过不同方程确立解变量，并互相迭代得出最终解。

算法 1　给定初始迭代解 u_1^0 与 u_2^0，对于 $j=1,2,3,\cdots$，进行以下迭代：

$$\text{给定 } u_2^{j-1}, \text{ 求 } u_1^j : A_1\left(u_1^j, u_2^{j-1}\right)(\psi_1) = F(\psi_1)$$

给定 u_1^j,求 $u_2^j : A_2\left(u_1^j, u_2^j\right)(\psi_2) = F(\psi_2)$

迭代停止条件为

$$\max\left(\|u_1^j - u_1^{j-1}\|, \|u_2^j - u_2^{j-1}\|\right) < \text{TOL}$$

若满足迭代停止条件,则停止迭代并输出解,若不满足,则继续迭代。

备注 20 或者可采用基于残差的停止条件:

$$\max\left(\left\|A_1\left(u_1^j, u_2^j\right)(\psi_1) - F(\psi_1)\right\|, \left\|A_2\left(u_1^j, u_2^j\right)(\psi_2) - F(\psi_2)\right\|\right) < \text{TOL}$$

3.3.3.2 变分全耦合方案

在全耦合方案中,我们通过设置好的求解公式一次求解全部解变量。为此,首先要设置共用的函数空间:

$$X = V_1 \times V_2$$

这样我们就可以在一个空间内定义一个包含两个解变量的半线性形式的解。

公式 7 令 $U := (u_1, u_2) \in X$,将式 (3.1) 与 (3.2) 两组方程结合:

$$\text{当 } U \in X \text{ 时, 有 } A(U)(\psi) = F(\psi), \quad \forall \psi \in X$$

这里 $\psi = (\psi_1, \psi_2) \in X$,且

$$A(U)(\psi) := A_1((u_1, u_2), \psi_1) + A_2((u_1, u_2), \psi_2)$$

在实际应用中,我们需要根据实际情况选择耦合方案。当 $A(U)(\psi)$ 可线性直接求解时,可采用直接求解法、共轭梯度法 (CG)、广义最小残差法 (GMRES) 或多重网格法等其他求解方案,当然这需要提前设置好适当的预处理器。当 $A(U)(\psi)$ 是非线性时,可以采用分割迭代、定点迭代或牛顿迭代之类的方法,相关方案可参考本书第 7 章。当然,在数值模型构建过程中,预处理器与高效并行计算方案往往是设计难点。

3.3.3.3 变分全耦合实例

我们耦合了两个椭圆型偏微分方程,其中一个用来描述位移量,另一个用来描述传热问题。

当 $u : \Omega \to \mathbb{R}$ 时

$$-\nabla \cdot (\vartheta \nabla u) = f, \quad \text{在 } \Omega \text{ 中}$$

$$u = 0, \quad \text{在 } \partial\Omega \text{ 上}$$

3.3 耦合问题与多物理场偏微分方程

给定 $\alpha > 0$, 当 $\vartheta : \Omega \to \mathbb{R}$ 时, 有

$$-\nabla \cdot (\alpha \nabla \vartheta) = g(u), \quad \text{在 } \Omega \text{ 中}$$

$$\vartheta = 0, \quad \text{在 } \partial\Omega \text{ 上}$$

根据定义 23, 在求解第一个 PDE 方程时, 我们通过解变量进行非线性耦合, 在第二个 PDE 方程中, 通过右手项耦合。

单个 PDE 系统函数空间为

$$V_1 := H_0^1(\Omega)$$

$$V_2 := H_0^1(\Omega)$$

共用解变量空间为

$$X = V_1 \times V_2$$

我们定义 $U := (u, \vartheta) \in X$, 后续的半线性形式为

$$A(U)(\psi) := (\vartheta \nabla u, \nabla \psi_1) + (\alpha \nabla \vartheta, \nabla \psi_2) - (g(u), \psi_2) \tag{3.3}$$

右手项为

$$F(\psi) := (f, \psi_1) \tag{3.4}$$

由此可得:

公式 8 当 $U \in X$ 时, 有

$$A(U)(\psi) = F(\psi)$$

3.3.3.4 变分界面全耦合方案

界面耦合问题中, 存在两个或多个分布于不同子域的 PDE/CVIS 方程, 本节介绍变分界面全耦合方案。在这里我们首先定义空间:

$$X = \{v \in H_0^1(\Omega) | v_1 = v|_{\Omega_1}, v_2 = v|_{\Omega_2}\}$$

这里函数 v 在整个空间 Ω 上有定义, 令 $U := (u_1, u_2) \in X$ 为解变量集。

命题 5 界面全耦合问题用以下形式表示:

当 $U \in X$ 时,

$$A(U)(\psi) = F(\psi), \quad \forall \psi \in X$$

其中 $\psi = (\psi_1, \psi_2) \in X$, 且

$$A(U)(\psi) := A_1((u_1, u_2), \psi_1) + A_2((u_1, u_2), \psi_2)$$

3.3.3.5 变分界面全耦合实例

令 $u_1 : \Omega_1 \to \mathbb{R}$ 与 $u_2 : \Omega_2 \to \mathbb{R}$，此时泊松问题表达如下：

$$-\nabla \cdot (\alpha \nabla u_1) = f_1, \quad 在 \ \Omega_1 \ 中$$

$$u_1 = 0, \quad 在 \ \partial\Omega_1 \backslash \Gamma \ 上$$

$$-\nabla \cdot (\beta \nabla u_2) = f_2, \quad 在 \ \Omega_2 \ 中$$

$$u_2 = 0, \quad 在 \ \partial\Omega_2 \backslash \Gamma \ 上$$

参考定义 32，有以下两个界面条件：

$$u_1 = u_2$$

$$\alpha \nabla u_1 n = \beta \nabla u_2 n$$

接下来我们回顾推导弱形式的步骤。

步骤 1：定义一个包含边界条件与界面数据的函数空间 X。

步骤 2：将强形式与测试函数相乘并整合。

这里全局函数空间定义为

$$X = \{ v \in H^1(\Omega) \mid v_1 = v_2, \quad 在 \ \Gamma \ 上;$$

$$v_1 = 0, \quad 在 \ \partial\Omega_1 \backslash \Gamma \ 上; \quad v_2 = 0, \ 在 \ \partial\Omega_2 \backslash \Gamma \ 上 \}$$

此处 $\Omega = \Omega_1 \cup \Omega_2$，在函数空间 X 中，函数 v 定义在域 Ω 中

$$v_1 = v|_{\Omega_1}, \quad v_2 = v|_{\Omega_2}$$

同时根据 Dirichlet 边界条件，可得：

命题 6 当 $U = (u_1, u_2)$ 时，有

$$A(U, \psi) := F(\psi), \quad \forall \psi \in X$$

此时 $\psi = (\psi_1, \psi_2) \in X$，且 $F(\psi) = (f_1, \psi_1) + (f_2, \psi_2)$

$$A(U, \psi) := A_1(u_1, \psi_1) + A_2(u_2, \psi_2)$$

其中

$$A_1(u_1, \psi_1) = (\alpha \nabla u_1, \nabla \psi_1), \quad A_2(u_2, \psi_2) = (\beta \nabla u_2, \nabla \psi_2)$$

3.4 弹性力学 Biot 方程耦合问题

本节将介绍一个简单案例，用以帮助读者理解相关概念。该问题在多孔介质问题中有着广泛应用。该问题有两个特点：第一，模型详细的边界条件未知；第二，该模型需要在现场实现大规模运用。

首先，我们根据 Biot 方程完成多孔弹性介质建模的工作[292]，将周边介质视为静态弹性固体[197]。实际上 Biot 方程问题[57,58,408]是一个均质化的问题，在微观尺度上需要考虑流固耦合[315]。通过均质化设置，我们在宏观尺度上导出 Biot 系统。关于该问题最新的算法与研究进展可参考文献 [314, 45, 233, 66, 103]。本节向读者介绍了一个变分全耦合框架。

3.4.1 符号与方程

首先需要描述一个纯孔隙弹性的空间，其中 Ω_B 是研究区域，$\partial\Omega_B$ 是它的边界。

$$\partial\Omega_B = \Gamma_u \cap \Gamma_t = \Gamma_p \cup \Gamma_f$$

其中 Γ_u 代表位移边界 (Dirichlet)，Γ_t 代表应力边界 (Neumann)，Γ_p 代表孔压边界 (Dirichlet)，Γ_f 代表流量边界 (Neumann)。对于位移变量 u 与柯西应力张量 σ 有

$$u = \bar{u}, \quad 在 \Gamma_u 上$$

$$\sigma(u)n = \bar{t}, \quad 在 \Gamma_t 上$$

给定 $\bar{u}$ 与 $\bar{t}$，法向量 n，渗透率 K，流体黏度 η_f，存在以下平衡条件：

$$p = \bar{p}, \quad 在 \Gamma_p 上$$

$$-\frac{K}{\eta_f}(\nabla p - \rho_f g) \cdot n = \bar{q}, \quad 在 \Gamma_f 上$$

给定 $\bar{p}$ 与 $\bar{q}$，密度 ρ_f，重力加速度 g。对于初始时间 $\tau = 0$ 时，我们规定：

$$p(\tau = 0) = p_0, \quad 在 \Omega_B 中$$

$$\sigma(u)(\tau = 0) = \sigma_0, \quad 在 \Omega_B 中$$

对以上方程进行扩展 (将岩石作为弹性介质考虑)，如图 1.4 和图 3.1 (右子图)。此时 $\partial\Omega_B := \Gamma_i = \bar{\Omega}_B \cap \bar{\Omega}_S$

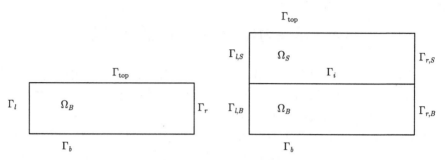

图 3.1 Mandel 问题模型
Ω_B 为 Biot 方程域，Ω_S 为线弹性域

3.4.2 控制方程

系统控制方程如下：压力 $p_B : \Omega \times I \to \mathbb{R}$，位移 $u_B : \Omega \times I \to \mathbb{R}^d$。初始状态下 $p_B(0) = p_B^0, \mathrm{div} u_B(0) = \mathrm{div} u_B^0$。有

$$\partial_t \left(c_B p_B + \alpha_B \mathrm{div} u_B \right) - \frac{1}{\eta_f} \mathrm{div} K \left(\nabla p_B - \rho_f g \right) = q, \quad 在 \Omega_B \times I 中$$

$$-\mathrm{div}\left(\sigma_B(u)\right) + \alpha_B \nabla p_B = f_B, \quad 在 \Omega_B \times I 中$$

其中

$$\sigma_B(u_B) := \mu_B \left(\nabla u_B + \nabla u_B^{\mathrm{T}} \right) + \lambda_B \mathrm{div} u_B I$$

其中压缩系数 $c_B \geqslant 0$，Biot-Willis 常数 $\alpha_B \in [0,1]$，渗透率 K，流体黏度与密度分别为 η_f 与 ρ_f，重力加速度 g，源项 q (参考采油井与注水井等实际情况设置)，Lamé 系数 $\lambda_B > -\frac{2}{3}\mu_B$，并且 $\mu_B > 0$，f_B 为体力。

根据达西定律可获得流速 v_B[121]:

$$v_B = -\frac{1}{\eta_f} K \left(\nabla p_B - \rho_f g \right)$$

假设其为线弹性模型，位移 $u_S : \Omega_S \times I \to \mathbb{R}^d$

$$-\mathrm{div}\left(\sigma_S(u_S)\right) = f_S, \quad 在 \Omega_S \times I 中$$

此时：

$$\sigma_S(u_S) := \mu_S \left(\nabla u_S + \nabla u_S^{\mathrm{T}} \right) + \lambda_B \mathrm{div} u_S I$$

λ_S 与 u_S 为 Lamé 系数，f_S 为体力。

3.4.2.1 界面条件

在两个子系统之间存在界面 Γ_i。系统中有两个子域，其中一个子域中有两个 PDE 方程 (位移与压力)，在另一个子域中只有一个压力方程。因此，我们得到了界面 Γ_i 上的三个耦合条件：

$$u_B = u_S$$
$$\sigma_B(u_B)n_B - \sigma_S(u_S)n_S = \alpha p_B n_B \qquad (3.5)$$
$$-\frac{1}{\eta_f}K(\nabla p_B - \rho_f g)\cdot n_S = 0$$

3.4.3 典型示例

模型配置与参数设置可参考文献 [183, 207]，模型域为 $(0, 100\text{m}) \times (0, 20\text{m})$，边界条件为

$$\partial_B n = \bar{t} - \alpha_B p_B n, \quad 在 \ \Gamma_{\text{top}} \ 上$$

$$p = 0, \quad 在 \ \Gamma_r \ 上$$

$$u_y = 0, \quad 在 \ \Gamma_b \ 上$$

$$u_x = 0, \quad 在 \ \Gamma_l \ 上$$

对应参数可参考文献：$M_B = 2.5 \times 10^{12}\text{Pa}$；$c_B = \dfrac{1}{M_B}\text{Pa}^{-1}$；$\alpha_B = 1.0$；$v_F = 1.0 \times 10^{-3}\text{m}^2/\text{s}$；$K_B = 100\text{md}$；$\rho_F = 1.0\text{kg/m}^3$；$\bar{t} = F = 1.0 \times 10^7\text{Pa}\cdot\text{m}$；$\mu_S = 10^8\text{Pa}$；$v_S = 0.2$。时间步长 $k = 1000\text{s}$，总时间为 $5 \times 10^6\text{s}$。

3.4.4 变分公式

变分全耦合系统包含以下方面：

(1) 域耦合：解变量作为线性项进入另一个方程。需要耦合这两个方程产生一个线性方程。

(2) 界面耦合：孔隙弹性模型和线性化弹性问题之间存在界面 Γ_i 耦合。

参考 3.4.3 节，我们首先定义函数空间：

$$V_P := \left\{p \in H^1(\Omega_B) \mid p = 0, \quad \Gamma_r \ 上\right\}$$

$$V_S := \left\{u \in H^1(\Omega_B) \mid u_y = 0, \quad 在 \ \Gamma_b \ 上; \quad u_x = 0, \ 在 \ \Gamma_l \ 上\right\}$$

整个系统中的变分弱形式如下。

公式 9 有 $(p,u) \in V_P \times V_S$,其中 $p(0) = p^0$,此时有

$$c_B(\partial_t p, \phi^p) + \alpha_B(\nabla \cdot u, \phi^p) + \frac{K}{v_F}(\nabla p, \nabla \phi^p) - \rho_F(g, \phi^p) - (q, \phi^p) = 0, \quad \forall \phi^p \in V_P$$

$$(\sigma_B, \nabla \phi^u) - \alpha_B(pI, \nabla \phi^u) - \int_{\Gamma_{\text{top}}} (\bar{t}_S - \alpha pn)\phi^u \mathrm{d}s - (f_B, \phi^u) = 0, \quad \forall \phi^u \in V_S$$

在这里我们需要引入一个弹性方程,即

$$V_S := \{u \in H^1(\Omega_B \cup \Omega_S) \mid u_y = 0, \quad \text{在 } \Gamma_b \text{ 上}; \quad u_x = 0, \quad \text{在 } \Gamma_{l,B} \cup \Gamma_{l,S} \text{ 上}\}$$

其中隐含动力学条件,在 Γ_i 上,$u_B = u_S$。

公式 10 有 $(p,u) \in V_P \times V_S$,其中 $p(0) = p^0$,此时有

$$c_B(\partial_t p, \phi^p) + \alpha_B(\nabla \cdot u, \phi^p) + \frac{K}{v_F}(\nabla p, \nabla \phi^p) - \rho_F(g, \phi^p) - (q, \phi^p) = 0, \quad \forall \phi^p \in V_P$$

$$(\sigma_B, \nabla \phi^u) - \alpha_B(pI, \nabla \phi^u) - (f_B, \phi^u) = 0, \quad \forall \phi^u \in V_S(\Omega_B)$$

$$(\sigma_S, \nabla \phi^u) - \int_{\Gamma_{\text{top}}} (\bar{t}_S)\phi^u \mathrm{d}s - (f_S, \phi^u) = 0, \quad \forall \phi^u \in V_S(\Omega_S)$$

3.4.5 离散,建模,结果分析

前文算例可参考文献 [203,134],以及 www.dopelib.net 中的相关案例 (目录为:dopelib/Examples/PDE/InstatPDE/Example6)。读者也可参考第 13 章中的相关软件。时间 $N = 100$ 时所得解如图 3.2 所示。此外,我们注意到压力随时间变化的过程中存在 Mandel-Cryer 效应 [292,203]。

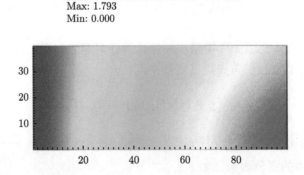

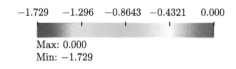
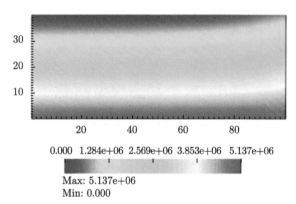
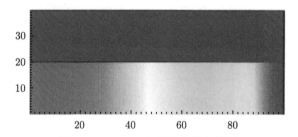

图 3.2 Mandel 问题数值模拟结果

u_x 与 u_y 代表位移。下面是压力值分布图。两个子域的位移场都是连续的。压力在界面 Γ_i 处是不连续的

3.5 原型准静态模型问题

在介绍了 PDE/CVIS 分类后,需要解决从相关公式到准静态模型建模的问题:

(1) 确定 $d+1$ 个自变量。

(2) 需要处理一个不包括时间导数的准静态问题,其中包括位移 PDE 方程与相场变分不等式。加载过程通过变量实现。在热力学背景下,每一个时间步均处于平衡状态。

(3) PDE 方程为椭圆型,整个过程满足能量守恒。相场变分不等式是不可逆的。

(4) 整个系统处于二维空间内部。

(5) 整个模型中包括 $d+1$ 个方程: d 个固体方程与 1 个相场变量方程。

(6) 模型非线性问题:

(a) 该耦合问题是非线性的，其中第一个方程是准线性的，第二个方程是半线性的。
　　(b) 该模型的本构关系是线性的。
　　(c) 函数集 K 并不是向量空间，因为我们还需要处理一个变分不等式。
　　(7) 即使假设给定变量后位移方程变为线性，但由于包含不等式约束，整个系统仍然是非线性的。

3.6　原型动态模型问题

在搭建动态相场裂缝模型时，我们需解决以下问题：
　　(1) 确定 $d+1$ 个自变量。
　　(2) 位移问题与相场问题在这里均是时间相关的。
　　(3) PDE 方程为双曲型，整个过程满足能量守恒。相场变分不等式是不可逆的。
　　(4) 位移 PDE 方程在时间上与空间上均是二阶的。相场变分不等式在时间上是一阶的，在空间上是二阶的。
　　(5) 整个模型中包括 $d+1$ 个方程：d 个固体方程与 1 个相场变量方程。
　　(6) 非线性问题：
　　(a) 该问题是准线性的，相场参数 φ 以系数的形式进入位移方程的高阶算子中。
　　(b) 该模型的本构关系是线性的。
　　(c) 函数集 K 并不是向量空间，因为我们还需要处理一个变分不等式。
　　(7) 即使假设给定变量后位移方程变为线性，但由于包含不等式约束，整个系统仍然是非线性的。

第 4 章 相场裂缝模型

本章主要研究脆性材料准静态变分相场裂缝模型。基础研究工作可参考文献 [171, 70, 71, 312]。

4.1 连续介质力学介绍与本构关系

本节首先介绍所用到的连续介质力学部分内容。

定义 34 (位移) 令 $B \subset \mathbb{R}^d, I \subset \mathbb{R}$，位移变量表示为 $u := u(x,t) : B \times I \to \mathbb{R}^d$，即

$$u := u(x,t) = (u_1(x,t), \cdots, u_d(x,t)) : B \times I \to \mathbb{R}^d$$

备注 21 在相场模型中，通常研究拉格朗日坐标系和欧拉坐标系下的无穷小位移 (见 11.6.1 节命题 108)，因此不需要考虑坐标系统之间的转换。当然其他文献 [435, 437] 中考虑了流固耦合下的固体损伤，因此需要对拉格朗日坐标系与欧拉坐标系进行区分转换。

定义 35 线性应变张量表示为

$$e = e(u) = \frac{1}{2}\left(\nabla u + \nabla u^{\mathrm{T}}\right)$$

例 7 (三维空间中的位移梯度) 在 $\mathbb{R}^3$ 中，位移梯度是一个 $\mathbb{R}^{3 \times 3}$ 矩阵，表示为

$$\nabla u = \begin{pmatrix} \partial_x u_x & \partial_y u_x & \partial_z u_x \\ \partial_x u_y & \partial_y u_y & \partial_z u_y \\ \partial_x u_z & \partial_y u_z & \partial_z u_z \end{pmatrix}$$

定义 36 (线性应力应变关系) 线性应力应变关系 [112,360] 可表示为

$$\sigma := \sigma(u) = Ce(u)$$

其中张量 $C = (C_{ijkl})_{i,j,k,l=1}^{3}$ 是一个正定的四阶张量，即存在一个正常数 $\alpha > 0$，使得对于任意对称矩阵 ξ 以及任意一点 $x \in B$ 均有

$$\alpha |\xi|^2 \leqslant C(x)\xi : \xi = \sum_{i,j,k,l=1}^{d} C_{ijkl} \xi_{ij} \xi_{kl}$$

基于各向同性，均匀对称的假设条件[360]可简化为

$$\sigma = Ce(u) = 2\mu e(u) + \lambda \mathrm{tr}(e(u))I \tag{4.1}$$

此处 I 为单位张量，$\mathrm{tr}(\cdot)$ 为迹。

定义 37 $W(e(u))$ 表示能量密度函数。

例 8 这里我们给出两个弹性能密度函数的实例。

(1) 在 Laplacian 方程或 Poisson 方程中，能量密度函数为

$$W(e(u)) = \frac{\mu}{2}|\nabla u|^2$$

此处 $\mu > 0$, 是材料参数，其能量定义为

$$E_b(u) = \frac{1}{2}\int_B \mu |\nabla u|^2 \mathrm{d}x - \int_B f(u)\mathrm{d}x \tag{4.2}$$

PDE 方程为

$$E_b'(u)(\psi) = \frac{1}{2}\int_B \mu \nabla u \cdot \nabla \psi \mathrm{d}x - \int_B f\psi \mathrm{d}x \tag{4.3}$$

(2) 对于一般弹性模型，我们有

$$W(e(u)) = Ce(u) : e(u)$$

这里用 Frobenius 标量积计算能量密度，具体细节可参考文献 [232]。

定义 38 在工程实践中，常采用杨氏模量 E_Y，泊松比 ν_s 来定义材料性质，拉梅系数 μ_s 与 λ_s 关系如下

$$\nu_s = \frac{\lambda_s}{2(\lambda_s + \mu_s)}, \quad E_Y = \frac{\mu_s(\lambda_s + 2\mu_s)}{\lambda_s + \mu_s}$$

$$\mu_s = \frac{E_Y}{2(1+\nu_s)}, \quad \lambda_s = \frac{\nu_s E_Y}{(1+\nu_s)(1-2\nu_s)}$$

备注 22 根据以上公式可知

$$-1 \leqslant \nu_s \leqslant 0.5, \quad E_Y, \mu_s > 0, \quad \lambda_s > -\frac{2}{3}\mu_s \tag{4.4}$$

根据泊松比的定义可知，其表示横向正应变与轴向正应变的绝对值的比值。因此当 $\nu_s \to 0.5$ 时，材料变为不可压缩材料。对于绝大多数材料，$0.5 \geqslant \nu_s > 0$。实际工程中某些聚合物材料的泊松比为负值，代表着拉伸时这些材料会变厚。我们有时会采用体积模量表示材料参数：$k = \lambda_s + \frac{2}{3}\mu_s$。

定义 39 给定参数 $\kappa > 0$, 采用退化函数来表示相场函数 $\varphi : B \times I \to [0,1]$。

$$g(\varphi) := (1-\kappa)\varphi^2 + \kappa$$

4.2 简单脆性断裂建模

临界能量释放率 G_c 是一个与断裂韧性相关的物理量[455],是相场裂缝模型中的重要参数。

定义 40 (临界能量释放率) 临界能量释放率 G_c 是用来描述材料断裂强度的一个物理量,其物理含义为在某一点 $x \in B$ 上产生一个无限小的裂纹所需的能量,其中 B 代表完好材料区域。

定义 41 (当前能量释放率) 当前能量释放率用 G 表示,定义为

$$G := G(x) = -\frac{\partial(U-V)}{\partial A}, \quad x \in B \tag{4.5}$$

此处 U 为模型 $B \subset \mathbb{R}^d$ 中可用于形成裂缝的全部能量,V 为外力所做的功,A 为模型区域。当前能量释放率的单位为 J/m^2。

裂缝扩展模型中存在以下三种情况:
(1) $G < G_c$: 此时当前能量释放率小于临界能量释放率,裂缝不扩展。
(2) $G = G_c$: 此时当前能量释放率等于临界能量释放率,裂缝稳定扩展。
(3) $G > G_c$: 此时当前能量释放率大于临界能量释放率,裂缝扩展不稳定。

4.3 Griffith 模型

根据格里菲斯 (Griffith) 准则,当单位表面上当前弹性能 G 等于 (或大于) 临界能量释放率 G_c 时,裂缝扩展。表面能这个物理量是微观层面上晶格脱落现象的宏观描述。因此,裂缝扩展问题其实就是模型内部的能量问题。

命题 7 Griffith 理论假设:
(1) 表面能与变形不连续的物体表面有关。
(2) 这些表面当中存在裂缝扩展准则。
(3) 裂缝扩展过程中不可逆。

根据 Griffith 的假设,断裂的过程是准静态的,该破裂准则适用于玻璃、混凝土以及钢铁等相关材料。

定义 42 (裂缝) 在相场模型中,裂缝定义如下 (图 4.1):

$$C(l) = C_0 \cup \{x(s); \quad 0 \leqslant s \leqslant 1\}$$

缝长等相关参数可通过积分计算该充分光滑的曲线长度获得 $L_0 = \int_{x_0}^{x_1} \mathrm{d}s$。

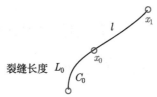

图 4.1 相场裂缝示意图
预制裂缝 C_0，新裂缝 $x(s)$，裂缝整体为 $C(l)$

定义 43 (体积势能与表面能) 我们分别定义了表面能与体积势能：
体积势能 (potential energy)：$P(t,l) := E_b\left(C(l), u_0\right)$ 单调递减；
表面能 (surface energy)：$Q(t,l) := E_s(C(l)) - E_s(C_0)$ 单调递增。
其中，u_0 是初始给定位移场。我们将研究随时间变化的裂缝变化情况：

$$t \to l(t)$$

这里 t 为时间步或加载步。当前能量释放率定义为

$$G := -\frac{\partial P}{\partial l}$$

命题 8 (Griffith 准则) 令 $l(t)$ 在时间 t 上连续，则裂缝扩展的 Griffith 准则可数学表示为：
(1) $l(0) = 0$ (初始裂缝)。
(2) $\partial_t l(t) \geqslant 0$ (裂缝扩展不可逆)。
(3) $G \leqslant G_c$ (能量释放率以临界能量释放率为界)。
(4) $(-G + G_c)\,\partial_t l = 0$ (裂缝只有在能量释放率达到临界时才会扩展)。

4.4 Francfort-Marigo 脆性断裂变分模型

1998 年，Francfort 与 Marigo 率先提出了一个基于 Griffith 准则的裂缝模型[171]。本节将对该模型进行简要介绍。

4.4.1 表面能与体积势能

接下来我们重点介绍裂缝的表面能 E_s。该物理量与临界能量释放率 G_c 有关。

定义 44 (表面能) 裂缝 $C \subset B$ 的表面能 E_s 定义为

$$E_s(C) = \int_c G_c(s)\mathrm{d}s = G_c H^{d-1}(C)$$

4.4 Francfort-Marigo 脆性断裂变分模型

这里 H 表示 Hausdorff 测度[213]。在二维 (2D) 空间的光滑表面中, Hausdorff 测度是裂缝 C 的长度; 三维 (3D) 空间中 Hausdorff 测度是裂缝 C 的面积。

备注 23 表面能 $E_s(C)$ 的值越高, 则表明越多的材料断裂。$G_c > 0$ 的值越大, 则表明起裂所需的能量也就越高。当 $G_c = \infty$ 时, 材料无法产生断裂。

定义 45 (体积势能) 体积势能定义为

$$E_b(C, u) = \int_\Omega W(e(u)) \mathrm{d}x - \int_\Omega f u \mathrm{d}x$$

在裂缝问题中, 通常不考虑体力 (如重力等) 的变化, 即 $f \equiv 0$。此时有

$$E_b(C, u) = \int_\Omega W(e(u)) \mathrm{d}x$$

命题 9 表面能与体积势能应满足以下特征:
(1) E_s 在裂缝 C 中是单调递增的。
(2) E_b 在裂缝 C 中任意 $u \in V_u^D$ 处都是单调递减的。

命题 10 (系统总能量) 给定裂缝 C, u_D 是边界 $\partial\Omega_D$ 上的给定载荷。总能量 $E_T(C, u)$ 定义为

$$E_T(C, u) = E_b(C, u) + E_s(C)$$

此时裂缝扩展问题转变为能量最小化问题:

$$\min_{u \in V_u^D} E_T(C, u) \tag{4.6}$$

总能量 $E_T(C, u)$ 是非线性的, 相关细节可参考文献 [71]。

4.4.2 模型准则

命题 11 (Francfort 和 Marigo 准则) 令 $u(x, t)$ 是一个 $\partial\Omega_D$ 上单调递增的载荷, $u(x, t) = t\bar{u}(x)$, 给定 $\bar{u}$, 对于 $t \geqslant 0$ 时, 应满足:
(1) 不可逆准则: $C(t)$ 是一个随时间 t 单调递增的函数, 其中 $C(0) = C_0$, 为初始时刻裂缝。
(2) 能量最小化原理: 对于 $t \in I, \tau \in I$, 有

$$E_T(C(t), u(t)) \leqslant E_T(C, u(t)), \quad \forall U_{\tau<t} C(\tau) \subset C$$

(3) 解变量约束:

$$E_T(C(t), u(t)) \leqslant E_T(C(\tau), u(t)), \quad \forall \tau < t$$

最后的解变量约束用以减少过多断裂点的情况，相关细节请参考文献 [171]。

基于以上准则，能量最小化的不可逆准静态裂缝扩展问题可以定义为一个满足以下三个条件的映射 $t \to (u(t), C(t))$，该映射应满足：

(1) 全局稳定条件：对于任意时刻 $t \in I$，解集 $(u(t), C(t))$ 应符合最小能量准则。

(2) 不可逆条件：当 $0 \leqslant \tau \leqslant t \leqslant T$ 时，$C(\tau) \subset C(t)$。

(3) 能量平衡条件：体积势能与裂缝表面能的增加等于外力所做的功 (热力学第一定律)。

备注 24　随着 $t \to \infty$，载荷单调增加，试样出现机械破坏 (即完全开裂)，热力学解释请参考文献 [308]。

备注 25 (严格熵增)　熵表征能量转移的方向，一般来说，机械能可以耗散成其他形式的能量。例如，在本书讨论的裂缝扩展问题中，机械能转化为裂缝表面能，同时裂缝一旦产生就不会愈合。从物理学角度来讲，当严格意义上的熵增出现时，裂缝表面不能再转化为机械能。这一原理也可用生活中的例子来说明。当你拿剪刀剪开纸张时，纸张上的裂口是无法修复的。

4.4.3　系统总能量的具体表现形式

命题 10 中系统总能量计算公式如下：

(1) 给定位移 $u : \Omega \to \mathbb{R}$ 为标量函数，系统总能量为

$$E_T(C, u) = \frac{1}{2} \int_\Omega |\nabla u|^2 \mathrm{d}x + \int_c G_c(x) \mathrm{d}s$$

(2) 给定位移 $u : \Omega \to \mathbb{R}^d$ 为向量值函数，系统总能量为

$$E_T(C, u) = \frac{1}{2} \int_\Omega \sigma(u) : e(u) \mathrm{d}x + \int_c G_c(x) \mathrm{d}s$$

4.4.3.1　裂缝表面积分的计算方法

由于裂缝在模型中是一个界面 (1.2 节)，因此 E_b 定义在 $\Omega \to \mathbb{R}^d$ 空间上，而 E_s 定义在 $\Omega \to \mathbb{R}^{d-1}$ 空间上，这就导致了数值离散与建模过程中的一系列问题。广义有限元法[323,181,305,32] 或边界元法[108,341,377,299] 中将裂缝界面处理为网格线条或者网格区域。在相场模型中，我们通过对整个域 B 进行积分获得 E_s 的近似值。

4.4.3.2　表面能计算：椭圆泛函的界面处理

相场裂缝问题是自由间断问题中的一种，在这里，我们假设未知数 (u, Γ)，其中 Γ 为在一个固定开集 $B \subset \mathbb{R}^d$ 的闭子集中的变量。然后计算 $E_T(C, u)$ 的最小值，从而获得 $u : B \backslash \Gamma \to \mathbb{R}^{d[195]}$。

4.4 Francfort-Marigo 脆性断裂变分模型

使用椭圆泛函逼近求解相关方程的思想可以追溯到 Modica 和 Mortola [322]。他们采用这样的方法近似求解了周长函数。Ambrosio 与 Tortorelli[20,21] 介绍了求解变量 φ 的近似过程。除了裂缝问题，该椭圆泛函近似方法也常用于图像分割领域 [325,105,107,79] 与材料科学领域 [350] 中。

命题 12 (椭圆泛函逼近)　为了近似求解界面 C 上的表面能，提出以下椭圆泛函：

$$\frac{1}{2}\int_B G_c\left(\frac{1}{\varepsilon}(1-\varphi)^2 + \varepsilon|\nabla\varphi|^2\right)\mathrm{d}x \to \int_c G_c \mathrm{d}s,\quad \varepsilon \to 0$$

在式 $\frac{1}{2}\int_B \left(\frac{1}{\varepsilon}(1-\varphi)^2 + \varepsilon|\nabla\varphi|^2\right)\mathrm{d}x$ 中，φ 的值决定了这一点是基质还是裂缝。实际上由于 $\varepsilon \to 0$，第一项占主导地位，而 φ 则会在 0 与 1 之间迅速切换。为了研究裂缝扩展的规律，我们引入了补偿项使 0 到 1 的转化更为平滑。根据上式可知，ε 的值越大，过渡也就越平滑，相关证明可参考 5.5.2 节与文献 [331,170]。

命题 13 (正则化相场参数)　非正则化能量泛函如下：

$$E_T(u,C) = \frac{1}{2}\int_\Omega |\nabla u|^2 \mathrm{d}x + G_c H^{d-1}(C)$$

在这里我们利用 Ambrosio-Tortorelli 椭圆泛函，结合定义 39 将其正则化

$$E_{T,\varepsilon}(u,\varphi) = \frac{1}{2}\int_B g(\varphi)|\nabla u|^2 \mathrm{d}x + \frac{1}{2}\int_B G_c\left(\frac{1}{\varepsilon}(1-\varphi)^2 + \varepsilon|\nabla\varphi|^2\right)\mathrm{d}x$$

其中 $u \in H^1(B)$，$u\varphi \in H^1(B)$ 且 $0 \leqslant \varphi \leqslant 1$。

备注 26 (正则化参数 κ)　当采用椭圆泛函近似时，需要引入退化函数 $g(\varphi)$ 进行修正。在这个过程中我们采用正则化参数 κ 将体积势能从 Ω 域扩展到 B。该参数是用来在 $\varphi \to 0$ 时使离散矩阵在数值处理中保持正则化。

当 $\varphi = 0$ 时，

$$\left[(1-\kappa)\varphi^2 + \kappa\right]|\nabla u|^2 \to \kappa|\nabla u|^2$$

在数值方面，$\kappa = 0$ 时，$\phi^2|\nabla u|^2$ 同样有效 [78,21,22]。

命题 14　当 $\varepsilon \to 0$ 时，有

$$E_{T,\varepsilon}(u,\varphi) = \begin{cases} E_T(u,C), & \varphi = 1 \\ +\infty, & \varphi = 0 \end{cases}$$

相关证明请参考文献 [70] 与 2.1 节。

命题 15 (正则化线弹性相场公式) 正则化能量函数可写作：

$$E_{T,\varepsilon}(u,\varphi) = \frac{1}{2}\int_B g(\varphi)\sigma(u):e(u)\mathrm{d}x + \frac{1}{2}\int_B G_c\left(\frac{1}{\varepsilon}(1-\varphi)^2 + \varepsilon|\nabla\varphi|^2\right)\mathrm{d}x$$

其中 $\tau \in H^1(B)^d, \varphi \in H^1(B)^d$ 且 $0 \leqslant \varphi \leqslant 1$。

4.4.3.3 无不等式约束的 Euler-Lagrange 方程

为了求解 $E_{T,\varepsilon}(u,\varphi)$ 的最小值，我们对公式进行变分处理，首先从能量最小化开始：

$$\min_{u \in V_u^D, \varphi \in K} E_{T,\varepsilon}(u,\varphi)$$

我们通过微分处理得到一阶变分形式：

$$E'_{T,\varepsilon}(u,\varphi)(w, \psi - \varphi) \geqslant 0$$

此时我们引入 φ 参数，根据不可逆约束，$0 \leqslant \varphi^n \leqslant \varphi^{n-1} \leqslant 1$。因此，假定每个时间步中

$$E'_{T,\varepsilon}(u,\varphi)(w,\psi) = 0$$

这样我们有：

引理 1 $(u,\varphi) \in V_u^D \times H^1(B)$ 变形如下：

$$E'_{T,\varepsilon,u}(u,\varphi)(w) = \int_B g(\varphi)\nabla u \cdot \nabla w \mathrm{d}x$$

$$E'_{T,\varepsilon,\varphi}(u,\varphi)(\psi) = \int_B (1-\kappa)\varphi|\nabla u|^2 \psi \mathrm{d}x + \int_B G_c\left(\frac{1}{\varepsilon}(1-\varphi)(-\psi) + \varepsilon \nabla\varphi \cdot \nabla\psi\right)\mathrm{d}x$$

证明 根据 2.2.4 节中的微分法则。以链式法则求解 (2.2.5 节) $f(u) := |\nabla u|^2$ 在方向 w 上的方向导数：

$$f'(u)(w) = \lim_{s \to 0}\frac{f(u+sw)-f(u)}{s} = \left[|\nabla u|^2\right]'(w) = \left(\sum_{i=1}^d (\partial_i u)^2\right)'(w)$$

$$= 2\sum_{i=1}^d (\partial_i u)(\partial_i w)$$

$$= 2\nabla u \cdot \nabla w$$

4.4 Francfort-Marigo 脆性断裂变分模型

对于能量泛函本身，有

$$E'_{T,\varepsilon,u}(u,\varphi)(w) = \lim_{s\to 0}\frac{E_{T,\varepsilon}(u+sw,\varphi)-E_{T,\varepsilon}(u,\varphi)}{s}$$

固定 φ，其中，

$$\lim_{s\to 0}\frac{1}{2}\int_B g(\varphi)s|\nabla w|^2 \mathrm{d}x = 0$$

则有

$$\lim_{s\to 0}\frac{E_{T,\varepsilon}(u+sw,\varphi)-E_{T,s}(u,\varphi)}{s} = \int_B g(\varphi)\nabla u\cdot\nabla w \mathrm{d}x$$

对于 φ 的方向导数，当固定 u 时，有

$$g'(\varphi)(\psi) = 2(1-\kappa)\varphi\psi$$

因此，

$$E'_{T,\varepsilon,\varphi}(u,\varphi)(\psi) = \lim_{s\to 0}\frac{E_{T,\varepsilon}(u,\varphi+s\psi)-E_{T,\varepsilon}(u,\varphi)}{s}$$

然后我们可以获得与前面相似的微分形式：

$$\begin{aligned}
& E'_{T,\varepsilon,\varphi}(u,\varphi)(\psi) \\
&= \frac{1}{2}\int_B g'(\varphi)(\psi)|\nabla u|^2\mathrm{d}x + \frac{1}{2}\int_B G_c\left(2\frac{1}{\varepsilon}(1-\varphi)(-\psi)+2\varepsilon\nabla\varphi\cdot\nabla\psi\right)\mathrm{d}x \\
&= \int_B (1-\kappa)\varphi\psi|\nabla u|^2\mathrm{d}x + \int_B G_c\left(-\frac{1}{\varepsilon}(1-\varphi)\psi+\varepsilon\nabla\varphi\cdot\nabla\psi\right)\mathrm{d}x
\end{aligned}$$

因此我们可得到：

公式 11 引入半线性形式 $A(\cdot)(\cdot)$，将以上公式改写为

$$A_1(U)(\psi) = E'_{T,\varepsilon,u}(u,\varphi)(w) = (g(\varphi)\nabla u,\nabla w)$$

$$\begin{aligned}
A_2(U)(\psi) &= E'_{T,\varepsilon,u,\varphi}(u,\varphi)(\psi) \\
&= ((1-\kappa)\varphi|\nabla u|^2,\psi) - \frac{1}{\varepsilon}(G_c(1-\varphi),\psi) + \varepsilon(G_c\nabla\varphi,\nabla\psi)
\end{aligned}$$

这些公式与我们之前在式 (2.4) 与 (2.5) 中所讨论的在结构上相似。

4.4.3.4 障碍问题相关证明：带不等式约束的一阶变分形式

我们在本节讨论以下障碍问题。

公式 12 令 $g \in L^2(\Omega)$，障碍问题的能量最小化形式为

$$\min_{u \in V, u \geqslant g} E(u)$$

函数空间为

$$K := \{v \in H^1(\Omega) \mid v = u_D, \text{ 在 } \partial\Omega \text{ 上}; \quad v \geqslant g, \text{ 在 } \Omega \text{ 中}\}$$

此时可写作：

$$\min_{u \in K} E(u)$$

定义 46 (凸集) 当 $0 \leqslant s \leqslant 1$，$u, v \in K$ 时，满足：

$$su + (1-s)v \in K$$

则称 K 为凸集。

定义 47 (凸函数) 当且仅当函数满足：

$$E(su + (1-s)v) \leqslant sE(u) + (1-s)E(v)$$

时，函数 $E: K \to \mathbb{R}$ 是凸的。

公式 13

$$E(u) = \min_{v \in K} E(v)$$

此时 $u \in K$，是能量函数的最小值。

下面我们导出一阶变分形式。令 $v, u \in K, sv + (1-s)u = u + s(v-u) \in K$，其中 $s \in [0,1]$，此时通过前文定义可知：

$$E(u + s(v-u)) \geqslant E(u)$$

由此我们有

$$\left.\frac{\mathrm{d}}{\mathrm{d}s}E(u + s(v-u))\right|_{s=0} \geqslant \frac{\mathrm{d}}{\mathrm{d}s}E(u) \Leftrightarrow \left.\frac{\mathrm{d}}{\mathrm{d}s}E(u + s(v-u))\right|_{s=0} \geqslant 0$$

此外我们有

$$\left.\frac{\mathrm{d}}{\mathrm{d}s}\int_\Omega \left(\mu\frac{1}{2}|\nabla(u + s(v-u))|^2 - f(v-u)\right)\mathrm{d}x\right|_{s=0} \geqslant 0$$

4.4 Francfort-Marigo 脆性断裂变分模型

$$\Rightarrow \int_\Omega (\mu\nabla(u+s(v-u))\cdot\nabla(v-u) - f(v-u))\mathrm{d}x \Big|_{s=0} \geqslant 0$$

$$\Rightarrow \int_\Omega (\mu\nabla u\cdot\nabla(v-u) - f(v-u))\mathrm{d}x \geqslant 0$$

$$\Leftrightarrow \int_\Omega \mu\nabla u\cdot\nabla(v-u)\mathrm{d}x - \int_\Omega f(x-u)\mathrm{d}x \geqslant 0$$

$$\Leftrightarrow \int_\Omega \mu\nabla u\cdot\nabla(v-u)\mathrm{d}x - \int_\Omega f(v-u)\mathrm{d}x \geqslant 0$$

简而言之，参考 2.1.9 节，最后一行可改写为

$$(\mu\nabla u, \nabla(v-u)) \geqslant (f, v-u)$$

通过计算我们可以得到以下变分形式。

命题 16 (障碍问题的变分不等式形式) 障碍问题以变分不等式写为如下形式。

给定 $u \in K$, 有

$$(\mu\nabla u, \nabla(v-u)) \geqslant (f, v-u), \quad \forall v \in K$$

其中，

$$K = \{v \in H^1 \mid v = u_D, \text{ 在 } \partial\Omega \text{ 上}; \quad v \geqslant g, \text{ 在 } \Omega \text{ 中}\}$$

4.4.3.5 考虑裂缝不可逆性的 Euler-Lagrange 系统

基于前文，有

$$E'_{T,\varepsilon}(u,\varphi)(w, \psi-\varphi) \geqslant 0$$

命题 17 对于 $(u,\varphi) \in V_u^D \times K$, 考虑裂缝不可逆性的 Euler-Lagrange 系统为

$$E'_{T,\varepsilon,u}(u,\varphi)(w) = \int_B g(\varphi)\nabla u \cdot \nabla w \mathrm{d}x$$

$$E'_{T,\varepsilon,\varphi}(u,\varphi)(\psi-\varphi) = \int_B (1-\kappa)\varphi|\nabla u|^2(\psi-\varphi)\mathrm{d}x$$
$$+ \int_B G_c \left(-\frac{1}{\varepsilon}(1-\varphi)(\psi-\varphi) + \varepsilon\nabla\varphi\cdot(\nabla(\psi-\varphi))\right)\mathrm{d}x$$

命题 18 研究位移向量值 $u: B \times I \to \mathbb{R}^d$, $(u,\varphi) \in V_u^D \times K$ 时，考虑裂缝不可逆性 Euler-Lagrange 系统的一阶变分形式为

$$E'_{T,\varepsilon,u}(u,\varphi)(w) = \int_B g(\varphi)\sigma(u) : \nabla w \mathrm{d}x$$

$$E'_{T,\varepsilon,\varphi}(u,\varphi)(\psi-\varphi) = \int_B (1-\kappa)\varphi\sigma(u) : e(u)(\psi-\varphi)\mathrm{d}x$$
$$+ \int_B G_c \left(-\frac{1}{\varepsilon}(1-\varphi)(\psi-\varphi) + \varepsilon\nabla\varphi\cdot(\nabla(\psi-\varphi))\right)\mathrm{d}x$$

4.4.4 准静态脆性相场裂缝问题弱形式

在前人研究的基础上，我们可以初步建立起一个相场裂缝模型作为本书的基本模型。

命题 19 (相场裂缝模型耦合系统) 对于加载步 $n=1,2,3,\cdots,N$，给定 $\varphi^{n-1}, K := K(\varphi^{n-1})$，令 $U=(u,\varphi) := U^n = (u^n,\varphi^n) \in V_u^D \times K$，其中 $\varphi(0)=\varphi_0$。此时我们有

$$A(U)(\Psi-\phi) \geqslant 0, \quad \forall \Psi = (w,\psi) \in V_u^0 \times K$$

$\phi = (0,\psi)$，将上式分解为半线性形式可得

$$A(U)(\Psi-\phi) = A_1(U)(\Psi) + A_2(U)(\Psi-\phi)$$

其中

$$A_1(U)(\Psi) = (g(\varphi)\sigma(u), \nabla w)$$

$$A_2(U)(\Psi-\phi) = ((1-\kappa)\varphi\sigma(u):e(u), \psi-\varphi)$$
$$+ \left(-\frac{G_c}{\varepsilon}(1-\varphi), \psi-\varphi\right) + (G_c\varepsilon\nabla\varphi, \nabla(\psi-\varphi))$$

证明：我们根据命题 18 对每一个加载步进行计算并得到 t_n 时刻的相关方程。

4.5 热力学扩展

本节将讨论两个热力学问题：
(1) 裂缝拉伸或压缩的分解；
(2) 采用最大应变能代替不等式约束的相关模型。

4.5.1 裂缝拉伸或压缩的能量问题

命题 19 所介绍的模型在应用于压缩断裂问题时是不适用的。基于热力学原理，作者在文献 [23] 与 [308] (以及 4.5.1.3 节) 中进行了相关扩展。关于这两种受力方式，我们在 Miehe 等研究的基础上对应力张量进行了分解，相关内容也可参考文献 [17, 65]。

4.5 热力学扩展

(1) 拉应力 σ^+；

(2) 压应力 σ^-。

从工程的角度而言，I 型裂缝仅由拉应力引起，因此方程进行了以下修正：

$$\sigma(u) = \sigma^+(u) + \sigma^-(u) \rightarrow g(\varphi)\sigma^+(u) + \sigma^-(u)$$

其中能量退化函数应仅考虑 σ^+，相关的数值模拟可参考文献 [312]。图 4.2 提供了相关数模案例。

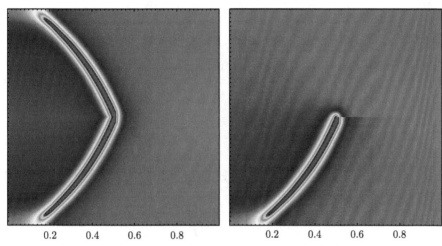

图 4.2 单边缺口剪切实验 (详见第 8 章)[312]
左图，无分裂的结果；右图，裂缝扩展路径

4.5.1.1 Miehe 应力分解法

首先我们将应力分解为拉应力 σ^+ 与压应力 σ^-，并且仅考虑拉应力造成的裂缝 (I 型裂缝)，则

$$\frac{1}{2}\int_B g(\varphi)\sigma(u) : e(u)\mathrm{d}x$$

可分解为

$$\frac{1}{2}\int_B g(\varphi)\sigma^+(u) : e(u)\mathrm{d}x + \frac{1}{2}\int_B \sigma^-(u) : e(u)\mathrm{d}x$$

其中拉应力与压应力为

$$\sigma^+(u) := 2\mu e^+ + \lambda[\mathrm{tr}(e)]^+ I$$

$$\sigma^-(u) := 2\mu\left(e - e^+\right) + \lambda\left(\mathrm{tr}(e) - [\mathrm{tr}(e)]^+\right)I$$

其中,
$$e := P\Lambda P^{\mathrm{T}}, \quad e^+ := P\Lambda^+ P^{\mathrm{T}}, \quad e^- := P\Lambda^- P^{\mathrm{T}}$$

在此基础上我们引入截止函数:
$$[x]^+ := \begin{cases} x, & x > 0 \\ 0, & 0 \leqslant 0 \end{cases}$$

备注 27 本书符号标注时,存在 $[x]_+ := [x]^+$ 的情况。

矩阵 P 与 Λ 可以根据特征值进行分解:
$$\Lambda = \begin{pmatrix} \lambda_1 & & \\ & \lambda_2 & \\ & & \lambda_3 \end{pmatrix}, \quad \Lambda^+ = \begin{pmatrix} [\lambda_1]^+ & & \\ & [\lambda_2]^+ & \\ & & [\lambda_3]^+ \end{pmatrix}, \quad \Lambda^- = \Lambda - \Lambda^+$$

4.5.1.2 Amor 应力分解法

此外还存在另一种更为简单的应力分解法,但该方法在数值模拟与力学解释中存在一定瑕疵。该方法认为
$$\sigma^+(u) := k\mathrm{tr}^+(e)I + 2\mu e_D$$

$$\sigma^-(u) := k\mathrm{tr}^-(e)I$$

其中体积模量 $k = \dfrac{2}{d}\mu + \lambda$,应变张量的分量 $e(u)$ 为
$$e_D := e - \frac{1}{d}\mathrm{tr}(e)I, \quad d = 2, 3$$

同时:
$$\mathrm{tr}^+(e) = [\mathrm{tr}(e)]^+, \quad \mathrm{tr}^-(e) = \mathrm{tr}(e) - \mathrm{tr}^+(e)$$

4.5.1.3 其他应力分解法

其他应力分解法可参考 Amor[23], Miehe[312], Zhang[458], Strobl 和 Seelig[394]、Steinke 和 Kaliske[393]、Bryant 和 Sun[87]、Freddi 和 Royer-Carfagni[173]、Bilgen、Homberger 和 Weinberg[56] 等的研究成果。当然作者研究团队同样在该方面开展了相关研究[158]。

4.5.2 基于应力分解的相场裂缝模型

在这里我们介绍用于描述脆性相场裂缝扩展模型的另一种思路。

命题 20 (采用应力分解法的相场裂缝模型耦合系统) 对于加载步 $n = 1, 2, 3, \cdots, N$,给定 φ^{n-1}, $K := K(\varphi^{n-1})$, 令 $U = (u, \varphi) := U^n = (u^n, \varphi^n) \in V_u^D \times K$, 其中 $\varphi(0) = \varphi_0$。此时我们有

$$A(U)(\Psi - \phi) \geqslant 0, \quad \forall \Psi = (w, \psi) \in V_u^0 \times K$$

将上式分解为半线性形式可得

$$A(U)(\Psi - \phi) = A_1(U)(\Psi) + A_2(U)(\Psi - \phi)$$

其中

$$A_1(U)(\Psi) = \left(g(\varphi)\sigma^+(u), \nabla w\right) + \left(\sigma^-(u), \nabla w\right)$$

$$A_2(U)(\Psi - \phi) = \left((1-\kappa)\varphi\sigma^+(u) : e(u)(\psi - \varphi), \psi - \varphi\right)$$
$$+ \left(-\frac{G_c}{\varepsilon}(1-\varphi), \psi - \varphi\right) + (G_c\varepsilon\nabla\varphi, \nabla(\psi - \varphi))$$

证明:将应力分解为拉应力与压应力并代入命题 19 即可得。

4.5.3 利用应变场替换不可逆约束的相场裂缝模型

Miehe 等从热力学角度出发对裂缝总能量进行了讨论,并建立了基于应变场的相场裂缝模型[308]。该模型从热力学角度解释了裂缝表面能 $E_s(\varphi)$ 并放松了模型中的不等式约束。

命题 21 (裂缝表面能 $E_s(\varphi)$ 的热力学性质) 给定

$$E_s(\varphi) = \frac{1}{2}\int_B G_c\left(\frac{1}{\varepsilon}(1-\varphi)^2 + \varepsilon|\nabla\varphi|^2\right)\mathrm{d}x$$

并提出以下假设:
(1) 裂缝扩展是一个能量耗散过程;
(2) 整个系统是一个熵增的过程;
(3) 能量耗散的物理表现为裂缝扩展的不可逆现象。
则在裂缝扩展过程中,有

$$\frac{\mathrm{d}}{\mathrm{d}t}E_s(\varphi) = \int_B G_c\left(-\frac{1}{\varepsilon}(1-\varphi)\partial_t\varphi + \varepsilon\nabla\varphi \cdot \nabla\partial_t\varphi\right)\mathrm{d}x$$

$$= \int_B G_c \left(-\frac{1}{\varepsilon}(1-\varphi) + \varepsilon \nabla \varphi \right) \partial_t \varphi \mathrm{d}x$$

证明：考虑边界齐次 Neumann 条件的情况下采用链式法则进行微分处理即可。

接下来讨论热力学第二定律 [232,142]，该理论认为裂缝扩展是一个不可逆的熵增过程。

$$\frac{\mathrm{d}}{\mathrm{d}t} E_s(\varphi) \geqslant 0$$

即

$$\int_B G_c \left(\frac{1}{\varepsilon}(1-\varphi) + \varepsilon \nabla \varphi \right) (-\partial_t \varphi) \, \mathrm{d}x \geqslant 0$$

第一项即变分导数 [312]，非负，第二项同样非负，因此该不等式恒成立。

接下来我们引入应力场 β，并将不等式约束正则化：

$$\frac{1}{\varepsilon}(1-\varphi) + \varepsilon \nabla \varphi \geqslant \beta \geqslant 0$$

我们可以看到变量 β 与位移方程 (公式 1，命题 19 和 20) 中的 φ 的导数相关。

Laplacian：

$$\beta := (1-\kappa)\varphi |\nabla u|^2$$

不考虑应力分解的弹性模型：

$$\beta := (1-\kappa)\varphi \sigma(u) : e(u)$$

考虑应力分解的弹性模型：

$$\beta := (1-\kappa)\varphi \sigma^+(u) : e(u)$$

基于第三项我们可得

$$G_c \left(\frac{1}{\varepsilon}(1-\varphi) + \varepsilon \nabla \varphi \right) \geqslant \beta := (1-\kappa)\varphi \sigma^+(u) : e(u) \tag{4.7}$$

命题 22 在命题 20 的 Euler-Lagrange 系统中存在以下条件：

$$\left[(1-\kappa)\varphi \sigma^+(u) : e(u) + G_c \left(-\frac{1}{\varepsilon}(1-\varphi) - \varepsilon \nabla \varphi \right) \right] \cdot \partial_t \varphi = 0 \tag{4.8}$$

证明：请读者参考 2.4.1 节中的障碍问题以及公式 1。

4.5 热力学扩展

对于扩展裂缝，我们参考式 (4.7) 与 (4.8)。当 $(1-\kappa)\varphi\sigma^+(u):e(u)$ 取最大时，

$$(1-\kappa)\varphi\sigma^+(u):e(u) + G_c\left(\frac{1}{\varepsilon}(1-\varphi) + \varepsilon\nabla\varphi\right) = 0$$

参考式 (4.7) 可知

$$G_c\left(\frac{1}{\varepsilon}(1-\varphi) + \varepsilon\nabla\varphi\right) = \beta = (1-\kappa)\varphi\sigma^+(u):e(u)$$

当 $\partial_t\varphi = 0$ 时，我们有

$$(1-\kappa)\varphi\sigma^+(u):e(u) - G_c\left(\frac{1}{\varepsilon}(1-\varphi) + \varepsilon\nabla\varphi\right) < 0$$

即

$$(1-\kappa)\varphi\sigma^+(u):e(u) < G_c\left(\frac{1}{\varepsilon}(1-\varphi) + \varepsilon\nabla\varphi\right)$$

定义 48 应变储能函数为

$$\psi_e^+ := \sigma^+(u):e(u)$$

命题 23 令 $\varepsilon > 0, \psi_e^+ \gg 0$，则 $\varphi \approx 0$。

证明 首先

$$\frac{1}{\varepsilon}(1-\varphi) + \varepsilon\nabla\varphi \geqslant \beta \geqslant 0$$

我们已知

$$\beta := (1-\kappa)\varphi\sigma^+(u):e(u) = (1-\kappa)\varphi\psi_e^+$$

当 $\varphi > 0$，且值较小时，有

$$\frac{G_c\left(\frac{1}{\varepsilon}(1-\varphi) + \varepsilon\Delta\varphi\right)}{(1-\kappa)\varphi} \geqslant \psi_e^+$$

即

$$\frac{G_c\left(\frac{1}{\varepsilon}(1-\varphi) + \varepsilon\Delta\varphi\right)}{(1-\kappa)\varphi} = \max_u \psi_e^+ \tag{4.9}$$

由于 $\max_u \psi_e^+ \gg 0$，相场方程的左手项趋向于 0，即 $\varphi \approx 0$。

推理 1 以上命题还可推出，当考虑最大应变能 ψ_e^+ 时可实现裂缝不可逆条件。根据式 (4.7) 与 (4.8)，当 $\partial_t\varphi < 0$ 在任意 t 均成立时，裂缝不可逆条件可忽略。

定义 49 (应变场函数) 应变场函数为

$$\mathcal{H} := \max_{s\in[0,t]} \psi_e^+(u(s))$$

命题 24 (基于应变场的相场裂缝模型) 对于加载步 $n = 1, 2, 3, \cdots, N$，给定 $\mathcal{H}(u^{n-1})$，令 $U := U^n = (u^n, \varphi^n) \in V_u^D \times V_\varphi$，其中 $\varphi(0) = \varphi_0$。此时有

$$A(U)(\Psi) = 0, \quad \forall \Psi = (w, \psi) \in V_u^0 \times V_\varphi$$

其中

$$A(U)(\Psi) = A_1(U)(\Psi) + A_2(U)(\Psi)$$

$$A_1(U)(\Psi) = (g(\varphi)\sigma^+(u), \nabla w) + (\sigma^-(u), \nabla w)$$

$$A_2(U)(\Psi) = ((1-\kappa)\varphi\mathcal{H}, \psi) + \left(-\frac{G_c}{\varepsilon}(1-\varphi), \psi\right) + (G_c\varepsilon\nabla\varphi, \nabla\psi)$$

备注 28 该模型并非 CVIS 耦合系统而是 PDE 耦合系统，因为该模型直接对相场方程进行处理而不是对变分不等式进行处理。基于应变场函数 $\mathcal{H}$ 进行数值建模的过程中需要注意数据储存问题。

4.6 变分能量公式与 Euler-Lagrange 系统总结

正如本章所介绍的，在弹性力学中通过求解能量公式的局部最小值来模拟相场裂缝问题[71]，即

$$\min_{u\in V_u^D} E_T(U)$$

通过微分，可以获得一个 CVIS 耦合系统，即所谓的 Euler-Lagrange 耦合系统：

$$A_1(U)(w) = E'_u(U)(w) = 0$$

$$A_2(U)(\psi) = E'_\varphi(U)(\psi - \varphi) \geqslant 0$$

通过对该系统进行求解，我们不仅需要获得局部最小值，还需要得到所有的驻点。换而言之，我们所求得的解有可能是鞍点。

另外，PDE 方程允许我们在系统中耦合更多的物理场，例如我们可以用能量泛函的形式来表示 Stokes 方程，但针对不可压缩流体的 Navier-Stokes 方程并没

有适用的能量方程。此外，简单的抛物型方程无法引入能量的概念，例如热方程、多孔介质流动方程等。读者可参考文献 [317,435,419,319] 中关于相场裂缝的成因以及形成机理等内容。

对于某些模型，基于部分假设，我们仍然可以获得适用的能量公式。例如，在流固耦合多孔介质准静态裂缝中，一些研究人员将整个系统设定为一个时间离散模型，并在给定的时间步上得到椭圆型偏微分方程，如文献 [310,319]。而在时间连续系统中，该方法不再适用。另外在相场-裂缝-孔隙相互耦合的模型中，仍未找到适用的能量公式。

第 5 章 数值建模 I：正则化与离散化

在第 4 章中我们推导了时空连续的变分相场裂缝模型。分别在命题 19 介绍了不等式约束；命题 20 介绍了应力分解法，将应力分解为拉应力与压应力进行处理；命题 24 建立了基于应变场的相场裂缝模型。而在接下来的章节中，我们将讨论相场裂缝模型的正则化与离散化 (第 5 章)，线性与非线性解问题 (第 7 章) 以及相场裂缝模型的自适应方法 (第 9 章)。我们在介绍这几个模块的过程中穿插补充了一些数值模拟的章节供读者参考学习。

离散与求解是有限元建模过程中的一个重要部分。另外还可以引入并行计算与自适应模块来加快求解速度。本章提供了两个经典离散案例：

(1) 弹性动力学中的时间离散；
(2) 多孔介质中的空间离散。

5.1 裂缝不可逆性约束

裂缝扩展不可逆是相场裂缝模型中的一个重要约束，可以通过固定裂缝节点[70,264]、引入应变场函数[308]、引入补偿项[190,313,424]、原始-对偶活动集方法[217]等方式实现。

5.1.1 应变场函数

基于应变场施加不可逆约束请参考 4.5.3 节。

5.1.2 简单补偿法与增广 Lagrangian 补偿法

命题 19 和命题 20 中通过向后差分实现不等式约束 $\partial_t \varphi \leqslant 0$，

$$\partial_t \varphi \approx \frac{\varphi - \varphi^{n-1}}{k} \leqslant 0 \Rightarrow \varphi \leqslant \varphi^{n-1} \Rightarrow \varphi - \varphi^{n-1} \leqslant 0$$

这意味着在 $\varphi > \varphi^{n-1}$ 时需要进行补偿。以障碍问题为例，将解空间扩展到线性空间 V_φ[256]。一种简单补偿项可写为 $\gamma \left[\varphi - \varphi^{n-1} \right]_+$，其中补偿参数 $\gamma > 0$，$[x]_+ = [x]^+ := \max(0, x)$，时间步 $k' := t_n - t_{n-1}$，$\varphi := \varphi^n := \varphi(t_n)$。当补偿参数过小时，补偿可能无效。当补偿参数过大时，可能会形成病态 Jacobian 矩阵，增大非线性程度。

5.1 裂缝不可逆性约束

命题 25 (引入简单补偿项的相场模型) 给定 $\gamma > 0$，解变量 $U = (u, \varphi)$。引入简单补偿项后，相场模型改写为

$$A_2(U)(\Psi) + \left(\gamma\left[\varphi - \varphi^{n-1}\right]_+, \psi\right) = 0 \tag{5.1}$$

具体而言，解变量 φ 不再局限于凸集 $K := K\left(\varphi^{n-1}\right)$，而是分布于线性空间 $V_\varphi := H^1(B)$。

此时我们通过补偿法放松不等式约束，得到了新的相场方程。

命题 26 (引入简单补偿项的命题 19) 在加载步 $n = 1, 2, 3, \cdots, N$ 中，给定 φ^{n-1}，令 $U := U^n = (u^n, \varphi^n) \in X$，其中 $\varphi(0) = \varphi_0$

$$A(U)(\Psi) = 0, \quad \forall \Psi = (w, \psi) \in X^0$$

其半线性形式为

$$A(U)(\Psi) = A_1(U)(\Psi) + A_2(U)(\Psi)$$

其中

$$A_1(U)(\Psi) = (g(\varphi)\sigma(u), \nabla w)$$

$$A_2(U)(\Psi) = ((1-\kappa)\varphi\sigma(u) : e(u), \psi) + \left(-\frac{G_c}{\varepsilon}(1-\varphi), \psi\right) + (G_c\varepsilon\nabla\varphi, \nabla\psi)$$
$$+ \left(\gamma\left[\varphi - \varphi^{n-1}\right]_+, \psi\right)$$

命题 27 (引入简单补偿项的命题 20) 在加载步 $n = 1, 2, 3, \cdots, N$ 中，给定 φ^{n-1}，令 $U := U^n = (u^n, \varphi^n) \in X$，其中 $\varphi(0) = \varphi_0$

$$A(U)(\Psi) = 0, \quad \forall \Psi = (w, \psi) \in X^0$$

其半线性形式为

$$A(U)(\Psi) = A_1(U)(\Psi) + A_2(U)(\Psi)$$

其中

$$A_1(U)(\Psi) = (g(\varphi)\sigma^+(u), \nabla w) + (\sigma^-(u), \nabla w)$$

$$A_2(U)(\Psi) = ((1-\kappa)\varphi\sigma^+(u) : e(u), \psi) + \left(-\frac{G_c}{\varepsilon}(1-\varphi), \psi\right) + (G_c\varepsilon\nabla\varphi, \nabla\psi)$$
$$+ \left(\gamma\left[\varphi - \varphi^{n-1}\right]_+, \psi\right)$$

备注 29 关于牛顿法的相关细节请读者参考第 7 章。

增广 Lagrangian 补偿法是数值优化的一种常见方法[223,349,202,168]。我们将该方法引入相场裂缝模型中。首先引入增广 Lagrangian 补偿系数 $\Xi \in L^2(B)$，此时我们有增广补偿项

$$\left[\Xi + \gamma\left(\varphi - \varphi^{n-1}\right)\right]_+ \tag{5.2}$$

命题 28 (基于增广 Lagrangian 补偿法的相场变分模型) 给定 $\gamma > 0$，解变量 $U = (u, \varphi)$，增广 Lagrangian 补偿系数 $\Xi \in L^2(B)$，此时有

$$A_2(U)(\Psi) + \left(\left[\Xi + \gamma\left(\varphi - \varphi^{n-1}\right)\right]_+, \psi\right) = 0 \tag{5.3}$$

与之前相似，解变量 φ 不再局限于凸集 $K := K(\varphi^{n-1})$，而是分布于线性空间 $V_\varphi := H^1(B)$ 中，求解细节请参考 7.11 节。

我们需要注意的是增广 Lagrangian 补偿项可以拆分为

$$\left(\left[\Xi + \gamma\left(\varphi - \varphi^{n-1}\right)\right]_+, \psi\right) + (\min(0, \Xi + \gamma\varphi), \psi) \tag{5.4}$$

在模拟过程中，我们发现后一项对数模结果影响较小。

5.1.3 原始-对偶活动集法

原始-对偶活动集是一类特殊方法，可以视为一种半光滑的牛顿法[240,227,228]。在障碍问题中，整个域被划分为活动集 (约束) 与非活动集 (PDE)，这种方法被称为活动集法。该方法在相场裂缝模型中的应用请参考文献 [217] 与 7.9.5 节。

5.1.3.1 有限维情况

本节我们将考虑以下问题。

公式 14 给定 $\varphi^{n-1} \in \mathbb{R}^d$，解变量 $\varphi : B \to \mathbb{R}^d$，增广 Lagrangian 补偿系数 $\Xi : B \to \mathbb{R}$，有

$$\begin{aligned} A\varphi + \Xi &= f \\ \varphi &\leqslant \varphi^{n-1} \\ \Xi &\geqslant 0 \\ (\Xi, \varphi - \varphi^{n-1}) &= 0 \end{aligned}$$

这里 $A \in \mathbb{R}^{d \times d}$ 是正定矩阵，且 $f \in \mathbb{R}^d$。

若 A 为对称正定矩阵，则上述系统为下式的最优解：

$$\min_{\varphi} E(\varphi) = \frac{1}{2}(\varphi, A\varphi) - (f, \varphi)$$

5.1 裂缝不可逆性约束

其中 $\varphi \leqslant \varphi^{n-1}$。根据互补条件：

$$\varphi \leqslant \varphi^{n-1}, \quad \Xi \geqslant 0, \quad (\Xi, \varphi - \varphi^{n-1}) = 0 \Leftrightarrow C(\varphi, \Xi) = 0$$

$$C(\varphi, \Xi) = \Xi - \left[\Xi + c\left(\varphi - \varphi^{n-1}\right)\right]^+ \tag{5.5}$$

其中 $c > 0$。且

备注 30 式 (5.5) 与式 (5.2)、式 (5.4) 有密切联系。

公式 15 有解 $\varphi : B \to \mathbb{R}^d$ 以及增广 Lagrangian 补偿系数 $\Xi : B \to \mathbb{R}$，则

$$A\varphi + \Xi = f$$
$$C(\varphi, \Xi) = 0$$

现在我们将指定一个原始-对偶活动集方案，对于给定组 (φ, Ξ)，我们可采用约束 $C(\varphi, \Xi) = 0$ 来决定活动集与非活动集：

$$\mathcal{N} = \left\{i : \Xi_i + c\left(\varphi - \varphi^{n-1}\right)_i \leqslant 0\right\}, \quad \text{求解 PDE 方程}$$

$$\mathcal{A} = \left\{i : \Xi_i + c\left(\varphi - \varphi^{n-1}\right)_i > 0\right\}, \quad \text{激活约束方程}$$

算法 2 原始-对偶活动集算法如下：
(1) 初始化 φ^0, Ξ^0，令 $j = 0$。
(2) 确定活动集与非活动集 $\mathcal{N}^j, \mathcal{A}^j$。
(3) 求解

$$A\varphi^{j+1} + \Xi^{j+1} = f$$
$$\varphi^{j+1} = \varphi^{n-1}, \quad \text{在 } \mathcal{A}^j \text{ 上}$$
$$\Xi^{j+1} = 0, \quad \text{在 } \mathcal{N}^j \text{ 上}$$

(4) 若满足终止准则就停止计算，否则令 $j = j + 1$ 并转到步骤 (2)。

备注 31 在文献 [227] 中，该算法也被称为半光滑牛顿法。

5.1.3.2 基于增广 Lagrangian 补偿法的相场裂缝模型

首先我们令 Q^* 为 Banach 空间 Q 的对偶空间：

$$V_\Xi = \{\chi \in Q^* | \chi \geqslant 0\}$$

随后我们得到以下等价公式。

命题 29 令 $(u, \varphi, \Xi) \in V_u^D \times V_\varphi \times V_\Xi$，有

$$A_1(U)(w) = 0, \quad \forall w \in V_u^0$$

$$A_2(U)(\psi) + \langle \Xi, \psi \rangle = 0, \quad \forall \psi \in V_\varphi$$

$$\langle \Xi - \chi, \varphi - \varphi^{n-1} \rangle \geqslant 0, \quad \forall \chi \in V_\Xi$$

备注 32　相关模型请参考文献 [296]。

命题 29 中的不等式可改写为

$$\varphi \leqslant \varphi^{n-1}, \quad 在 B \times I 中$$

$$\Xi \geqslant 0, \quad 在 B \times I 中$$

$$\langle \Xi, \varphi - \varphi^{n-1} \rangle = 0, \quad 在 B \times I 中$$

根据式 (5.5) 可知：

$$C(\varphi, \Xi) = 0$$

其中

$$C(\varphi, \Xi) = \Xi - \left[\Xi + c\left(\varphi - \varphi^{n-1}\right)\right]^+ \tag{5.6}$$

5.2　时间离散化

到目前为止介绍的都是准静态相场裂缝模型。本节将介绍时间离散化方面的研究。

5.2.1　单次 θ 分解法与分步 θ 分解法

定义 50　给定 $U \in X$，$X = V_u^D \times V_\varphi$，对于 $t \in I$，有

$$\int_0^T A(U)(\Psi)\mathrm{d}t = \int_0^T F(\Psi)\mathrm{d}t, \quad \forall \Psi \in X^0 \tag{5.7}$$

这里半线性形式的 $A(U)(\Psi)$ 包含时间导数、显式与隐式项，我们假定该式可分解为

$$A(U)(\Psi) := A_T(U)(\Psi) + A_E(U)(\Psi) + A_I(U)(\Psi)$$

需要注意的是 A_T, A_E, A_I 仍是时空连续的，且有可能是非线性的。

备注 33　在增量准静态相场裂缝模型中，给定 $U^{n-1}, U^n \in X, n = 1, 2, 3, \cdots, N$

$$A(U^n)(\Psi) = F^n(\Psi), \quad \forall \Psi \in X^0$$

这里 $F^n(\Psi) := (f(t_n), \Psi)$，为给定右手项的数据。

5.2 时间离散化

例 9 在之前的命题中有

$$A(U)(\Psi) = A_1(U)(\Psi) + A_2(U)(\Psi)$$

由于我们规定：

$$A(U)(\Psi) := A_E(U)(\Psi) = A_1(U)(\Psi) + A_2(U)(\Psi)$$

此时涉及多个 PDE/CVIS 耦合，请参考文献 [428]。

定义 51 (半线性分类) 我们在这里正式定义了以下几种半线性形式：
(1) 带时间导数的形式；
(2) 显式形式；
(3) 隐式形式。

定义 52 (单次 θ 分解法) 给定 $U^{n-1}, F^n, F^{n-1}, \theta \in [0,1]$，我们利用单次 θ 分解法求解：

$$A_T\left(U^{n,k}\right)(\Psi) + \theta A_E\left(U^n\right)(\Psi) + A_I\left(U^n\right)(\Psi)$$
$$= -(1-\theta)A_E\left(U^{n-1}\right)(\Psi) + \theta F^n(\Psi) + (1-\theta)F^{n-1}(\Psi) \tag{5.8}$$

$\theta = 0$ 时为一阶显式欧拉差分形式，$\theta = 1$ 时为一阶隐式向后差分形式，$\theta = 0.5$ 时为二阶隐式差分形式，$\theta = 0.5 + k_{\text{char}}$ 时为二阶隐式 Crank-Nicolson 差分形式 (这里 $k_{\text{char}} \ll 0.5$)。

定义 53 (分步 θ 分解法) 我们令 $\theta = 1 - \frac{\sqrt{2}}{2}$，$\theta' = 1 - 2\theta$，$\alpha = \frac{1-2\theta}{1-\theta}$，$\beta = 1 - \alpha$。将一个时间步分解为三个连续的子时间步。给定 U^{n-1}，时间步长为 $k := k_n = t_n - t_{n-1}$：

$$A_T\left(U^{n-1+\theta,k}\right)(\Psi) + \alpha\theta A_E\left(U^{n-1+\theta}\right)(\Psi) + A_I\left(U^{n-1+\theta}\right)(\Psi)$$
$$= -\beta\theta A_E\left(U^{n-1}\right)(\Psi) + \theta F^{n-1}(\Psi)$$
$$A_T\left(U^{n-\theta,k}\right)(\Psi') + \beta\theta' A_E\left(U^{n-\theta}\right)(\Psi) + A_I\left(U^{n-\theta}\right)(\Psi')$$
$$= -\alpha\theta' A_E\left(U^{n-1+\theta}\right)(\Psi) + \theta' F^{n-\theta}$$
$$A_T\left(U^{n,k}\right)(\Psi') + \alpha\theta A_E\left(U^n\right)(\Psi) + A_I\left(U^n\right)(\Psi)$$
$$= -\beta\theta A_E\left(U^{n-\theta}\right)(\Psi') + \theta F^n(\Psi) + \theta F^{n-\theta}(\Psi) \tag{5.9}$$

备注 34 在流固耦合问题中，本书中的分步 θ 分解法略有改动，具体请参考文献 [84, 201, 360, 414]。

5.2.2 例题

在本节中，我们基于传热学部分内容简单展示抛物型方程的差分方法。

$$\partial_t u - \Delta u = f$$

基于有限差分法，令 $u := u^n$

$$\frac{u - u^{n-1}}{k} - \theta \Delta u - (1-\theta)\Delta u^{n-1} = \theta f + (1-\theta)f^{n-1}$$

整理上式可得

$$\frac{u - u^{n-1}}{k} - \theta \Delta u = (1-\theta)\Delta u^{n-1} + \theta f + (1-\theta)f^{n-1}$$

整理时间项 A_T 以及定常项 A_E 后，该问题可转化为：给定 $U \in X$，$t \in (0, T)$，有

$$A(U)(\Psi) = F(\Psi), \ \Psi \in X$$

其中

$$A(U)(\Psi) = A_T(U)(\Psi) + A_E(U)(\Psi)$$

对于上述热方程而言：

$$A_T(U)(\Psi) = (\partial_t u, \Psi)$$
$$A_E(U)(\Psi) = (\nabla u, \nabla \Psi)$$
$$F(\Psi) = (f, \Psi)$$

$A_T(U)(\Psi)$ 的向后差分形式写作：

$$A_T(U)(\Psi) \approx A_T\left(U^{n,k}\right)(\Psi) = \left(\frac{u - u^{n-1}}{k}, \Psi\right)$$

5.2.3 非线性时间导数

当时间函数为非线性时，首先我们应用乘积法则，随后采用非线性系数 θ 进行加权差商。

令 $u \in V_u, v \in V_v$ 为非线性系统的解变量，$J := J(u)$，此时有强形式：

$$\partial_t(Jv) = J\partial_t v + v\partial_t J$$

5.2 时间离散化

弱形式：
$$A_T(U)(\psi) = (\partial_t(Jv), \psi) = (J\partial_t v + v\partial_t J, \psi)$$

接下来我们引入差商与参数 θ

$$J^\theta\left(\frac{v - v^{n-1}}{k}\right) + v^\theta\left(\frac{J - J^{n-1}}{k}\right)$$

其中

$$J^\theta = \theta J^n + (1-\theta)J^{n-1}$$
$$v^\theta = \theta v^n + (1-\theta)v^{n-1}$$

此时：

$$A_T(U)(\psi) \approx A_T\left(U^{n,k}\right)(\psi) = \left(J^\theta\left(\frac{v - v^{n-1}}{k}\right), \psi\right) + \left(v^\theta\left(\frac{J - J^{n-1}}{k}\right), \psi\right)$$

例 10 我们令

$$\partial_t(Ju) - \Delta u = f$$

其中 $J := J(u)$, J 取决于 u 的值，进行时间离散处理后：

$$A_T\left(U^{n,k}\right)(\psi) = \left(J^\theta\left(\frac{u - u^{n-1}}{k}\right), \psi\right) + \left(u^\theta\left(\frac{J - J^{n-1}}{k}\right), \psi\right)$$
$$A_E(u^n)(\psi) = (\nabla u, \nabla \psi)$$

这里与前文相同，$u := u^n = u(t_n)$。

5.2.4 准静态相场裂缝

在推导准静态相场裂缝模型的过程中，我们用到了参数 θ ($\theta = 1$ 为一阶隐式后向差分形式)，该不等式约束 $\partial_t \varphi \leqslant 0$ 可写为

$$A_T(\varphi)(\psi) = (\partial_t \varphi, \psi) \approx A_T\left(\varphi^{n,k}\right)(\psi) = \left(\frac{\varphi - \varphi^{n-1}}{k}, \psi\right) \leqslant 0$$

5.2.5 非稳态多物理场模拟过程中的时间步设置

在非稳态多物理场模型中，时间步的选择至关重要。为了保证时间离散化后的能量守恒，我们首选耗散较少的方案，如 $\theta = 0.5$, $\theta = 0.5 + k_n$ 或分步 θ 分解法。

通过不可压缩流体-弹性结构相互作用 (FSI) 流固耦合基准算例 [236]，我们向读者展示了各个时间步进方案的不同模拟结果 [428,374,157]。图 5.1 展示了绕弹性物体的不可压缩流体在层流状态下引起结构自激振荡的数值模拟。

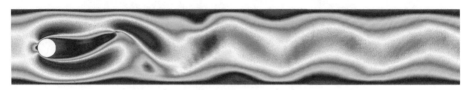

图 5.1 FSI 基准问题的速度场云图数值模拟

$\theta = 1$ 时的一阶隐式向后差分格式的模拟结果如图 5.2 所示。

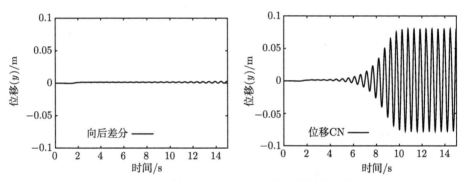

图 5.2 不同时间步方案下 FSI-2 数值模拟收敛情况
左：向后差分格式；右：Crank-Nicolson (CN) 格式

在模拟过程中 Crank-Nicolson 格式也出现数值不稳定性，由此可知，选择正确的时间步方案对模型的求解至关重要 [236,428,354,287]。在求解过程中仅有 Rannacher 方案以及分步 θ 分解法求解稳定 (图 5.3)。

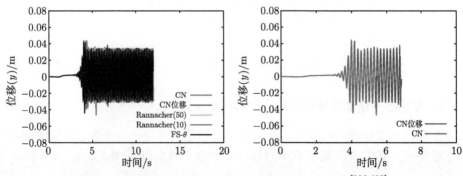

图 5.3 不同时间步方案下 FSI-3 数值模拟收敛情况 [236,432]

基于以上讨论可知，在显式求解方案中，较小的时间步长能够增强模型的数值稳定性；而在隐式求解方案中，我们则可采取相对较大的时间步长，这也正是隐式方案的优势之一。

5.3 空间离散化

在前文中我们完成了时间离散化的介绍，但方程此时仍然包含连续的空间 V_u^D 与 V_φ，为此我们需要继续对空间进行离散化操作并构建有限维子空间 $V_{u,h} \subset V_u^D, V_{\varphi,h} \subset V_\varphi$。本节我们将简要介绍 Galerkin 空间离散化方法，细节部分请读者参考相关文献 [99, 123, 124, 126, 249, 113, 386, 77, 80, 237, 359]。

5.3.1 有限元空间

计算域 B 根据空间维度 d 可被分割为开集 K。其网格由四边形或者六面体单元组成，这些网格对计算域 $B \subset \mathbb{R}^d$ 实现了非重复覆盖，$\mathcal{T}_h = \{K\}$ 由 B 中的所有有限元节点组成。空间离散化参数 h 为常数，是网格的单元长度。

定义 54 (规则网格) 网格 $\mathcal{T}_h = \{K\}$ 若满足以下条件，则为规则网格。
(1) $\overline{B} = U_{K \in \mathcal{T}_h} \overline{K}$；
(2) $K_1 \cap K_2 = \varnothing$，对于任意 K_1，K_2 均成立；
(3) 任意一个单元 $K_1 \in \mathcal{T}_h$，它的任意一个面或边均为另外一个单元 $K_2 \in \mathcal{T}_h$ 的面或边。

但在建模过程中，条件 (3) 对模型限制很大。为适配自适应网格等技术，我们引入悬挂节点的概念，允许网格单元具有相邻网格单元面或边的中点上的节点。当然每个面或者边的中点上最多允许悬挂一个节点。

我们定义连续 H_1 协调有限元空间 V_h^l 如下：

$$V_h^l := \left\{ v_h \in C(\overline{B}), \ |v_h|_K \in \mathcal{Q}(K), \quad \forall K \in \mathcal{T}_h \right\} \subseteq H^1(B)$$

其中 $\mathcal{Q}(K)$ 代表类多项式函数空间。接下来我们引入一阶以下张量积多项式空间 $\mathcal{Q}_1(K)$：

$$\mathcal{Q}_l(K_{\text{unit}}) := \text{span} \left\{ \prod_{i=1}^d x_i^{\alpha_i} | \alpha_i \in \{0, 1, \cdots, l\} \right\}$$

考虑到 $K \in \mathcal{T}_h$ 中的双线性变换 $\sigma_K : K_{\text{unit}} \to K$，我们定义 Q_1^c：

$$Q_1^c(K) = \left\{ q \circ \hat{\sigma}_K^{-1} : q \in \text{span} \langle 1, x, y, xy \rangle \right\} \quad (d = 2)$$

$$Q_1^c(K) = \left\{ q \circ \sigma_K^{-1} : q \in \text{span} \langle 1, x, y, z, xy, xz, yz, xyz \rangle \right\} \quad (d = 3)$$

二维空间中 $Q_1^c = 4$，三维空间中 $Q_1^c = 8$。维数表示单个单元的局部自由度。

在二维空间中，Q_2^c 定义为

$$Q_2^c(K) = \left\{ q \circ \sigma_K^{-1} : q \in \text{span}\langle 1, x, y, xy, x^2, y^2, x^2y, y^2x, x^2y^2 \rangle \right\}$$

当 $Q_2^c = 9$ 时，P_1^{dc} 由以下线性函数定义而来:

$$P_1^{dc}(\hat{K}) = \left\{ q \circ \hat{\sigma}_K^{-1} : q \in \text{span}\langle 1, x, y \rangle \right\}$$

其中 $P_1^{dc}(\hat{K})$ 的维度为 3。

备注 35 如果 σ_K 变换后仍为 $Q_1(K)^d$ 中的一个元素，则相应的有限元空间是等参的。

将以上概念推广到有限元空间中，并讨论引入悬挂节点后的一些情况。为全局连续性 (全局一致性) 考虑，位于不同细化界面上的自由度必须满足一定的约束。它们由相邻自由度的插值确定。基于以上考虑，悬挂节点本身并不具备任何自由度，关于该方面的相关细节可参考文献 [99]。

为保证有限元空间的逼近性质，我们介绍两个经典假设 [80,77]。

定义 55 (准均匀性假设) 若存在常数 κ 满足以下条件:

(1) 对于任意映射 $K_{\text{unit}} \to K$，均有

$$\frac{\sup\{\|\nabla \sigma_K(x)x\| \,|\, x \in K, \|x\|=1\}}{\inf\{\|\nabla \hat{\sigma}_K(x)x\| \,|\, x \in K, \|x\|=1\}} \leqslant \kappa, \quad K \in \bigcup_h \mathcal{T}_h \quad (5.10)$$

(2) 当 ρ_k 为 K 的内圆半径时

$$\frac{h_K}{\rho_K} \leqslant \kappa, \quad \forall K \in \bigcup_h \mathcal{T}_h$$

则称网格集 $\{T_A | h \to 0\}$ 准均匀。

5.3.2 准静态脆性相场裂缝模型空间离散弱形式

命题 30 (有限元空间离散化) 令 $V_{u,h} \subset V_u^D, V_{\varphi,h} \subset V_\varphi$ 为有限元空间，加载步 $n = 1, 2, 3, \cdots, N$。给定 φ_h^{n-1}，当 $U_h^n = (u_h, \varphi_h) \in V_{u,h} \times V_{\varphi,h}$，$\varphi_h(0) = \varphi_{0,h}$ 时，有

$$\text{当 } \forall \Psi_h = (w_h, \psi_h) \in V_{u,h}^0 \times V_{\varphi,h} \text{ 时}, \quad A(U_h^n)(\Psi_h) = 0$$

其中

$$A(U_h^n)(\Psi) = A_1(U_h^n)(w_h) + A_2(U_h^n)(\psi_h)$$

$$A_1(U_h^n)(w_h) = (g(\varphi_h^n)\sigma(u_h^n), \nabla w_h)$$

$$A_2(U_h^n)(\psi_h) = ((1-\kappa)\varphi_h^n\sigma(u_h^n) : e(\psi_h^n), \psi_h) + \left(-\frac{G_c}{\varepsilon}(1-\varphi_h^n), \psi_h\right)$$

$$+ (G_c\varepsilon\nabla\varphi_h^n, \nabla\psi_h) + \left(\gamma[\varphi_h^n - \varphi_h^{n-1}]_+, \psi_h\right)$$

命题 31 (有限元半线性空间离散化形式) 基于前文，我们可得到两个半线性形式：

$$A_1(U_h^n)(w_h) = (g(\varphi_h^n)\sigma^+(u_h^n), \nabla w_h) + (\sigma^-(u_h^n), \nabla w_h)$$

$$A_2(U_h^n)(\psi_h) = ((1-\kappa)\varphi_h^n\sigma^+(u_h^n) : e(u_h^n), \psi_h) + \left(-\frac{G_c}{\varepsilon}(1-\varphi_h^n), \psi_h\right)$$

$$+ (G_c\varepsilon\nabla\varphi_h^n, \nabla\psi_h) + \left(\gamma[\varphi_h^n - \varphi_h^{n-1}]_+, \psi_h\right)$$

备注 36 上式中，须考虑增广 Lagrangian 补偿项中 Ξ_h，细节请参考 7.11 节。接下来我们讨论命题 24 的完全离散格式。

命题 32 对于加载步 $n = 1, 2, 3, \cdots, N$，给定 $\mathcal{H}_h^{n-1} := \mathcal{H}(u_h^{n-1})$，当 $U_h^n = (u_h^n, \varphi_h^n) \in V_{u,h}^D \times V_{\varphi,h}$，$\varphi_h(0) = \varphi_{0,h}$ 时，有

$$A(U_h^n)(\Psi_h) = 0, \quad \forall \Psi_h = (w_h, \psi_h) \in V_{u,h}^0 \times V_{\varphi,h}$$

其中

$$A(U_h^n)(\Psi_h) = A_1(U_h^n)(\Psi_h) + A_2(U_h^n)(\Psi_h)$$

$$A_1(U_h^n)(\Psi_h) = (g(\varphi_h^n)\sigma^+(u_h^n), \nabla w_h) + (\sigma^-(u_h^n), \nabla w_h)$$

$$A_2(U_h^n)(\Psi_h) = ((1-\kappa)\varphi_h^n\mathcal{H}_h^{n-1}, \psi_h) + \left(-\frac{G_c}{\varepsilon}(1-\varphi_h^n), \psi_h\right) + (G_c\varepsilon\nabla\varphi_h^n, \nabla\psi_h)$$

$\mathcal{H}_h^{n-1}$ 将在 7.13 节中详细介绍。

5.4 模型离散参数介绍

正则化参数 ε 与空间离散参数 h 的选择在 (裂缝、流体、材料等) 数值模拟过程中至关重要。我们在模拟过程中发现收敛性是一个非常 "微妙" 的性质，与我们的研究目标及领域息息相关。接下来我们将介绍相关参数的设定以及设定理由。

5.4.1 ε 与 h 的关系

本节我们将讨论如何对 ε 进行取值。根据图 5.4 和图 5.5，我们认为以下情况显然成立。

命题 33 (ε 的取值) 根据几何形状可知，在使用线性或双线性有限元模型时 $h \leqslant \varepsilon$，这意味着我们需要在有限元网格分辨率范围内捕获到过渡区 (即断裂边界)。

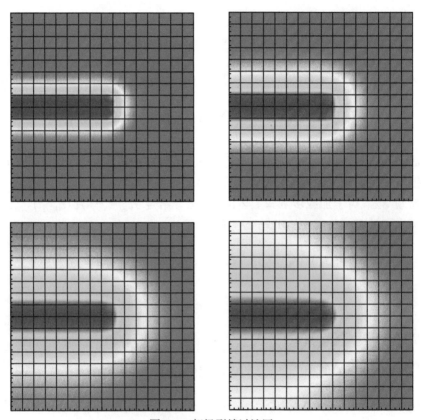

图 5.4 相场裂缝过渡区

左上至右下分别为 $\varepsilon = 0.5h, h, 2h, 4h$。随着 ε 增大，过渡区被捕获的越多，极限情况为图 2，此时 $\varepsilon = h$

备注 37 我们必须保证在空间离散化后还能得到梯度 $\nabla \varphi$ 的值，关于参数取值的讨论可参考相应文献 [70, 78, 67]。我们推荐读者参考关于 Allen-Cahn 方程的数值模拟部分 [109,457,162,161]。

命题 34 (高阶有限元) 基于前文，我们可以做出这样的假设：

$$\frac{h}{2^{l-1}} \leqslant \varepsilon$$

5.4 模型离散参数介绍

其中 l 为有限元多项式次数，当 $l=1$ 时，即为命题 33。

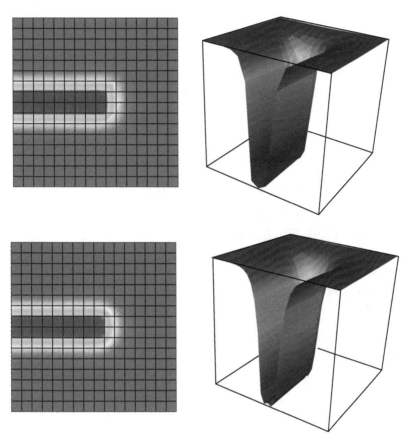

图 5.5 相场参数分布的比较 Q_1 (上一行) 和 Q_2 (下一行) 情况的近似值 $\varepsilon = 0.5h$
在右栏中，3D 曲面图更好地突出了相场参数 φ 的梯度

推理 2 当 $0 < \varepsilon \ll 1$ 时，h 在裂缝范围内的取值需要足够小才能保证 φ 具有足够的梯度。出于以上原因，局部自适应网格技术在相场裂缝模型中是不可或缺的。

备注 38 在多尺度数值建模中，ε 的取值同样至关重要[457]。

5.4.2 简化原型数值模型分析

本节我们提出了一个简化的原型数值模型来帮助读者理解相场正则化参数 k, ε 以及网格大小 h 之间的关系。

5.4.2.1 主要研究对象

主要研究对象有：

(1) $0 \leqslant \varphi \leqslant 1$ 的有界性及 H^1 范数；

(2) $|\nabla u|$ 与 ε 的关系；

(3) 位移 PDE 方程中，逼近结果与 k, h 对有限元误差的影响；

(4) 相场变分不等式中，逼近结果与 ε, h 对误差的影响。

部分结论可参考文献 [67, 78]。本节不设定 $k \to 0$ 或 $\varepsilon \to 0$，仅固定 $k > 0, \varepsilon > 0$。

5.4.2.2 问题陈述

首先我们介绍模型基本假设并给出控制方程。

假设 1 首先做出以下假设：

(1) 域 B 以及边界 ∂B 是正则的；

(2) 不考虑孔隙弹性；

(3) 模型不设置不等式约束；

(4) 方程均为经典椭圆型方程；

(5) 固定 $0 < \kappa \ll \varepsilon$；

(6) 模型满足文献 [170] 中有限元逼近及插值的基本假设。

接下来我们介绍两个原型问题。

公式 16 (稳态相场位移原型问题) 基于以上假设，给定 $\varphi: B \to \mathbb{R}$ 且为正，从而控制方程为椭圆型。另外根据定义 39，给定：

$$g(\varphi) = (1-\kappa)\varphi^2 + \kappa$$

此时 $u: B \to R$

$$-\nabla \cdot (g(\varphi)\nabla u) = f, \quad \text{在 } B \text{ 中}$$

$$u = 0, \quad \text{在 } \partial B \text{ 上}$$

且方程右手项 $f: B \to \mathbb{R}$。

命题 35 (κ 的取值) 退化函数符合以下条件：

$$\varphi = 1 \Rightarrow g(\varphi) = 1$$
$$\varphi = 0 \Rightarrow g(\varphi) = \kappa$$

由此可得：当 $\varphi = 1$ 时，$g(\varphi) = 1$ 恒成立；当 $\varphi = 0$ (裂缝区域) 时，κ 应当满足 $\kappa \ll 1$，如：$\kappa = 10^{-10}$。

5.4 模型离散参数介绍

例 11 (基于 κ 的模型误差分析) 在这里我们对三种情况分别进行模型误差分析。

(1) $\varphi = 1$，基质材料；
(2) $\varphi = 0.1$，裂缝边界过渡区域；
(3) $\varphi = 0$，裂缝区域；

尽管 $\varphi = 0.1, 0.2, 0.3$ 这样的取值在实际应用方面有待进一步的讨论，但在数模方面均表示裂缝边界过渡区域，因此不予深究。我们设定 $\kappa = 1, 0.1, 10^{-12}$ 并进行误差分析，结果如表 5.1 所示。结合定理 6 可知，当 $\varphi = 1$ 时，$g(1) = 1$ 独立于 κ；当 $\varphi = 0.1$ 时，由 κ 引起的误差绝对值为

$$|g_{\kappa=0.1}(0.1) - g_{\kappa=10^{-12}}(0.1)| = 0.099$$

表 5.1 不同 φ 与 κ 取值下 $g(\varphi)$ 的取值情况

κ	φ	$g(\varphi)$
1	1	1
1	0.1	1
1	10^{-12}	1
0.1	1	1
0.1	0.1	0.109
0.1	10^{-12}	0.1
10^{-12}	1	1
10^{-12}	0.1	0.01
10^{-12}	10^{-12}	10^{-12}

推理 3 基于前文的研究结果可知，κ 的影响域可定义为

$$\Omega_\kappa = \{x \in B | 0 \leqslant \varphi^2 < \kappa \ll 1\}$$

在实际应用中，κ 能够避免离散化产生奇异矩阵，当 $\varphi = 0$ 时，有

$$-\nabla \cdot (\kappa \nabla u) = f$$

实际上，我们甚至可以令 $\kappa = 0$。

公式 17 (稳态相场模型方程) 基于前文假设。令 $u := B \to \mathbb{R}$ 给定且 $|\nabla u|^2 > 0, \varphi := B \to \mathbb{R}$：

$$-\varepsilon \Delta \varphi - \frac{1}{\varepsilon}(1-\varphi) + |\nabla u|^2 \varphi = 0, \quad \text{在 } B \text{ 中}$$

$$\varepsilon \partial_n \varphi = 0, \quad \text{在 } \partial B \text{ 上}$$

接下来我们将进行弱形式推导。

命题 36 (位移方程弱形式) 令 $f \in L^2(B), g(\varphi) \in L^\infty(B)$。当 $u \in V := H_0^1(B)$ 时：
$$a(u,w) = l(w), \ \forall w \in V$$

其中
$$a(u,w) = (g(\varphi)\nabla u, \nabla w)$$
$$l(w) = (f,w)$$

我们对相场参数的变分不等式进行了简单重构，利用 $-\dfrac{1}{\varepsilon}$ 作为右手项，可得以下命题。

命题 37 (相场方程弱形式) 令 $\varepsilon > 0$，给定 $|\nabla u|^2 \in L^\infty(B)$，$\varphi \in H_0^1(B)$，有
$$a(\varphi, \psi) = l(\psi), \quad \forall \psi \in H^1(B)$$

其中
$$a(\varphi, \psi) = \varepsilon(\nabla\varphi, \nabla\psi) + \frac{1}{\varepsilon}(\varphi, \psi) + (|\nabla u|^2 \varphi, \psi)$$
$$l(\psi) = \frac{1}{\varepsilon}(1, \psi)$$

5.4.2.3 适定性

由于求解的两组方程均为椭圆型方程，可参考 Lax-Milgram 定理，采用 H^1 范数，有
$$\|v\|_V^2 := \|v\|_{H^1}^2 = \int v^2 + |\nabla v|^2 \mathrm{d}x$$

假设 2

(1) $l(\cdot)$ 为有界线性形式
$$|l(u)| \leqslant C\|u\|_V, \quad u \in V$$

(2) $a(\cdot, \cdot)$ 为 $V \times V$ 连续的双线性形式
$$|a(u,v)| \leqslant \gamma \|u\|_V \|v\|_V, \quad \gamma > 0, \quad u, v \in V$$

(3) $a(\cdot, \cdot)$ 为椭圆型形式
$$a(u,u) \geqslant \alpha \|u\|_V^2, \quad \alpha > 0, \quad u \in V$$

5.4 模型离散参数介绍

命题 38 命题 36 中推导了 $u \in V, H_0^1(B) \subset V \subset H^1(B)$ 时的位移方程弱形式，其适定性分析如下：

$$\|u\|_V \leqslant \frac{\|l\|_{V^*}}{\alpha(\kappa)} = \frac{\|f\|_{L^2}}{\alpha(\kappa)}$$

式中

$$\|l\|_{V^*} := \sup_{u \neq 0} \frac{|l(u)|}{\|u\|_V}, \quad \alpha(\kappa) := \frac{\kappa}{2} \min\left(1, \frac{1}{c_{\mathrm{PF}}}\right)$$

其中 $c_{\mathrm{PF}} > 0$ 为 Poincaré-Friedrichs 常数[77]。

命题 39 命题 37 中推导了 $\varphi \in H^1(B)$ 的弱形式，其稳定性分析如下：

$$\|\varphi\|_{H^1} \leqslant \frac{1}{\alpha \varepsilon} \|1\|_{L^2} \cong \frac{1}{\varepsilon^2} \|1\|_{L^2}$$

其中 $\alpha := \alpha(\varepsilon) \sim \varepsilon$。

5.4.2.4 数值近似

定理 5 根据前文可知：
(1) 相场参数的取值是有界的

$$0 \leqslant \varphi \leqslant 1$$

(2)

$$\|\nabla \varphi\|_{L^2}^2 + \frac{1}{2\varepsilon^2} \|\varphi\|_{L^2}^2 \leqslant \frac{1}{2\varepsilon^2} \|1\|_{L^2}^2$$

(3) φ 的梯度满足

$$\|\nabla \varphi\|_{L^2} \cong \frac{1}{\varepsilon}$$

(4) 位移梯度满足

$$\|\nabla u\|_{L^\infty} \cong \frac{1}{\varepsilon}$$

备注 39 (网格细分方案) 预测-校正网格细化方案是为了解决过渡区相场参数变化的问题，由于过渡区范围较小，数值变化相对较快，我们需要足够的离散化网格才能达到精度要求，因此提出了预测-校正自适应网格方法，后续章节将对其进行详细介绍。

5.4.2.5 Céa's 定理

命题 40 (位移问题 Céa's 定理) 令 V 为 Hilbert 空间，$V_h \subset U$ 为有限维子空间，假设 Lax-Milgram 定理成立。令 $u \in V, u_h \in V_h$ 为离散后的解变量，则

$$\|u - u_h\|_V = C_{\text{Cea}} \inf_{w_h \in V_h} \|u - w_h\|_V$$

其中

$$C_{\text{Cea}} = \frac{\gamma}{\alpha} = \frac{1}{\frac{\kappa}{2}\min\left(1, \frac{1}{C_{\text{PF}}}\right)} \simeq \frac{1}{\kappa}$$

且

(1) 对于 $g(\varphi)$ 尤其当 $\varphi^2 \approx 0$ 时，由于 κ 的影响，预测误差比近似误差大得多；

(2) 在裂缝周边 Ω_κ 区域会出现较大误差；

(3) 在远离裂缝区域，$g(\varphi) = 1$，与经典泊松问题结果相似。

命题 41 (相场问题 Céa's 定理) 令 V 为 Hilbert 空间，$V_h \subset U$ 为有限维子空间，假设 Lax-Milgram 定理成立。令 $\varphi \in V, \varphi_h \in V_h$ 为离散后的解变量，则

$$\|\varphi - \varphi_h\|_V = C_{\text{Cea}} \inf_{\psi_h \in V_h} \|\varphi - \psi_h\|_V$$

其中

$$C_{\text{Cea}} = \frac{\gamma}{\alpha} = \frac{\max\left(\varepsilon, \frac{1}{\varepsilon}, |\nabla u|^2\right)}{\min\left(\varepsilon, \frac{1}{\varepsilon}, |\nabla u|^2\right)} \simeq \frac{\max\left(\varepsilon, \frac{1}{\varepsilon}, \frac{1}{\varepsilon^2}\right)}{\min\left(\varepsilon, \frac{1}{\varepsilon}, \frac{1}{\varepsilon^2}\right)} \simeq \frac{1}{\varepsilon^3}$$

参考定理 5，有

$$|\nabla u|^2 \simeq \varepsilon^{-2}$$

5.4.2.6 有限元误差分析

我们首先假设模型标准正则并采用经典插值法。

引理 2 (插值误差估计) 在三角剖分中误差分析如下。

$$\|v - i_h v\|_{L^2} \leqslant ch^2 |v|_{H^2} = \mathcal{O}\left(h^2\right) \tag{5.11}$$

$$\|v - i_h v\|_{H^1} \leqslant ch |v|_{H^2} = \mathcal{O}(h) \tag{5.12}$$

5.4 模型离散参数介绍

命题 42 令 $\mathcal{T}_h$ 为域 B 的三角剖分，固定 $\varepsilon > 0$。令 $u_A \in V_h$ 为离散化解变量，则有

$$\|u - u_h\|_V \leqslant cC_{\text{Cea}}h\|f\|_{L^2}$$

在 Ω_κ 中，有

$$\|u - u_h\|_V = O\left(\frac{h}{\varepsilon}\right)$$

证明：参考命题 37 并根据 Céa's 引理处理插值结果。

命题 43 令 $\mathcal{T}_h$ 为域 B 的三角剖分，固定 $\varepsilon > 0$。令 $\varphi_h \in V_h \subset H^1(B)$ 为离散化解变量，则有

$$\|\varphi - \varphi_h\|_V \leqslant cC_{\text{Cea}}h\frac{1}{\varepsilon}\|1\|_{L^2} = O\left(\frac{h}{\varepsilon^4}\right)$$

5.4.3 正则化参数取值建议

基于文献资料与前文简化原型模型的误差分析，我们总结了相关参数的取值建议。

5.4.3.1 具体研究目标

在建模过程中最重要的一步是找出具体研究目标。基于以上讨论，我们可以给出三个方向：
(1) 当 $\kappa \to 0, \varepsilon \to 0, h \to 0$ 时的模型收敛性；
(2) 数值离散化后的网格与时间步研究；
(3) 工程应用时的损伤区域，裂缝扩展路径，位移与应力分布的预测以及证明。

5.4.3.2 数值分析

定理 6 基于文献 [70] 中提出的经典相场裂缝模型，在 Ambrosio-Tortorelli 近似下有

$$h \ll \kappa \ll \varepsilon$$

即

$$h = o(\varepsilon), \quad h = o(\kappa), \quad \kappa = o(\varepsilon)$$

定理 7 (k 的取值) 在裂缝区 Ω_κ 有 $\varphi^2 \ll 1, \kappa \approx \varphi^2$，且

$$h \ll \kappa$$

在 $B \backslash \Omega_\kappa, \varphi \leqslant 1$ 中，我们令

$$\kappa \ll 1$$

此时 h 取值相对独立，如命题 35。然而为了避免退化函数 $g(\varphi)$ 中的模型误差，κ 取值不宜过小。

定理 8 (ε 的取值)　在设定相场裂缝正则化参数 ε 时，我们可参考定理 5：

$$h \leqslant \varepsilon$$

其中 $\|\nabla\varphi\|_{L^2} \sim \dfrac{1}{\varepsilon}$ 在足够的精度下近似成立。参考命题 43 可知：$\|\varphi - \varphi_h\|_V = O\left(\dfrac{h}{\varepsilon^4}\right)$。

5.4.3.3　数值模拟

命题 44　h 为空间离散化参数，在实际数值模拟中我们常设置：

$$\kappa = 10^{-12}, \cdots, 10^{-18}$$

$$\varepsilon = 2h, \cdots, 8h$$

此时：

(1) 这两项取值在渐进极限上均与定理 6 存在出入。

(2) 根据定理 7，κ 的取值需要足够小。然而，材料在 Ω_κ 中是完全破碎的，该区域的数值模拟并不需要高精度。

(3) 设定 $\varepsilon = ch$，此时由于 $h \ll \kappa \ll \varepsilon, c > 1$。然而根据定理 6，$\varepsilon = ch$ 会影响模型收敛性，因此可令 $\varepsilon = ch^\beta, 0 < \beta < 1$，更多细节可参考命题 44。需要强调的是该设定仅在 $\varepsilon > \varepsilon_0 \geqslant 0$ 时成立。

备注 40　Allen-Cahn 相场模型的相关证明请参考文献 [457, 161, 109]。

5.4.3.4　模型离散误差分析

在进行误差分析的过程中，首先需要对误差进行分类。

引理 3　令 $U = (u, \varphi)$，$U := u$ 或 $U := \varphi$ 是极限情况 $\varepsilon = 0$ 下的连续解，而 U_ε 为 $\varepsilon > 0$ 情况下的连续解，$U_{\varepsilon,h}$ 为 $\varepsilon > 0$ 情况下的离散解，此时我们有

$$\|U - U_{\varepsilon,h}\| \leqslant \|U - U_\varepsilon\| + \|U_\varepsilon - U_{\varepsilon,h}\| \tag{5.13}$$

其中 $\|U - U_\varepsilon\|$ 为模型误差，$\|U_\varepsilon - U_{\varepsilon,h}\|$ 为离散误差。

备注 41 (U_ε 与 U)　除本节以外，当 $\varepsilon > 0$ 时，本书其他部分不需要特别备注 U_ε，仅用 U 指代即可。

备注 42　公式 (5.13) 中右手项的离散误差部分可参考前文的经典误差分析，模型误差部分则与模型收敛速度有关。

5.4.3.5 本节小结

命题 45 基于以上讨论我们可得：

(1) 在工程计算领域 $\kappa = 10^{-12}$，$\varepsilon = ch$，$c = 2$ 的取值均可适用。

(2) 在收敛性研究方面，可令 $\kappa = 10^{-12}$，并对以下设置进行对比分析：

(a) 固定 $\varepsilon = 2h_{\text{coarse}}$，其中 h_{coarse} 为最粗的网格尺寸，此时 $h \to 0$；

(b) 令 $\varepsilon = 2h$；

(c) 令 $\varepsilon = ch^\beta$，$0 < \beta < 1$。

(3) 在网格细分研究方面，如果可以的话至少要在三重网格上分别进行计算；对于瞬态模型，至少研究三种不同的时间步长。

(4) 全局变量的选择对模拟结果至关重要。

同时我们需要注意：

(1) ε 是一个有着物理单位的参数，代表着损伤尺度[70,312]而且其很有可能有自己的物理含义，我们在工程分析时可重点关注 ε 的工程意义。

(2) 另一方面，数模研究中可令 ε 不为 0，以多孔介质研究[318,317]为例，随着 $\varepsilon \to 0$，数值模型也会相应改变。

(3) 相场裂缝被认为是一个有一定厚度的物体时，相场变量被解释为有助于确定流体区域的水平集函数[316,282]。

基于以上讨论，相场模型数值分析简化为经典的 h (空间) 和 k (时间步长) 的离散研究。然而这并不意味着问题简单化了，此时仍然需要处理带有不等式约束的非平稳、非线性、多物理场问题，这依旧会给我们带来极大的挑战。

5.5 动态相场裂缝模型

本节我们将介绍一个动态相场裂缝模型及其对应的变分形式。

命题 46 令 $B \subset \mathbb{R}^d$ 为开集且 $I := (0, T)$，$T > 0$ 为结束时间，有 $u : B \times I \to \mathbb{R}^d$ 与 $\varphi : B \times I \to \mathbb{R}$：

$$\rho_s \partial_t^2 u - \nabla \cdot (g(\varphi)\sigma(u)) = f, \quad 在 B \times I 中$$

$$\partial_t \varphi + (1-\kappa)\varphi \sigma(u) : e(u) - G_c \varepsilon \Delta \varphi - \frac{G_C}{\varepsilon}(1-\varphi) \leqslant 0, \quad 在 B \times I 中$$

$$\partial_t \varphi \leqslant 0, \quad 在 B \times I 中$$

$$u = 0, \quad 在 \partial B \times I 上$$

$$\varepsilon \partial_n \varphi = 0, \quad 在 \partial B \times I 上$$

$$u = u_0, \quad 在 B \times \{0\} 时$$

$$\partial_t u = v_0, \quad 在 B \times \{0\} 时$$

$$\varphi = \varphi_0, \quad 在 B \times \{0\} 时$$

接下来我们重新定义了时间连续情况下的凸集：

$$K := \{\varphi \in V_\varphi | \partial_t \varphi \leqslant 0\}, \quad 在 B \times I 中$$

变分形式推导如下。

命题 47 有 $(u,\varphi) \in V_u^D \times K$，其中 $u(0) = u_0$，$\partial_t u(0) = v_0$，$\varphi(0) = \varphi_0$，则对 $t \in I$ 有

$$(\rho_s \partial_t^2 u, w) + (g(\varphi)\sigma(u), \nabla w) = (f, w), \quad \forall w \in V_u^0$$

$$(\partial_t \varphi, \psi - \varphi) + (1-\kappa)(\varphi \sigma(u) : e(u), \psi - \varphi)$$
$$+ \varepsilon (G_c \nabla \varphi, \nabla(\psi - \varphi)) - \frac{1}{\varepsilon}(G_c(1-\varphi), \psi - \varphi) \geqslant 0, \quad \forall \psi \in K$$

接下来采用单次或分步 θ 分解法处理位移公式。

命题 48 有 $(u, v, \varphi) \in V_u^D \times L^2(B) \times K$，其中 $u(0) = u_0, v(0) = v_0, \varphi(0) = \varphi_0$，则对 $t \in I$ 有

$$(\partial_t u, w^u) - (v, w^u) = 0, \quad \forall w^u \in L^2(B)$$

$$(\rho_s \partial_t v, w) + (g(\varphi)\sigma(u), \nabla w) = (f, w), \quad \forall w \in V_u^0$$

$$(\partial_t \varphi, \psi - \varphi) + (1-\kappa)(\varphi \sigma(u) : e(u), \psi - \varphi)$$
$$+ \varepsilon (G_c \nabla \varphi, \nabla(\psi - \varphi)) - \frac{1}{\varepsilon}(G_c(1-\varphi), \psi - \varphi) \geqslant 0, \quad \forall \psi \in K$$

与之前一样，我们将方程求和，可得到一个通用的半线性形式：

$$A(U)(\Psi - \Phi) := A_1(U)(w^u) + A_2(U)(w) + A_3(U)(\psi - \varphi)$$

其中 $\Phi = (0, 0, \varphi)$，A_1, A_2, A_3 为前三个弱形式的半线性形式。

时间离散化与变分不等式约束的增广 Lagrangian 正则化形式相结合可得

$$\gamma [\partial_t \varphi]^+ \to \gamma \left[\frac{\varphi - \varphi^{n-1}}{k}\right]^+$$

命题 49 (半离散动态相场裂缝模型) 令 $u^{n-1}, v^{n-1}, \varphi^{n-1}$ 为前一个时间步的解，对于 $n = 1, 2, 3, \cdots, n$，$\forall \Psi \in X^0$ 有

$$A(U^n)(\Psi) = F(\Psi)$$

且

$$A(U^n)(\Psi) := A_T(U^{n,k})(\Psi) + A_E(U^n)(\Psi)$$

$$= \left(\frac{u^n - u^{n-1}}{k}, w^u\right) - \theta\left(v^n, w^u\right) - (1-\theta)\left(v^{n-1}, w^u\right) + \left(\rho_s \frac{v^n - v^{n-1}}{k}, w\right)$$

$$+ \theta\left(g\left(\varphi^n\right)\sigma\left(u^n\right), \nabla w\right) + (1-\theta)\left(g\left(\varphi^{n-1}\right)\sigma\left(u^{n-1}\right), \nabla w\right) + \left(\frac{\varphi^n - \varphi^{n-1}}{k}, \psi\right)$$

$$+ \theta(1-\kappa)\left(\varphi^n \sigma\left(u^n\right) : e\left(u^n\right), \psi\right) + (1-\theta)(1-\kappa)\left(\varphi^{n-1}\sigma\left(u^{n-1}\right) : e\left(u^{n-1}\right), \psi\right)$$

$$+ \theta\varepsilon\left(G_c \nabla \varphi^n, \nabla \psi\right) + (1-\theta)\varepsilon\left(G_c \nabla \varphi^{n-1}, \nabla \psi\right)$$

$$- \theta\frac{1}{\varepsilon}\left(G_c\left(1-\varphi^n\right), \psi\right) - (1-\theta)\frac{1}{\varepsilon}\left(G_c\left(1-\varphi^{n-1}\right), \psi\right)$$

$$+ \left(\left[\Xi + \gamma\frac{\varphi^n - \varphi^{n-1}}{k}\right]^+, \psi\right)$$

$$F(\Psi) := \theta(f, w) + (1-\theta)\left(f^{n-1}, w\right)$$

由此我们得到了一系列准稳态问题 Galerkin 空间离散化的形式。

命题 50 (时空全离散化的动态相场裂缝模型) 令 $u_h^{n-1}, v_h^{n-1}, \varphi_h^{n-1}$ 为前一个时间步的解，对于 $\bar{n}_n = 1, 2, 3, \cdots, n$，$U_h^n \in X_h^D$，有

$$A\left(U_h^n\right)\left(\Psi_h\right) = F\left(\Psi_h\right), \quad \forall \Psi_h \in X_h^0$$

需要注意的是，这仍是一个非线性离散的问题。

第 6 章 数值模拟 I：相场裂缝域

第 5 章讨论了相场裂缝模型的正则化与离散化。本章提供了相关数值案例，帮助读者更好地理解建模过程中的相关细节。在与 Ivo Babuska 讨论中我们初步确立了模型的空间域 [63,436,24]，并在此基础上提出了相场建模方法。尽管本章并不涉及裂缝扩展的内容，但仍能帮助读者更好地理解相场裂缝的相关特征。

6.1 研究对象与特征

在文献 [24, 63] 中，研究人员提出了一种非齐次 Dirichlet 条件裂缝的模拟方案。我们在前文中推导了简化弹性相场裂缝模型的相关公式。在泊松问题与弹性问题的研究过程中存在以下注意点：

(1) 利用相场函数可以完成初始裂缝的设置，即在 B 中，$\varphi^0 = \varphi(t_0)$，但是我们需要重点关注如何处理相场参数 φ^0 和初始网格尺寸 h 的关系。

(2) 注意补偿项 γ 对模型的影响。

(3) 能够固定参数 κ 与 ε 的值并保证我们能够在一个足够小的网格尺寸 h 下计算数值解。

(4) 便于后续研究 κ 与 ε 取值的变化对数模的影响。

(5) 能够进行自适应算法中的局部网格细化与误差控制研究 (参考第 10 章)。

(6) 模型计算的同时能够输出相关参数，如全局误差范数、总裂缝体积、缝宽、局部应力等。

(7) 能够进行线性或非线性求解器性能研究。

6.2 建模方案

模型空间域为 $B := (-1,1)^2$，初始裂缝位于 $\{(x,y)|-1 \leqslant x \leqslant 0,\ y=0\}$，该裂缝在 $(-1,0)$ 处造成了边界的不连续，该缝的解析解可写为

$$\left(\lambda_{G_c} r^{\frac{1}{2}} \sin \frac{\phi}{2};\ \{(x,0)|-\infty \leqslant x \leqslant 0\}\right)$$

其中极坐标 $r^2 = x^2 + y^2$。

6.3 边界条件

将 ∂B 上的边界条件设置为 $u_D := g = \lambda_{G_c} \sin\dfrac{\phi}{2}$。假定该模型中均为非齐次 Dirichlet 条件，则可表达为

$$x \leqslant 0,\ y \geqslant 0 \text{时},\ g(x,y) = \frac{\lambda_{G_c}}{\sqrt{2}} * \sqrt{\sqrt{x^2+y^2}-x}$$

$$x \leqslant 0,\ y \leqslant 0 \text{时},\ g(x,y) = -\frac{\lambda_{G_c}}{\sqrt{2}} * \sqrt{\sqrt{x^2+y^2}-x}$$

$$x \geqslant 0,\ y \geqslant 0 \text{时},\ g(x,y) = \frac{\lambda_{G_c}}{\sqrt{2}} * \sqrt{\sqrt{x^2+y^2}-x}$$

$$x \geqslant 0,\ y \leqslant 0 \text{时},\ g(x,y) = -\frac{\lambda_{G_c}}{\sqrt{2}} * \sqrt{\sqrt{x^2+y^2}-x}$$

对于相场变量则采用齐次 Neumann 条件，即 $\varepsilon \partial_n \varphi = 0$。

6.4 初始条件

对于相场函数而言，需要首先在初始几何模型中定义裂缝位置。在该案例中我们采用：

$$\varphi^0 := \{(x,y) \in B \mid -1 \leqslant x \leqslant 0;\ -h \leqslant y \leqslant h\}$$

h 为网格大小参数，详见图 6.1。

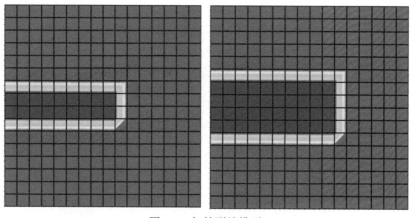

图 6.1 初始裂缝模型

左图，初始裂缝采用网格尺寸 h 进行设置；右图，初始裂缝采用 $2h$ 进行设置

6.5 模 型 参 数

在该模型中，我们引入材料参数 G_c，λ_{G_c}，μ，以及离散化参数 h。在本章模型中，$\kappa = 10^{-12}$，$G_c = \lambda_{G_c}^2 \times \sqrt{\pi/2}$，$\lambda_{G_c} = 1.0$，$\mu = 1.0$，$\gamma = 10^3$。相场正则化参数 ε 则根据模型实际情况调整。

6.6 解变量与相关参数

我们在建模过程中主要研究以下解变量及相关参数：
(1) φ 的 L^2 范数：$\|\varphi - \varphi_{\varepsilon,h}\|_{L^2(B)}$；
(2) u 的 L^2 范数：$\|u - u_{\varepsilon,h}\|_{L^2(B)}$；
(3) u 的 H^1 范数：$\|u - u_{\varepsilon,h}\|_{H^1(B)}$。

此外还有：
(1) 单点值取 $J^P(u) = u(0.75, -0.5)$；
(2) 半域函数取 $J^D(u) = \int_{\{x \geqslant 0\}} u \mathrm{d}x$；
(3) 裂缝能量函数为 $J^C(\varphi) = E_s(\varphi)$。

6.7 本 章 小 结

我们运行了几个算例来帮助读者更好地了解相场裂缝的数值模拟方法。
(1) $\varepsilon = 2h$ 算例图如下。
图 6.2 为位移场的 2D 与 3D 云图，图 6.3 为对应的相场裂缝扩展云图。

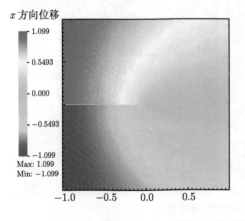 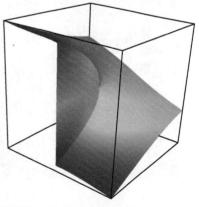

图 6.2 位移场云图

左图：标准 2D 空间内位移场分布云图；右图：3D 曲面图，可观察到位移值在裂缝区域存在突变

6.7 本章小结

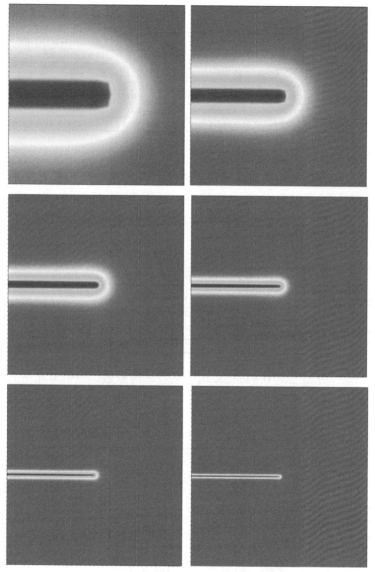

图 6.3 $\varepsilon = 2h$ 时相场参数分布云图 (全局细分 6 次)
在最终的网格中我们有 262144 个单元

(2) 令 $\varepsilon = 2h$, Q_1^c 离散时的范数、点值分布研究。

在第一组计算过程中,我们研究了 ε-h 的关系。如图 6.1 与图 6.2 所示,对于全局 L^2 范数与对应点值分析可知,模型收敛速度介于 0.5 与 1 之间。参考 5.5 节中假设可知,我们无法得到全局位移值的 H^1 收敛范数,然而当我们排除裂缝区域后可知 H^1 收敛范数约在 0.85 左右,详见表 6.1。

表 6.1　$\varepsilon = 2h$，Q_1^c 离散时参数表

h	ε	$u(0.75, -0.5)$	$\|\varphi - \varphi_{\text{eps},h}\|_{L^2}$	$\|u - u_{\text{eps},h}\|_{L^2}$	$\|u - u_{\text{eps},h}\|_{H^1}$	$\|u - u_{\text{eps},h}\|_{H^1(\tilde{B})}$
1.76776×10^{-1}	3.53553×10^{-1}	9.92766×10^{-3}	6.65147×10^{-1}	2.25360×10^{-1}	3.28350×10^{0}	1.28949×10^{-1}
8.83883×10^{-2}	1.76776×10^{-1}	5.33744×10^{-3}	4.07929×10^{-1}	1.57825×10^{-1}	4.39123×10^{0}	7.17441×10^{-2}
4.41941×10^{-2}	8.83883×10^{-2}	2.87674×10^{-3}	2.64571×10^{-1}	1.09320×10^{-1}	5.97844×10^{0}	3.98377×10^{-2}
2.20970×10^{-2}	4.41941×10^{-2}	1.56450×10^{-3}	1.77630×10^{-1}	7.57405×10^{-2}	8.25670×10^{0}	2.20526×10^{-2}
1.10485×10^{-2}	2.20970×10^{-2}	8.54456×10^{-4}	1.22886×10^{-1}	5.26920×10^{-2}	1.15146×10^{1}	1.21450×10^{-2}
5.52427×10^{-3}	1.10485×10^{-2}	4.66603×10^{-4}	8.66212×10^{-2}	3.68272×10^{-2}	1.61551×10^{1}	6.64832×10^{-3}

(3) 令 $\varepsilon = 2h$，Q_2^c 离散时的范数、点值分布研究。

在该模拟中我们再次观察到其收敛速度小于 1，对 H_1 范数而言并无变化，由此可知高阶离散并无明显优势。其中模拟结果证明了备注 46 的真实性。令 $\varepsilon = 2h$，此时 ε-h 之间为线性关系，此时有表 6.2。

表 6.2　$\varepsilon = 2h$，Q_2^c 离散时参数表

h	ε	$u(0.75, -0.5)$	$\|\varphi - \varphi_{\text{eps},h}\|_{L^2}$	$\|u - u_{\text{eps},h}\|_{L^2}$	$\|u - u_{\text{eps},h}\|_{H^1}$	$\|u - u_{\text{eps},h}\|_{H^1(\tilde{B})}$
1.76776×10^{-1}	3.53553×10^{-1}	8.30593×10^{-3}	6.77870×10^{-1}	2.12162×10^{-1}	3.35415×10^{0}	1.07176×10^{-1}
8.83883×10^{-2}	1.76776×10^{-1}	4.41495×10^{-3}	4.18879×10^{-1}	1.50870×10^{-1}	4.42787×10^{0}	5.89949×10^{-2}
4.41941×10^{-2}	8.83883×10^{-2}	2.35965×10^{-3}	2.73134×10^{-1}	1.05871×10^{-1}	5.99364×10^{0}	3.24509×10^{-2}
2.20970×10^{-2}	4.41941×10^{-2}	1.27261×10^{-3}	1.84253×10^{-1}	7.41070×10^{-2}	8.25971×10^{0}	1.78001×10^{-2}
1.10485×10^{-2}	2.20970×10^{-2}	6.89267×10^{-4}	1.28390×10^{-1}	5.19459×10^{-2}	1.1511×10^{1}	9.71886×10^{-3}

(4) 令 $\varepsilon = 2h_{\text{conrse}} = 0.353553$，$Q_1^c$ 离散时的范数、点值分布研究。

接下来我们根据粗网格的尺寸固定 ε，仅改变 h。参数如表 6.3 所示。

表 6.3　$\varepsilon = 2h_{\text{conrse}} = 0.353553$，$Q_1^c$ 离散时参数表

h	ε	$u(0.75, -0.5)$	$\|\varphi - \varphi_{\text{eps},h}\|_{L^2}$	$\|u - u_{\text{eps},h}\|_{L^2}$	$\|u - u_{\text{eps},h}\|_{H^1}$	$\|u - u_{\text{eps},h}\|_{H^1(\tilde{B})}$
1.76776×10^{-1}	3.53553×10^{-1}	9.92766×10^{-3}	6.65147×10^{-1}	2.25360×10^{-1}	3.28350×10^{0}	1.28949×10^{-1}
8.83883×10^{-2}	3.53553×10^{-1}	1.66914×10^{-2}	7.77520×10^{-1}	2.01387×10^{-1}	4.28395×10^{0}	1.71584×10^{-1}
4.41941×10^{-2}	3.53553×10^{-1}	1.75405×10^{-2}	8.30471×10^{-1}	1.81495×10^{-1}	5.98922×10^{0}	1.74764×10^{-1}
2.20970×10^{-2}	3.53553×10^{-1}	1.75858×10^{-2}	8.53918×10^{-1}	1.72314×10^{-1}	6.63652×10^{0}	1.74176×10^{-1}
1.10485×10^{-2}	3.53553×10^{-1}	1.75052×10^{-2}	8.63479×10^{-1}	1.68434×10^{-1}	6.08944×10^{0}	1.72781×10^{-1}
5.52427×10^{-3}	3.53553×10^{-1}	1.73946×10^{-2}	8.66369×10^{-1}	1.66089×10^{-1}	5.82924×10^{0}	1.71000×10^{-1}

(5) $\varepsilon = 2\sqrt{h}$，Q_1^c 离散时的范数、点值分布研究。

在该模拟中我们令 $\varepsilon = 2\sqrt{h}$，参考 5.5.3.6 节，$\beta = 0.5$。观察全局 L^2 范数，明显有高阶收敛，与之前的模拟一样，无全局 H^1 收敛范数。而局部 H^1 收敛范数约为 0.85，这说明在不考虑裂缝影响下 ε 的取值对数值模拟收敛速度影响较小。计算结果如表 6.4 所示。

(6) 点值、半域函数与裂缝能量离散误差分析。

固定 $\varepsilon = 2h_{\text{coarse}} = 0.353553$ 采用 Q_1^c 离散，参数如表 6.5 所示。在这里我们仅计算小网格下的解析解，进行收敛性方面的研究。

6.7 本章小结

我们在一个 9 次全局细分网格上计算得

$$J_{\text{ref}}^P(\varphi) = 0.40491845333765075$$
$$J_{\text{ref}}^D(u) = 1.1820392158614590$$
$$J_{\text{ref}}^C f(\varphi) = 1.8152453540665945$$

表 6.4 $\varepsilon = 2\sqrt{h}$，Q_1^c 离散时参数表

h	ε	$u(0.75, -0.5)$	$\|\varphi - \varphi_{\text{eps},h}\|_{L^2}$	$\|u - u_{\text{eps},h}\|_{L^2}$	$\|u - u_{\text{eps},h}\|_{H^1}$	$\|u - u_{\text{eps},h}\|_{H^1(\tilde{B})}$
1.76776×10^{-1}	8.40896×10^{-1}	9.92766×10^{-3}	1.20001×10^{0}	2.25360×10^{-1}	3.28350×10^{0}	1.28949×10^{-1}
8.83883×10^{-2}	5.94603×10^{-1}	5.33744×10^{-3}	9.21985×10^{-1}	1.57825×10^{-1}	4.39123×10^{0}	7.17441×10^{-2}
4.41941×10^{-2}	4.20448×10^{-1}	2.87674×10^{-3}	7.16836×10^{-1}	1.09320×10^{-1}	5.97844×10^{0}	3.98377×10^{-2}
2.20970×10^{-2}	4.20448×10^{-1}	1.56450×10^{-3}	5.77596×10^{-1}	7.57405×10^{-2}	8.25670×10^{0}	2.20526×10^{-2}
1.10485×10^{-2}	2.10224×10^{-1}	8.54456×10^{-4}	4.78872×10^{-1}	5.26920×10^{-2}	1.15146×10^{1}	1.21450×10^{-2}
5.52427×10^{-3}	1.48650×10^{-1}	4.66603×10^{-4}	4.01274×10^{-1}	3.68272×10^{-2}	1.61551×10^{1}	6.64832×10^{-3}

表 6.5 固定 $\varepsilon = 0.353553$，小网格下的数值模拟参数表

h	$J^P(u_h)$	$\|J^P(u_h) - J_{\text{ref}}^P(u_h)\|$	$J^D(u_h)$	$\|J^D(u_h) - J_{\text{ref}}^D(u_h)\|$	$J^c(\varphi_h)$	$\|J^c(\varphi_h) - J_{\text{ref}}^c(\varphi_h)\|$
1.76776×10^{-1}	4.00505×10^{-1}	4.41306×10^{-3}	1.13282×10^{0}	4.92143×10^{-2}	2.18138×10^{0}	3.66142×10^{-1}
8.83883×10^{-2}	4.04598×10^{-1}	3.19599×10^{-4}	1.17345×10^{0}	8.57989×10^{-3}	2.03920×10^{0}	2.23964×10^{-1}
4.41941×10^{-2}	4.05037×10^{-1}	1.18667×10^{-4}	1.18937×10^{0}	7.33318×10^{-3}	1.95764×10^{0}	1.42396×10^{-1}
2.20970×10^{-2}	4.05041×10^{-1}	1.22623×10^{-4}	1.19160×10^{0}	9.56847×10^{-3}	1.90215×10^{0}	8.69079×10^{-2}
1.10485×10^{-2}	4.04987×10^{-1}	6.90576×10^{-5}	1.18669×10^{0}	4.66065×10^{-3}	1.85612×10^{0}	4.08773×10^{-2}

随后我们在单次细分网格系统中进行模拟，模型收敛性相对较差。

(7) 非线性求解器。

本节我们将提供一些非线性求解器的计算过程，该算法将在第 7 章详细介绍。我们采用 7.9.1 节中基于残差的牛顿求解器，$\varepsilon = 2h, \gamma = 10^3$，并采用 Q_1^c 离散。初始细分时 $h = 0.176777$，有 256 个有限元，非线性求解器求解过程如图 6.4 所示。

```
:=================================================:
Refinement cycle 0:
Iter   Residual       Reduction       Mat.build   LS   CPU time
1      1.76271e+00    1.00000e+00     r           6    2.09250e-02
2      1.68046e+00    9.53344e+01     r           0    1.12610e-02
3      7.34867e-01    4.37300e-01     r           0    1.13100e-02
4      3.19290e-01    4.34487e-01     r           0    1.13130e-02
5      2.59225e-01    8.11878e-01     r           0    1.01670e-02
6      1.65960e-01    6.40218e-01     r           0    1.09050e-02
7      1.57344e-01    9.48083e-01     r           0    9.82000e-03
8      8.68526e-02    5.51991e-01     r           0    8.60100e-03
       2.91434e-16
:=================================================:
```

图 6.4 初始细分时非线性求解器求解结果

细分 5 次后 $h = 5.5242717 \times 10^{-3}$，有 262144 个有限元，非线性求解器求解过程如图 6.5 所示。

```
:==============================================================:
Refinement cycle 5:
Iter    Residual         Reduction        Mat.build    LS    CPU time
1       1.82899e+00      1.00000e+00      r            0     8.09725e+01
2       1.05220e+00      5.75289e-01      r            0     9.11733e+01
3       4.29467e-03      4.08161e-03                   0     9.14439e+01
4       3.82478e-04      8.90588e-02                   0     9.11363e+01
5       2.66439e-07      6.96612e-04                   0     9.10128e+01
        5.88071e-16
:==============================================================:
```

图 6.5　细分 5 次后非线性求解器求解结果

图 6.5 中，Iter 为牛顿迭代步；Residual 为 $A\left(U^j\right)(\Psi)_{l\infty}$；Reduction 为 $\theta_j = \dfrac{A\left(U^j\right)(\Psi)}{A\left(U^{j-1}\right)(\Psi)}$。

(8) 补偿项 γ 对求解的影响。

在图 6.6 中介绍了补偿项 γ 的影响，当补偿项 γ 太小时，不可逆约束 $\partial_t \varphi \leqslant 0$ 则不够强。在模拟中意味着裂缝有可能消失；如果补偿项 γ 太大，则会对模型求解造成影响。因此，我们在模拟的过程中需要从足够小的补偿项开始[355]，以扩展拉格朗日或其他方法进行迭代 (参考 5.2.2 节)。

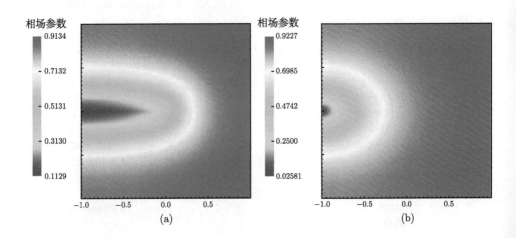

6.7 本章小结

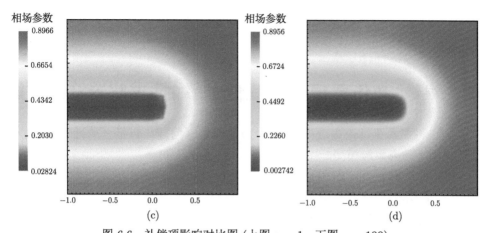

图 6.6 补偿项影响对比图 (上图 $\gamma = 1$；下图 $\gamma = 100$)

γ 越小，裂缝无法打开并会逐渐消失，在 (a)、(c) 中给出了裂缝初始细分状态，(b)、(d) 中给出了细分三次后的裂缝形态

第 7 章 数值建模 II：线性/非线性求解器及线性解

我们将在本章讨论离散模型的线性/非线性解问题，包括之前讨论的准静态裂缝模型与动态裂缝模型。主要有以下三个部分：
(1) 耦合 PDE/CVIS 系统求解；
(2) 非线性 PDE/CVIS 系统求解；
(3) 线性 PDE/CVIS 系统求解。

在求解耦合 PDE/CVIS 系统时主要有两种基本思路：
(1) 分部耦合求解：按照设定好的顺序对多个方程循环求解；
(2) 全耦合求解：一次求解出全部未知量。

正如第 3 章讨论的那样，大部分耦合问题是非线性的。在研究线性/非线性耦合问题时需要注意，当耦合问题 (整个耦合系统或其中的子系统) 是非线性时，常用到以下几种线性化方法：
(1) 解耦、时滞或外推处理；
(2) 固定点迭代法；
(3) 泛函迭代或算子分裂法；
(4) 牛顿迭代法。

关于这些方法的详细介绍可参考本书其他章节与相应参考文献 [361, 128, 335, 376, 55]。

本书提供了求解线性问题的线性求解器，迭代线性求解器开发时需要对预条件进行处理，由于内存以及计算成本问题，本书中介绍的求解器往往仅适用于原型问题 [376]。

7.1 分部耦合和全耦合

3.3 节对分部耦合与全耦合进行了简要介绍。全耦合方式一般是隐式的，被归类为强耦合方法。分部耦合 (分区耦合) 则是多个偏微分方程组组合而成的松散耦合方法。图 7.1 展示了两种方法流程上的区别。

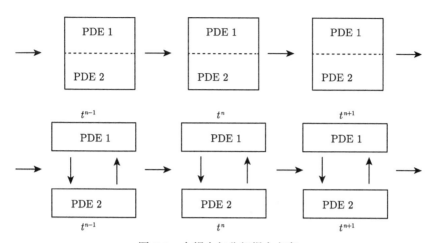

图 7.1 全耦合与分部耦合方案
在全耦合方案中,耦合系统一次求解;而在分部耦合方案中,经常需要内部多个方程迭代

7.2 线性化方案

本节我们将介绍几种常见的线性化方案,其中最简单的方法包括用已有的信息替换非线性项。这些信息可以来自前一个时间步的解 (时滞) 或者时插值、外推得来的相关变量。

以瞬态 PDE 方程为例:令 $u \in V$,当 $t \in I$ 时,有

$$a(u)(\phi) := (\partial_t u, \phi) + (u^2, \phi) = (f, \phi) =: l(\phi), \quad \forall \phi \in V$$

中间项可以按以下方法线性化:

$$(u^2, \phi) \to (uu^{n-1}, \phi)$$

其中 $u := u^n := u(t_n)$,是当前时间步未知量。$u^{n-1} := u(t_{n-1})$ 是上一个时间步的解变量。这是一个显式耦合方案。当我们令 $u^{n,0} := u^{n-1}$ 在 $j = 1, 2, 3, \cdots$ 时成立,则有望获得

$$\lim_{j \to \infty} uu^{n,j} = u^2$$

7.3 不动点迭代法

不动点迭代法是一种经典的迭代方案。

定义 56 (域 $\mathbb{R}$ 中不动点) 给定函数 f,不动点为

$$x = f(x)$$

接下来通过构造序列 $(x_j)_{j \in \mathbb{N}}$，$j = 1, 2, 3, \cdots$ 时可得到以下迭代格式：

$$x^j = f\left(x^{j-1}\right)$$

这一思想可被推广到线性方程组求解过程中。

命题 51 ($Ax = b$ **的不动点迭代法**) 给定 $Ax = b$，其中 $A \in \mathbb{R}^{n \times n}$，$x \in \mathbb{R}^n$，$b \in \mathbb{R}^n$。$j = 1, 2, 3, \cdots$ 时有

$$x^j = x^{j-1} + Cd^{j-1}$$

其中 $d^{j-1} = b - Ax^{j-1}$，并且 $C \in \mathbb{R}^{n \times n}$ 是可逆矩阵。

证明 给定 $Ax = b$ 有

$$0 = b - Ax \quad \Leftrightarrow \quad x = x + b - Ax \quad \Leftrightarrow \quad x = x + d \quad \Rightarrow \quad x^j = x^{j-1} + Cd^{j-1}$$

推理 4 (误差校正) 不动点迭代法可用于误差校正，给定初始猜测值 x^0，对于 $j = 1, 2, 3, \cdots$，有

$$\delta x^{j-1} = Cd^{j-1}$$
$$x^j = x^{j-1} + \delta x^{j-1}$$

备注 43 该推理可采用 Banach 不动点定理来证明，其收敛速度取决于矩阵 C，在牛顿迭代法中，C 即为雅可比矩阵。

接下来我们将这些概念扩展到半线性形式。

命题 52 给定 $U \in X$，则

$$\text{当 } \Psi \in X \text{ 时}, \quad A(U)(\Psi) = F(\Psi)$$

其中不动点迭代格式为

$$\left(U^j, \Psi\right) = \left(U^{j-1}, \Psi\right) + C\left(F(\Psi) - A\left(U^{j-1}\right)(\Psi)\right)$$

误差校正形式为

$$C^{-1}\left(\delta U^{j-1}, \Psi\right) = F(\Psi) - A\left(U^{j-1}\right)(\Psi)$$
$$U^j = U^{j-1} + \delta U^{j-1}$$

其中 $C = \omega I$，I 为单元矩阵，$w > 0$ 为松弛系数。

备注 44 (牛顿迭代法) 在牛顿法中，我们有

$$C^{-1}\left(\delta U^{j-1}, \Psi\right) := A'\left(U^{j-1}\right)\left(\delta U^{j-1}, \Psi\right)$$

其中 A' 为雅可比矩阵。

命题 53 (非线性迭代的终止条件) 在非线性迭代中我们常设置两个基本终止条件。

基于残差的终止条件：

$$\left\|A\left(U^j\right)(\Psi) - F(\Psi)\right\| < \text{TOL} \quad \text{或} \quad \left\|A\left(U^j\right)(\Psi) - F(\Psi)\right\| < \|F(\Psi)\|\text{TOL}$$

基于误差的终止条件：

$$\left\|U^j - U^{j-1}\right\| < \text{TOL} \quad \text{或} \quad \left\|U^j - U^{j-1}\right\| < \|U^j\|\text{TOL}$$

其中 TOL 常设置为 $\text{TOL} = 10^{-12}, \cdots, 10^{-8}$。

7.4 函数迭代法

函数迭代法的基本思想在于引入一个迭代序列 j，并利用前一个迭代步 $j-1$ 中已知的非线性项进行线性化处理，最终令

$$\lim_{j \to \infty} U^j = U$$

该方法既适用于稳态问题，也适用于瞬态问题。

公式 18 给定 $U \in X$：

$$\text{当 } \Psi \in X \text{ 时}, \quad A(U)(\Psi) = F(\Psi)$$

初始值 $U^0 \in X$，$j = 1, 2, 3, \cdots$，给定 U^{j-1}，则

$$\text{当 } \Psi \in X \text{ 时}, \quad A\left(U^j, U^{j-1}\right)(\Psi) - F(\Psi) = 0$$

半线性形式 $A\left(U^j, U^{j-1}\right)(\Psi)$ 中的部分量来自于当前迭代步，另外部分来自于上一迭代步。终止条件可参考命题 53。

接下来我们介绍一个简单算例：令

$$-\Delta u + u^2 = f$$

变分形式如下：令 $u \in V$，有

$$\text{当 } \psi \in V \text{ 时}, \quad A(u)(\psi) = F(\psi)$$

其中
$$A(u)(\psi) = (\nabla u, \nabla \psi) + (u^2, \psi)$$
$$F(\psi) = (f, \psi)$$

迭代格式构造如下。

算法 3 $j = 1, 2, 3, \cdots$，给定 u^{j-1}，则

$$\text{当 } \psi \in V \text{ 时}, \ A\left(u^j, u^{j-1}\right)(\psi) = F(\psi)$$

其中
$$A\left(u^j, u^{j-1}\right)(\psi) = \left(\nabla u^j, \nabla \psi\right) + \left(u^j u^{j-1}, \psi\right)$$
$$F(\psi) = (f, \psi)$$

终止条件可参考命题 53。

7.5 PDE 系统分部耦合模型

前文介绍的函数迭代概念仅考虑单个偏微分方程，本节我们将考虑多个 PDE 方程系统的迭代形式。

公式 19 假设 $U = (u, \varphi) \in X$，则

$$\text{当 } \Psi \in X \text{ 时}, \ A(U)(\Psi) = F(\Psi)$$

初始值 $U^0 \in X$，$j = 1, 2, 3, \cdots$，给定 U^j，则

$$\text{当 } \Psi \in X \text{ 时}, \ A\left(U^j, U^{j-1}\right)(\Psi) - F(\Psi) = 0$$

这里可拆解为单个运算符，即 $A_1\left(u^j, \varphi^{j-1}\right)(\psi_1)$ 与 $A_1\left(u^j, \varphi^j\right)(\psi_2)$，接下来按照以下顺序：

(1) 给定 φ^{j-1}，求解 $A_1\left(u^j, \varphi^{j-1}\right)(\psi_1) = F_1(\psi_1)$，得到 u^j；

(2) 给定 u^j，求解 $A_2\left(u^j, \varphi^j\right)(\psi_2) = F_2(\psi_2)$，得到 φ^j。

依照以上流程即可完成迭代。

7.6 相场裂缝分部耦合模型

接下来我们介绍相场裂缝分部耦合模型。其中准脆性断裂的交替最小化分部耦合方案已被广泛研究[70,71]，相关收敛性证明可参考相关文献 [68, 92, 306, 424, 318]。然而该方案迭代次数较多，效率较低，因此相关研究人员对模型进行了完善，在降低计算量的同时保证了足够的精度[86]。

7.6 相场裂缝分部耦合模型

我们参考命题 19 并求解相场裂缝模型的 CVIS 系统：

$$A_1((u,\varphi))(w) = (g(\varphi)\sigma(u), \nabla w)$$

$$A_2((u,\varphi))(\psi - \varphi) = ((1-\kappa)\varphi\sigma(u) : e(u), \psi - \varphi)$$

$$+ \left(-\frac{G_c}{\varepsilon}(1-\varphi), \psi - \varphi\right) + (G_c\varepsilon\nabla\varphi, \nabla(\psi - \varphi))$$

采用部分耦合方式，求解算法如下。

算法 4 给定时间 t_n，初始迭代值为

$$u^0 := u^{n,0} := u^{n-1}, \quad \varphi^0 := \varphi^{n,0} := \varphi^{n-1}$$

接下来我们按照以下顺序进行迭代：

(1) 给定 $\varphi^{j-1} \in K$，求解 $A_1\left((u^j, \varphi^{j-1})\right)(\omega) = 0$ 得到 $u^j \in V_u^D$；

(2) 给定 $u^j \in V_u^D$，求解 $A_2\left((u^j, \varphi^j)\right)(\psi - \varphi^j) \geqslant 0$ 得到 $\varphi^j \in K$。

备注 45 在算法实现过程中，由于非线性本构应力张量以及变分不等式的影响，A_1, A_2 两个子方程可能是非线性的，需要用到非线性求解器进行求解。

接下来我们介绍一种引入稳定项的分部耦合方案，基于非线性椭圆-抛物型 Richards 方程进行求解。在这里我们引入了两个稳定项：$L_u\left(u^j - u^{j-1}\right)$ 与 $L_\varphi\left(\varphi^j - \varphi^{j-1}\right)$。

固定 $L_u, L_p > 0$，L 为稳定项，j 为迭代序列。

显然，我们有

$$\lim_{j \to \infty} L_u \|u^j - u^{j-1}\| = 0$$

$$\lim_{j \to \infty} L_\varphi \|\varphi^j - \varphi^{j-1}\| = 0$$

稳定项 $L_u, L_\varphi > 0$ 且为常数，可在迭代过程中提高收敛效率。接下来我们以考虑增广 Lagrangian 补偿项的命题 31 为例介绍稳定项的实际应用。

命题 54 令稳定项 L_u, L_φ 为正，给定 $\gamma > 0$。命题 31 的半线性形式为 $A_1\left(U_h^n\right)(\omega_h)$，$A_2\left(U_h^n\right)(\psi_h)$，我们在时间 t_n 内迭代解决以下问题：

(1) 给定初始值 $u_h^{n,0} := u_h^{n-1}, \varphi_h^{n,0} := \varphi_h^{n-1}$，给定 $\left(u_h^{n,j-1}, \varphi_h^{n,j-1}\right) \in V_{t,h}^D \times V_{\varphi,h}$，令 $u_h^j \in V_{u,h}^D$，有

$$A_{L_u}\left(u_h^{n,j}, \varphi_h^{n,j-1}\right)(w_h) := L_u\left(u_h^{n,j} - u_h^{n,j-1}, w_h\right)$$

$$+ A_1\left(u_h^{n,j}, \varphi_h^{n,j-1}\right)(w_h) = 0, \quad \forall w_h \in V_{u,h}^0 \quad (7.1)$$

(2) 给定 $\left(\varphi_h^{n,j-1}, u_h^{n,j} \cdot \varphi_h^{n-1}\right) \in V_{\varphi,h} \times V_{u,h}^D \times V_{\varphi,h}$，令 $\varphi_h^j \in V_{\varphi,h}$，则

$$A_{L_\varphi}\left(U_h^{n,j}\right)(\psi_h) := L_\varphi\left(\varphi_h^{n,j} - \varphi_h^{n,j-1}, \psi_h\right) + A_2\left(U_h^{n,j}\right)(\psi_h) = 0, \quad \forall \psi_h \in V_{\varphi,h}$$

$$(7.2)$$

该结果可用于以下增广 Lagrangian 补偿项中 Ξ_A 与稳定项结合的算法设计。

算法 5 令 $\gamma > 0, a > 1$，同时 Ξ_h 与 L^0 均为正，设定 $L = L_u = L_\varphi$，令 $j = 0$。

循环：

$$j = j + 1$$

求解 (7.1) 中的非线性弹性问题

求解 (7.2) 中的非线性相场问题

$$L^j = aL^{j-1}$$

令

$$\Xi_h^j = \left[\Xi_h^{j-1} + \gamma\left(\varphi_h^{n,j} - \varphi_h^{n-1}\right)\right]^+$$

直到满足 $\max\left\{\left\|A_1\left(u_h^{n,j}, w_h\right)\right\|, \left\|A_2\left(\varphi_h^{n,j}, \psi_h\right)\right\|\right\} \leqslant \text{TOL}$ 时停止循环。

令

$$U_h^n := (u_h^n, \varphi_h^n) := \left(u_h^{n,j}, \varphi_h^{n,j}\right)$$

读者可进一步研究稳定项动态更新的相关问题 [13,152]。

7.7 相场裂缝全耦合模型

在公式 19 中我们引入一种常见的半线性形式：

$$A(U)(\Psi) = F(\Psi) \tag{7.3}$$

现在我们一次求解全部方程，为此需要采取不动点迭代或牛顿迭代法进行全耦合处理。

定义 57（全耦合求解方案） 全耦合求解方案包括非线性迭代、线性求解器以及相应的预条件。一般而言全耦合方案需要一个全耦合公式以及对应的非线性求解器。如果对非线性迭代进行解耦，则会得到一个分部耦合方案。

备注 46（隐式耦合） 在全耦合方案中，所有项均采取隐式耦合方案，这意味着我们只需要处理当前 U^j 的迭代。

7.7.1 模型挑战

在相场裂缝模型中采取全耦合求解面临着各种挑战。在命题 15 的能量方程中，E_b 是一个四阶项：

$$\frac{1}{2}\int_B g(\varphi)\sigma(u) : e(u)\mathrm{d}x$$

7.7 相场裂缝全耦合模型

在一阶变分形式中,位移方程的第一项 $A_1(U)(w)$ 为

$$\int_B \left((1-\kappa)\varphi^2 + \kappa\right)\sigma(u) : e(w)\mathrm{d}x$$

这里相场变量 φ 与 u 均为未知量。参考命题 19,20 及 24,此时变分形式为

$$\int_B \left((1-\kappa)\varphi^2 + \kappa\right)\sigma^+(u) : e(w)\mathrm{d}x$$

相关的能量项在这里并不是严格的凸形式。正如之前所述,解耦处理将会使之变为分部耦合形式,其操作如下:

$$\left((1-\kappa)\varphi^2 + \kappa\right)\sigma(u) : e(u) \sim \varphi^2|\nabla u|^2$$

此时将 φ 替代为 $\bar\varphi$,则上式为

$$\bar\varphi^2|\nabla u|^2$$

此时该式为一个给定系数 $\bar\varphi^2$ 的经典泊松形式。另一方面可将 u 替代为 $\bar u$,则上式为

$$\varphi^2|\nabla \bar u|^2$$

此时该式为一个给定系数 $|\nabla \bar u|$ 的二阶加权项。

鉴于模型过于复杂,至今准静态脆性相场裂缝全耦合模型的研究仍然较少 [217,189,438,439,439,261]。

7.7.2 准全耦合方案

在这里我们首先介绍一个不常见的定义。

定义 58 (准全耦合函数迭代) 当一次得到全部解,但某些 (低阶) 项被线性化时,我们将其称之为准全耦合方案。在这里我们放松了隐式耦合的限制,但仍保证符合全耦合的特点。

这里我们采取部分线性化方案,但不涉及运算符拆分。

命题 55 (前时间步时滞法) 在命题 30、31、32 的第一项 $A_1(U)(\Psi)$ 中利用 $\tilde\varphi := \varphi^{n-1}$ 的时滞处理进行线性化,可得

$$\left((1-\kappa)\tilde\varphi^2 + \kappa\right)\sigma(u)$$

然而这会带来较大的时间离散化误差,在单边缺口剪切模拟等扩展裂缝模型中会非常明显。与其相比,线性外推法是另外一种适用性更好的方法,其在准确性与求解效率方面有较大优势。

命题 56 (线性外推法) 在命题 30、31、32 的第一项 $A_1(U)(\Psi)$ 中利用外推形式

$$\tilde{\varphi} := \tilde{\varphi}\left(\varphi^{n-1}, \varphi^{n-2}\right) = \varphi^{n-2}\frac{t_n - t_{n-1}}{t_{n-2} - t_{n-1}} + \varphi^{n-1}\frac{t_n - t_{n-2}}{t_{n-1} - t_{n-2}}$$

进行线性化, 可得

$$\left((1-\kappa)\tilde{\varphi}^2 + \kappa\right)\sigma(u)$$

线性外推是一个探索性的过程 (如图 7.2 所示)。由于我们研究的是一个准静态问题, 这也就意味着相场解 φ 可能随时间产生跳跃 [376,318]。

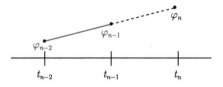

图 7.2 $\bar{\varphi}$ 的线性时间外推

研究人员对由外推法引入的误差分析进行了相关研究。我们假设随着时间步长 $k \to 0$, 解变量收敛于某一稳态极限值。对于特定假设下的解耦版本, 可参考文献 [318] (3.3 节)。对于全耦合或准全耦合公式, 这个假设没有数学证明。时滞误差的计算验证见文献 [438](图 7.3, 图 7.11, 以及图 7.12), 并且时滞误差与裂缝尖端扩展速度有关。加载步长细化的进一步计算结果可参考第 8 章。在稳态裂缝模型中, 时滞误差可以忽略不计, 因此在第 6 章与第 10 章中我们简单采用了 φ^{n-1} 进行处理。

最后我们介绍 7.4 节中的函数迭代法。

命题 57 (滞后迭代法) 我们首先给定前一个迭代步的解 U^{j-1} 并求解 U^j。在命题 30、31、32 的第一项 $A_1(U)(\Psi)$ 中利用 $\tilde{\varphi} := \varphi^{j-1}$ 进行迭代滞后处理, 可得

$$\left((1-\kappa)\tilde{\varphi}^2 + \kappa\right)\sigma(u)$$

7.7.3 外推迭代算法

本节我们介绍了外推迭代法, 该算法在外推法中引入了一个额外的迭代步骤, 流程如下。

算法 6 (迭代外推) 假设在时间步 t_n 中:
(1) 给定前两个时间步的解 φ^{n-1} 与 φ^{n-2}。
(2) 令 $\varphi^{n,-2} := \varphi^{n-2}, \varphi^{n,-1} := \varphi^{n-1}$。

(3) 构造线性外推

$$\tilde{\varphi}^{n,0} = \varphi^{n,-2}\frac{t_n - t_{n-1}}{t_{n-2} - t_{n-1}} + \varphi^{n,-1}\frac{t_n - t_{n-2}}{t_{n-1} - t_{n-2}}$$

(4) 令 $u^{n,0} := u^{n-1}$，$\varphi^{n,0} := \varphi^{n-1}$。

(5) 对于 $i = 1, 2, \cdots, N$。

(a) 根据 $u^{n,-1}, \varphi^{n,-1}, \varphi^{-n,-1}$ 求解相场位移方程，得到 $(u^{n,i}, \varphi^{n,i})$；

(b) 构造一个新的外推

$$\tilde{\varphi}^{n,i} = \varphi^{n,i-2}\frac{t_n - t_{n-1}}{t_{n-2} - t_{n-1}} + \varphi^{n,i-1}\frac{t_n - t_{n-2}}{t_{n-1} - t_{n-2}}$$

(c) $i \to i + 1$。

(6) 令 $u^n := u^{n,N}$，$\varphi^n := \varphi^{n,N}$。

相关计算请参考第 8 章。

7.8　牛顿迭代法概述

牛顿迭代法是一种有效的求解非线性方程组的方法，关于牛顿法在抽象空间中收敛性的分析可追溯到 Newton-Kantorovich 定理 [251,252]。在本节中，我们简要介绍了一些相关算法 [128,335,361]。

(1) 一维空间中 Newton-Raphson 法：$x \in \mathbb{R}$，按 $j = 0, 1, 2, 3, \cdots$ 迭代，使 $x^j \to x$

$$令\ \delta x \in \mathbb{R}, f'\left(x^j\right) \delta x = -f'\left(x^j\right)$$

$$x^{j+1} = x^j + \delta x$$

(2) $\mathbb{R}^d$ 空间中的 Newton 法：当 $x \in \mathbb{R}^d$，按 $j = 0, 1, 2, 3, \cdots$ 迭代，使 $x^j \to x$

$$令\ \delta x \in \mathbb{R}, F'\left(x^j\right) \delta x = -F\left(x^j\right)$$

$$x^{j+1} = x^j + \delta x$$

(3) Banach 空间中：令 $u \in V$，$\dim(V) = \infty$，按 $j = 0, 1, 2, 3, \cdots$ 迭代，使 $u^j \to u$

$$令\ \delta u \in V,\ F'\left(u^j\right) \delta u = -F\left(u^j\right)$$

$$u^{j+1} = u^j + \delta u$$

这里我们需要进行离散化处理并继续求解线性方程组。

(4) Banach 空间中的变分法：令 $U \in X$，$\dim(X) = \infty$，按 $j = 0, 1, 2, 3, \cdots$，迭代，使 $u^j \to u$

$$\text{令 } \delta U \in R, \ A'\left(U^j\right)\left(\delta U^j, \Psi\right) = -A\left(U^j\right)(\Psi) \tag{7.4}$$

$$U^{j+1} = U^j + \delta U \tag{7.5}$$

这里我们同样需要进行离散化处理并继续求解线性方程组。

7.8.1 单调性测试

为了确认迭代法的有效性，我们需要对式 (7.4) 和式 (7.5) 进行单调性测试。比较经典的方案是检查迭代步中是否满足 $\left\|A\left(U^{j+1}\right)(\Psi)\right\| < \left\|A\left(U^j\right)(\Psi)\right\|$。另外我们还可以选择采用自然单调性测试的方案，即观察 $\left\|\delta U_{\text{simp}}^{j+1}\right\| < \left\|\delta U^j\right\|$ 是否成立。相关细节请读者参考文献 [335, 128]。

7.8.2 基于残差的牛顿法基本算法

该类算法的关注点之一在于残差随迭代步减小 (即收缩)。

算法 7 (基于残差的牛顿迭代法算法) 给定初始值 $x^0, j = 0, 1, 2, 3, \cdots$，

$$\text{令 } \delta x \in \mathbb{R}^d, \ F'\left(x^j\right)\delta x^j = -F\left(x^j\right)$$

$$x^{j+1} = x^j + \lambda^j \delta x^j$$

其中 $\lambda^j \in (0, 1]$，当为一个完整牛顿迭代步时 $\lambda^j = 1$。收敛准则为

$$F\left(x^{j+1}\right) < F\left(x^j\right)$$

为简化计算过程，在逼近解 x^* 时，可改变 λ 取值。当 $\lambda^j = 1$ 时，有

$$\theta^j = \frac{F\left(x^{j+1}\right)}{F\left(x^j\right)} < 1$$

若 $\theta^j < \theta_{\max}$，如 $\theta_{\max} = 0.1$ 时，则保留 $F'\left(x^j\right)$ 并在 $k+1$ 迭代步中继续迭代。若 $\theta^j > \theta_{\max}$，则组装新的 $F'(x^j)$，迭代终止准则如下：

$$F\left(x^{j+1}\right) \leqslant \text{TOL}_{\text{New}} \quad (\text{绝对迭代终止准则})$$

$$F\left(x^{j+1}\right) \leqslant \text{TOL}_{\text{New}} F(x_0) \quad (\text{相对迭代终止准则})$$

当满足终止准则时，$x^* := x^{j+1}$ 为 $F(x) = 0$ 的近似解。

7.8.3 全耦合方案概述及其数值解-图片公式化

本节我们提供了变分方程的一个简要方案，可用来完成非线性方程的离散。首先假设给定 S 个方程以及相应的不等式已经完成了正则化处理。

(1) 将给定方程写成变分形式：

$$A_1(U)(\Psi) = F_1(\Psi)$$
$$A_2(U)(\Psi) = F_2(\Psi)$$
$$\vdots$$
$$A_S(U)(\Psi) = F_S(\Psi)$$

(2) 改写为半线性形式：令 $U \in X$，$\Psi \in X$

$$A(U)(\Psi) = \sum_{j=1}^{S} A_j(U)(\Psi)$$

此时该系统为时空连续的非线性系统。

(3) 时间离散化处理。

(a) 将 A_j 分为瞬态与静态：

$$\text{静态：} A_j \Rightarrow A_S \tag{7.6}$$

$$\text{瞬态：} A_j \Rightarrow A_T, A_E, A_I \tag{7.7}$$

其中 A_S 是平稳项，A_T 是时间导数项，A_E 是时间步中的显式项，A_I 是在时间步中的隐式项。

(b) 用向后差商法处理连续时间导数 $A_T\left(U^{n,k}\right) \approx A_T\left(U^n\right)$。

(c) 单次 θ 分解法：对于 $n = 1, 2, \cdots, N$

$$\underbrace{A_T\left(U^{n,k}\right)(\Psi) + \theta A_E\left(U^n\right)(\Psi) + A_I\left(U^n\right)(\Psi) + A_S\left(U^n\right)(\Psi)}_{=: A(U^n)(\Psi)}$$
$$= -\underbrace{(1-\theta)A_E\left(U^{n-1}\right)(\Psi)}_{=: A(U^{n-1})(\Psi)} + \underbrace{\theta F^n(\Psi) + (1-\theta)F^{n-1}(\Psi)}_{=: F^{n,n-1}(\Psi)}$$

备注 47 这里同样可以采用分步 θ 分解法。

(4) t_n 内的空间离散化处理：令 $U_h^n \in X_h \subset X$

$$A\left(U_h^n\right)(\Psi_h) = -A\left(U_h^{n-1}\right)(\Psi_h) + F_h^{n,n-1}(\Psi_h)$$

(5) 非线性处理 (如牛顿迭代法):

令 $M_X = \dim(X_h)$，$\delta U_h = \sum_{i=1}^{M_X} \delta u_i \Psi_i$，$\delta U_h \in X_A (j = 0, 1, 2, 3, \cdots)$

(i) $\underbrace{A'\left(U_h^{n,j}\right)(\delta U_h, \Psi_h)}_{=:AU} = \underbrace{-A\left(U_h^{n,j}\right)(\Psi_h) - A\left(U_h^{n-1,j}\right)(\Psi_h) + F_h^{n,n-1}(\Psi_h)}_{=:B}$。

(ii) $U_h^{n,j+1} = U_h^{n,j} + \omega \delta U_h, \omega \in [0,1]$。

(6) 求解线性方程

$$AU = B$$

其中 $U = (\delta u_1, \cdots, \delta u_{M_X})^{\mathrm{T}} \in \mathbb{R}^{M_X}$，当使用迭代法时，需构建预处理器 P^{-1}，令

$$P^{-1}AU = P^{-1}B$$

当系统中某些项受不等式约束时，可以应用以下基于补偿法的改进算法。

(1) 将给定方程写成变分形式：

$$A_1(U)(\Psi) = F_1(\Psi)$$
$$A_2(U)(\Psi) = F_2(\Psi)$$
$$\vdots$$
$$A_s(U)(\Psi) = F_s(\Psi)$$
$$A_{s+1}(U)(\Psi - U) \geqslant F_{s+1}(\Psi - U)$$
$$\vdots$$
$$A_S(U)(\Psi - U) \geqslant F_S(\Psi - U)$$

(2) 利用补偿项 $\gamma_{s+1}, \cdots, \gamma_S$ 实现不等式约束的正则化：

$$A_{s+1}(U)(\Psi) - A_{\gamma_{s+1}}(U)(\Psi) = F_{s+1}(\Psi)$$
$$\vdots$$
$$A_S(U)(\Psi) - A_{\gamma_S}(U)(\Psi) = F_S(\Psi)$$

(3) 整理半线性形式

$$A(U)(\Psi) = \sum_{j=1}^{s} A_j(U)(\Psi) - \sum_{j=s+1}^{S} A_{\gamma_j}(U)(\Psi)$$

此时我们获得一个时空连续的非线性正则化系统。

(4) 根据上一节步骤 (3)~(6) 完成后续离散化等处理。

7.9 牛顿迭代法在相场裂缝问题中的运用

本节我们将介绍几种不同类型的牛顿迭代法。在相场裂缝模型中规定 $U = (u, \varphi)$ 为解变量，并为此编写了以下大部分算法。本节提供的算法同样可应用于其他多物理场问题。例如，可利用经典牛顿迭代法实现流固耦合问题 $U = (v, u, p)$。同时文献 [435, 437] 中我们进一步完善了相场裂缝流固耦合数值模型 $U = (v, u, p, \varphi)$。具体细节请读者参考相关文献 [432, 134, 203, 217, 218]。

7.9.1 面向残差的牛顿迭代法算法

本节我们将提出一个面向残差的牛顿迭代法算法，并通过一种基于线性搜索算法的阻尼策略实现全局化处理。

这里我们将使用离散范数 $\|\cdot\| := \|\cdot\|_{l_\infty}$ 处理残差以及监测函数，在给定时间步中求解算法如下。

算法 8 (面向残差的牛顿迭代法算法) 这类方法的主要判别依据是每个迭代步中残差值减小与否，首先选取一个初始值 U^0，$U^0 := U^{n,0} := U^{n-1}$，按 $j = 0, 1, 2, 3, \cdots$ 的顺序迭代。

(1) 令 $\delta U^j := (\delta u, \delta \varphi) \in X^D$，$\lambda_j = 1$：

$$A'(U^j)(\delta U^j, \Psi) = -A(U^j)(\Psi), \quad \forall \Psi \in X^0 \tag{7.8}$$

$$U^{j+1} = U^j + \lambda_j \delta U^j \tag{7.9}$$

(2) 基于残差值判断收敛情况

$$\|A(U^{j+1})(\Psi)\| < \|A(U^j)(\Psi)\| \tag{7.10}$$

(3) 若式 (7.10) 不成立，则令 $\lambda_j^l = 0.5$ 重新计算式 (7.9) 获得新的 U^{j+1}，$l = 1, \cdots, l_M$

$$U^{j+1} = U^j + \lambda_j^l \delta U^j$$

若 $l^* = l_M$ 时仍未收敛，则终止程序计算。

(4) 当 $l^* < l_M$ 时，我们检验是否满足以下终止准则

$$\|A(U^{j+1})(\Psi)\| \leqslant \text{TOL}_{\text{New}}$$

若满足，则令 $U^n := U^{j+1}$，同时令 $j \to j+1$ 并重复步骤 (1)。

7.9.1.1 Jacobian 矩阵

为使用牛顿迭代法求解 $A(U_h)(\Psi_h) = 0$，我们需要计算相应的偏导，构造 Jacobian 矩阵 (2.2.4 节)：

$$A'(U)(\delta U, \Psi) := \lim_{s \to 0} \frac{A(U + s\delta U)(\Psi) - A(U)(\Psi)}{s}$$

其中 $\delta U := (\delta u, \delta \varphi) \in X^0$，参考命题 31：

$$\begin{aligned}\mathcal{A}'(U)(\delta U, \Psi) =\ & \left(2\delta\varphi(1-\kappa)\varphi\sigma^+(u) + \left((1-\kappa)\varphi^2 + \kappa\right)\sigma^+(\delta u), e(w)\right) \\ & + \left(\sigma^-(\delta u), e(w)\right) \\ & + (1-\kappa)\left(\delta\varphi\sigma^+(u) : e(u) + 2\varphi\sigma^+(\delta u) : e(u), \psi\right) \\ & + \left(\frac{1}{\varepsilon}(G_c\delta\varphi, \psi) + \varepsilon(G_c\nabla\delta\varphi, \nabla\psi)\right) \\ & + \gamma(\delta\varphi, \psi)_{\mathcal{A}(\varphi)}, \quad \forall \Psi := (w, \psi) \in X^0\end{aligned} \quad (7.11)$$

其中

$$\mathcal{A}(\varphi) = \left\{x \in B \mid \Xi + \gamma\left(\varphi(x) - \varphi(x)^{n-1}\right) > 0\right\}$$

备注 48 由于采用了截止值，其中 $\gamma(\delta\varphi, \psi)_{\mathcal{A}(\varphi)}$ 项并不平滑。

命题 58 (Miehe 应力分解偏导) 参考 4.5.1.1 节，拉应力可分解为

$$\sigma^+(u) := 2\mu e^+ + \lambda[\text{tr}(e)]^+ I$$

其方向导数为

$$\sigma^+(\delta u) := \sigma^+(u)'(\delta u) := 2\mu e^+(u)'(\delta u) + \lambda\left[\text{tr}\left(e(u)'(\delta u)\right)\right]^+ I$$

其中

$$\left[\text{tr}\left(e(u)'(\delta u)\right)\right]^+ = \begin{cases} \text{tr}(e(\delta u)), & \text{tr}(e) > 0 \\ 0, & \text{tr}(e) \leqslant 0 \end{cases}$$

且

$$e^+(\delta u) = P(\delta u)\Lambda^+ P^\mathrm{T} + P\Lambda^+(\delta u)P^\mathrm{T} + P\Lambda^+ P^\mathrm{T}(\delta u)$$

压应力的偏导求解过程与上述过程相似，算法可参考文献 [245]。

7.9.1.2 Jacobian 矩阵，解向量与右手项的块结构

本节我们将深入研究迭代过程中式 (7.8) 线性系统的结构问题。在空间离散化方面，我们采用前文介绍过的 X_h^D 与 X_h^0 处理。

$$\{\psi_i | i = 1, \cdots, M_X\}$$

其中 $M_X = \dim\left(X_h^D\right)$。我们可将基函数分为位移基函数以及相场基函数，并进行相应排序。

$$\psi_i = \begin{pmatrix} \chi_i^u \\ 0 \end{pmatrix}, \quad i = 1, \cdots, M_{X,u}$$

$$\psi_{(M_{X,u}+i)} = \begin{pmatrix} 0 \\ \chi_i^\varphi \end{pmatrix}, \quad i = 1, \cdots, M_{X,\varphi}$$

这里 $M_{X,u} + M_{X,\varphi} = M_X$，这样我们就可以将式 (7.8) 中的系统改写为

$$M\delta U = F \tag{7.12}$$

式中 M 为 Jacobian 矩阵，F 为右手项，解变量为 δU，块结构为

$$M = \begin{pmatrix} M^{uu} & M^{u\varphi} \\ M^{\varphi u} & M^{\varphi\varphi} \end{pmatrix}, \quad F = \begin{pmatrix} F^u \\ F^\varphi \end{pmatrix}, \quad \delta U = \begin{pmatrix} \delta U^u \\ \delta U^\varphi \end{pmatrix}$$

参考式 (7.11)

$$M_{i,j}^{uu} = \left(\left((1-\kappa)\varphi^2 + \kappa\right)\sigma^+\left(\chi_j^u\right), e\left(\chi_i^u\right)\right) + \left(\sigma^-\left(\chi_j^u\right), e\left(\chi_i^u\right)\right)$$

$$M_{i,j}^{u\varphi} = \left(2\chi_j^\varphi(1-\kappa)\varphi\sigma^+(u), e\left(\chi_i^u\right)\right)$$

$$M_{i,j}^{\varphi u} = 2(1-\kappa)\left(\varphi\sigma^+\left(\chi_j^u\right) : e(u), \chi_i^\varphi\right)$$

$$M_{i,j}^{\varphi\varphi} = -(1-\kappa)\left(\sigma^+(u) : e(u)\chi_j^\varphi, \chi_i^\varphi\right)$$
$$+ \left(\frac{1}{\varepsilon}\left(G_c\chi_j^\varphi, \chi_i^\varphi\right) + \varepsilon\left(G_c\nabla\chi_j^\varphi, \nabla\chi_i^\varphi\right)\right) + \gamma\left(\chi_j^\varphi, \chi_i^\varphi\right)_{A(\varphi)}$$

右手项为

$$F_i^u = -A\left(U^j\right)\left(\chi_i^u\right)$$
$$= \left(\left((1-\kappa)\varphi^{j^2} + \kappa\right)\sigma^+\left(u^j\right), e\left(\chi_i^u\right)\right) + \left(\sigma^-\left(u^j\right), e\left(\chi_i^u\right)\right)$$

$$F_i^\varphi = -A\left(U^j\right)\left(\chi_i^\varphi\right) = (1-\kappa)\left(\varphi^j \sigma^+\left(u^j\right) : e\left(u^j\right), \chi_i^\varphi\right)$$
$$+ \left(-\frac{1}{\varepsilon}\left(G_c\left(1-\varphi^j\right), \chi_i^\varphi\right) + \varepsilon\left(G_c \nabla \varphi^j, \nabla \chi_i^\varphi\right)\right)$$
$$+ \left(\left[\Xi^j + \gamma\left(\varphi^j - \varphi^{n-1}\right)\right]^+, \chi_i^\varphi\right)$$

7.9.1.3 块结构设计方案

当采用 7.7.2 节中的方案时，我们可得

$$M_{i,j}^{u\varphi} = 0$$

在对 φ 求导后，我们得到三角块结构

$$M = \begin{pmatrix} M^{uu} & 0 \\ M^{\varphi u} & M^{\varphi\varphi} \end{pmatrix}$$

该方案有利于线性求解时迭代求解器的设计。相关研究已经证明了该方案在二维与三维问题求解过程中的有效性[217,281]。

由于 M 是对称的，因此 $M^{\varphi u}$ 和 $M^{u\varphi}$ 具有相同的块结构，可以尝试构建简化块矩阵：

$$\begin{pmatrix} M^{uu} & 0 \\ 0 & M^{\varphi\varphi} \end{pmatrix} \tag{7.13}$$

7.9.2 面向误差的牛顿迭代法

接下来我们讨论另一种牛顿迭代法，基于该方法的相场裂缝模型请参考文献 [128, 438]。该方法基于自然单调性测试 (7.8.1 节)，在残差处理方面更具灵活性。如前所述，这里我们继续使用离散范数 $\|\cdot\| := \|\cdot\|_{l^\infty}$，算法细节如下。

算法 9 (面向误差的牛顿迭代法) 该方法的判断依据主要为 δU^j 的减少与否。令 $\lambda_{\min} = 10^{-10}$，选取一个初始值 $U^0 \in X^D$，令 $j = 1, 2, 3, \cdots$。

(1) 求解 $\delta U^j := (\delta u, \delta \varphi) \in V_u^0 \times V_\varphi$

$$A'\left(U^j\right)\left(\delta U^j, \Psi\right) = -A\left(U^j\right)(\Psi), \quad \forall \Psi \in V_u^0 \times V_\varphi \tag{7.14}$$

(2) 检查是否满足 $\|\delta U^j\| \leqslant \text{TOL}_{\text{New}}$，若成立，则解为

$$U^* := U^j + \delta U^j$$

若不成立且 $j > 0$，则：

$$\lambda_j := \min(1, \mu_j), \quad \mu_j := \frac{\|\delta U^{j-1}\| \cdot \|\delta U^j_{\text{simp}}\|}{\|\delta U^j_{\text{simp}} - \delta U^j\| \cdot \|\delta U^j\|}$$

(3) 若

$$\lambda_j < \lambda_{\min} \tag{7.15}$$

则收敛失败，程序终止。

(4) 若 $\lambda_j > \lambda_{\min}$，则继续迭代计算

$$U^{j+1} = U^j + \lambda_j \delta U^j \tag{7.16}$$

并计算新的残差 $A(U^{j+1})(\Psi)$。基于"旧" Jacobian 矩阵求解简化的线性系统：令

$$\delta U^{j+1}_{\text{simp}} := (\delta u_{\text{simp}}, \delta\varphi_{\text{simp}}) \in V_u^0 \times V_\varphi$$

$$A'(U^j)(\delta U^{j+1}_{\text{simp}}, \Psi) = -A(U^{j+1})(\Psi), \quad \forall \Psi \in V_u^0 \times V_\varphi \tag{7.17}$$

(5) 计算监测函数

$$\theta_j := \frac{\|\delta U^{j+1}_{\text{simp}}\|}{\|\delta U^j\|}, \quad \mu'_j := \frac{0.5\|\delta U^j\| \cdot \lambda_j^2}{\|\delta U^{j+1}_{\text{simp}} - (1-\lambda_j)\delta U^j\|}, \quad \lambda'_j := \min\left(\mu'_j, \frac{1}{2}\lambda_j\right) \tag{7.18}$$

(6) 若 $\theta_j \geqslant 1$（迭代后无收敛性），则令

$$\lambda_j := \lambda'_j$$

并返回步骤 (3) 重新计算。

(7) 若 $\theta_j < 1$，则此时收敛，继续计算：

(a) 若 $\lambda'_j = \lambda_j = 1$，检查是否满足 $\|\delta U^{j+1}_{\text{simp}}\| \leqslant \text{TOL}_{\text{New}}$，若满足，则此时解为

$$U^* := U^{j+1} + \delta U^{j+1}_{\text{simp}}$$

(b) 否则以 U^{j+1} 为新的迭代值，$j \to j+1$，返回步骤 (1) 重新计算。

7.9.3 修正 Jacobian 矩阵的牛顿迭代法

本节我们将介绍第三种牛顿迭代法，其特点是实现了 Jacobian 矩阵的内部动态更新，使其能在牛顿迭代与不动点迭代之间自由切换。相关研究成果请读者参考文献 [439, 234, 291, 301]。

根据 7.9.1.3 节相关内容可知，由于相场变量的影响，Jacobian 矩阵的一个块结构为 0，这就引出了一个全新方案。与算法 8 中的步骤 (3) 不同，在这里我们引入了一个控制参数 $\omega \in [0,1]$，用来确定是以全牛顿迭代法 ($\omega = 1$) 还是不动点迭代法 ($0 < \omega < 1$) 进行求解 ($\omega = 0$ 后续单独讨论)。

如 7.9.1.2 节所示，每个牛顿迭代步的 Jacobian 矩阵均可表示为

$$M = \begin{pmatrix} M^{uu} & M^{u\varphi} \\ M^{\varphi u} & M^{\varphi\varphi} \end{pmatrix}$$

其中块矩阵 $M^{u\varphi}$ 可表示为

$$\left(2\chi_j^\varphi(1-\kappa)\varphi\sigma^+(u), e\left(\chi_i^u\right)\right) \tag{7.19}$$

接下来我们引入参数 ω 控制块矩阵激活与否

$$M = \begin{pmatrix} M^{uu} & \omega M^{u\varphi} \\ M^{\varphi u} & M^{\varphi\varphi} \end{pmatrix} = \begin{pmatrix} M^{uu} & 0 \\ M^{\varphi u} & M^{\varphi\varphi} \end{pmatrix} + \omega \begin{pmatrix} 0 & M^{u\varphi} \\ 0 & 0 \end{pmatrix} \tag{7.20}$$

7.9.3.1 控制参数 ω

控制参数 ω 的值依赖于前两步的残差值，在每一个牛顿迭代步中参数 $\omega := \omega_j$ 均动态更新。其残差比与残差比倒数的定义为

$$Q_{j+1} = \frac{\|A(U^{j+1})(\Psi)\|}{\|A(U^j)(\Psi)\|}, \quad Q_{j+1}^{\text{rec}} = \frac{\|A(U^j)(\Psi)\|}{\|A(U^{j+1})(\Psi)\|} \tag{7.21}$$

若 $Q_{j+1} < 1$，则说明新的残差值较小，此时该迭代步中满足基于残差的单调性准则 (7.8.1 节)。若 $Q_{j+1} \to 0$ 则说明该步迭代效果好。相反，若 $Q_{j+1} > 1$，则说明新的残差值较上一步更大，此时采用基于误差的牛顿迭代法可能效果会更好。

为此，我们总结了控制参数 ω 的关键思想以及构造方法如下。

定义 59 在第 j 牛顿迭代步中，给定 $0 \leqslant \omega_j \leqslant 1$ 并令 $S \in \mathbb{R}_+ \cup \{0\}$。我们有

$$\omega := \omega_{j+1} = S\omega_j \tag{7.22}$$

命题 59 换算参数 S 应满足如下要求：
(1) S 必须满足 $\omega_{j+1} \in [0,1]$；
(2) $S \gg 1$ 时，$\omega_{j+1} \to 1$ (全牛顿迭代)；
(3) $S \to 0$ 时，$\omega_{j+1} \to 0$ (不动点迭代)。

命题 60 参数 S 可由式 (7.21) 中的 Q_{j+1} 与 Q_{j+1}^{rec} 通过以下流程计算得到：

7.9 牛顿迭代法在相场裂缝问题中的运用

(1) 当 $Q_{j+1} \to 0 (Q_{j+1}^{\text{rec}} \to \infty)$ 时，矩阵 M（式 (7.20)）非病态，此时可采用全牛顿迭代法，即 $S \gg 1$ 且 $\omega_{j+1} = 1$；

(2) 相应地，当 $Q_{j+1} \to \infty (Q_{j+1}^{\text{rec}} \to 0)$ 时，矩阵 M（式 (7.20)）病态，此时 $S \to 0$ 且 $\omega_{j+1} \ll 1$；

(3) 根据以上规律可得到 S 的计算公式

$$S := \left(\frac{a}{\exp\left(Q_{j+1}^{\text{rec}}\right)} + \frac{b}{\exp\left(Q_{j+1}\right)} \right) \tag{7.23}$$

(4) 其中控制参数 a 与不动点迭代步有关，因此 $a < 1$；

(5) 其中控制参数 b 与全牛顿迭代步有关，因此 $b \geqslant 1$；

(6) 这两个控制参数的取值在 7.9.3.3 节中进一步指定。

推理 5 (参数 S 的特性) 参数 S 具有以下特性：

(1) S 下界为 0，由于 $a, b, Q_{j+1}, Q_{j+1}^{\text{rec}} \geqslant 0$，由此可知 $S \geqslant 0$ 且 $\omega_{j+1} \geqslant 1$；

(2) S 无上界。因此可能存在式 (7.22) 中 $\omega_{j+1} > 1$ 的情况。为此我们设定，若 $\omega_{j+1} > 1$，则强制 $\omega_{j+1} = 1$。

7.9.3.2 改进后的牛顿迭代法算法

接下来参考命题 31，修正后的 Jacobian 矩阵如下：

$$\begin{aligned}
A'_\omega(U)(\delta U, \Psi) &= \left(\omega 2 \delta\varphi (1-\kappa) \varphi \sigma^+(u) + \left((1-\kappa)\varphi^2 + \kappa\right) \sigma^+(\delta u), e(w) \right) \\
&\quad + \left(\sigma^-(\delta u), e(w) \right) \\
&\quad + (1-\kappa) \left(\delta\varphi \sigma^+(u) : e(u) + 2\varphi \sigma^+(\delta u) : e(u), \psi \right) \\
&\quad + \left(\frac{1}{\varepsilon}(G_c \delta\varphi, \psi) + \varepsilon(G_c \nabla\delta\varphi, \nabla\psi) \right) \\
&\quad + \gamma(\delta\varphi, \psi)_{A(\varphi)}
\end{aligned} \tag{7.24}$$

算法 10 (修正 Jacobian 矩阵的牛顿迭代法) 给定初始值 U^0 以及初始控制参数 $\omega_0 = 1$。按 $j = 0, 1, 2, \cdots$ 迭代。

(1) 令 $\delta U^j := (\delta u, \delta\varphi) \in X^0$

$$A'_\omega(U^j)(\delta U^j, \Psi) = -A(U^j)(\Psi), \quad \forall \Psi \in X^0 \tag{7.25}$$

$$U^{j+1} = U^j + \delta U^j \tag{7.26}$$

(2) 根据式 (7.23) 计算：

$$\omega := \omega_{j+1} = S\omega_j \tag{7.27}$$

(3) 检查终止标准

$$\|A(U^{j+1})(\Psi)\| \leqslant \text{TOL}_{\text{New}}$$

若成立，则令 $U^n := U^{j+1}$，否则，令 $j \to j+1$ 并返回步骤 (1) 重新计算。

备注 49 在该算法中，我们未引入收敛监测函数，由此需要在步骤 (3) 中检验是否发散：

$$\|A(U^{j+1})(\Psi)\| < \text{TOL}_{\text{New}}^{\text{up}}, \quad \text{TOL}_{\text{New}}^{\text{up}} = 10^{12}$$

若不成立，则说明该算法发散，需要终止程序。相关收敛性研究请读者参考文献 [439, 29]。

7.9.3.3 控制参数取值

根据命题 60 可知，参数 a 控制 $M^{u\varphi}$，与不动点迭代步有关；参数 b 则与全牛顿迭代步有关，为此我们首先划定取值范围：

$$0 \leqslant a < 1, \quad 1 \leqslant b < \infty$$

接下来我们进一步分析。若 $Q_{j+1} \ll 1$，则说明残差较小，收敛情况好。此时下一迭代步可使用更高阶 ω_{j+1}，根据式 (7.23)

$$\lim_{Q_{j+1} \to \infty} \lim_{Q_{j+1} \to 0} S \to b \Rightarrow \omega_{j+1} = b\omega_j \Rightarrow \omega_{j+1} \geqslant \omega_j$$

反正，若 $Q_{j+1} > 1$ (或 $Q_{j+1} \gg 1$，即 $Q_{j+1}^{\text{rec}} \to 0$)，则说明残差较大，此时我们希望消除 Jacobian 矩阵中的不规则项或使用不动点迭代之类的方法减小这些项的影响

$$\lim_{Q_{j+1} \to 0} \lim_{Q_{j+1} \to \infty} S \to a \Rightarrow \omega_{j+1} = a\omega_j \Rightarrow \omega_{j+1} < \omega_j$$

备注 50 有关 a, b 更多取值细节以及相关研究请参考文献 [439]。

7.9.4 基于修正 Jacobian 矩阵的面向误差牛顿迭代法

接下来我们介绍第四种基于修正 Jacobian 矩阵的面向误差牛顿迭代法，适用于收敛性较差的情况。首先我们进行以下处理：

(1) 将式 (7.18) 中的 Q_{j+1} 替换为 θ_j，相应的 Q_{j+1}^{rec} 替换为 $\dfrac{1}{\theta_j}$。

(2) 通过式 (7.23) 反推 ω 的值。

(3) 在面向误差的牛顿迭代法中采用式 (7.24) 中的修正 Jacobian 矩阵进行计算。

接下来算法 9 中的步骤 (6) 需要修改为

(6) 若 $\theta_j \geqslant 1$：
$$\lambda_j := \lambda_j'$$
随后通过式 (7.22) 计算 ω，返回步骤 (3) 并继续计算。

在该方案中我们提高了迭代不收敛后的稳定性，其次我们在下一个循环中采用了新的控制参数以提升计算效率。

7.9.5 基于原始-对偶活动集的牛顿迭代法

本节我们介绍第五种牛顿迭代法，在这里我们采用原始-对偶活动集法处理裂缝不可逆性约束。我们将该方法与牛顿迭代法相结合。其处理算法与 7.9.1 节中的算法相似。

该方法主要用于处理非线性问题，主要流程如下：首先给定初始值 U^0，令 $j = 0, 1, 2, 3, \cdots,$

$$A'\left(U^j\right)(\delta U, \Psi) = -A\left(U^j\right)(\Psi), \quad \forall \Psi \in X^0$$

$$U^{j+1} = U^j + \omega \delta U$$

接下来我们定义

$$G := A'\left(U^j\right)(\cdot, \Psi)$$

$$F := -A\left(U^j\right)(\Psi)$$

为实现线性不可逆约束 $\varphi \leqslant \varphi^{n-1}$。我们规定 $\delta\varphi \leqslant 0$，为此，在 $0 \times V_\varphi$ 中，有 $\delta U \leqslant 0$。这样我们就可将相场部分方程改写为

$$\min_{\delta U \in \tilde{V}_\varphi} \frac{1}{2}(\delta U, G\delta U) - (F, \delta U)$$

其中

$$\tilde{V}_\varphi := \{\delta U \in 0 \times V_\varphi \mid \delta U \leqslant 0\}$$

命题 61 (基于增广 Lagrange 补偿项的牛顿法) 令 $\delta U = (\delta u, \delta\varphi) \in V_u^0 \times V_\varphi$，$(\delta U, \Xi) \in X^0 \times V_\Xi$ 有

$$A'(U)(\delta U, \Psi) + \langle \Xi, \psi \rangle = -A(U)(\Psi), \quad \forall \Psi = (w, \psi) \in X^0$$

$$\langle \Xi - \chi, \delta\varphi \rangle \geqslant 0, \quad \forall \chi \in V_\Xi$$

对其中的不等式有

$$\delta\varphi \leqslant 0, \quad 在 B \times I 中$$
$$\Xi \geqslant 0, \quad 在 B \times I 中$$
$$\langle \Xi, \delta\varphi \rangle = 0, \quad 在 B \times I 中$$

根据 5.2.3 节可知
$$C(\delta\varphi, \Xi) = 0$$
其中
$$C(\delta\varphi, \Xi) = \Xi - [\Xi + c\delta\varphi]^+ \tag{7.28}$$

根据 5.2.3 节，我们可以初步制定一个原始-对偶活动集算法，该算法在相场裂缝模型中的应用可参考文献 [217]。命题 61 中的线性方程组具有以下结构：

$$\begin{pmatrix} G & B \\ B^{\mathrm{T}} & 0 \end{pmatrix} \begin{pmatrix} \delta U \\ \Xi \end{pmatrix} = \begin{pmatrix} F \\ 0 \end{pmatrix}$$

其中矩阵 B 代表 $\langle \Xi, \psi \rangle$，采用 Gauss-Lobatto 法在空间离散化模型中对 $\langle \Xi_h, \psi_h \rangle$ 求积。牛顿迭代步中单元 i 是否属于活动集 $\mathcal{A}^j$ 的判断准则如下：

$$\left(B^{-1}\right)_{ii} \left(F - G\delta U^j\right)_i + c\left(\delta U^j\right)_i > 0 \tag{7.29}$$

若：
$$\left(B^{-1}\right)_{ii} \left(F - G\delta U^j\right)_i + c\left(\delta U^j\right)_i \leqslant 0 \tag{7.30}$$

则表面单元 i 属于非活动集 $\mathcal{N}^j$。

完成以上准备工作后，处理算法如下：

算法 11 给定初始迭代值 U^0，迭代顺序 $j = 1, 2, 3, \cdots$。
(1) 组装残差 $R\left(U_h^j\right) := -A\left(U_h^j\right)(\Psi)$；
(2) 计算活动集 $\mathcal{A}_j = \{i| \left(B^{-1}\right)_{ii}(R_j)_i + c\left(\delta U_h^j\right)_i > 0\}$；
(3) 组装矩阵 $G = A'\left(U_h^j\right)\left(\delta U_h^j, \Psi_h\right)$ 与右手项 $F = -A\left(U_h^j\right)(\Psi_h)$；
(4) 去除 G 与 F 中的活动集 $\mathcal{A}^j$ 相关项，获得 $\tilde{G}$ 与 $\tilde{F}$；
(5) 求解线性方程组 $\tilde{G}\delta U_j = \tilde{F}$，令 $\delta U_h^j \in X_h^0$

$$A'\left(U_h^j\right)\left(\delta U_h^j, \Psi_h\right) = -A\left(U_h^j\right)(\Psi_h), \quad \forall \Psi_h \in X_h^0 \tag{7.31}$$

(6) 取 $0 < \lambda \leqslant 1$，此时 $U_h^{j+1} = U_h^j + \lambda \delta U_h^j$ 且 $\tilde{R}\left(U_h^{j+1}\right) < \tilde{R}\left(U_h^j\right)$；
(7) 检查活动集与非活动集的终止准则是否同时满足

$$\mathcal{A}_{j+1} = \mathcal{A}_j \text{ 且 } \tilde{R}\left(U_h^j\right) < \mathrm{TOL}_{\mathrm{New}}$$

备注 51 我们需要注意 $R\left(U_h^j\right)$ 与 $\tilde{R}\left(U_h^j\right)$ 是不同的残差，后者指非活动集上的残差，可通过忽略活动集部分的约束来进行计算。该算法的 C++ 实现部分请读者参考第 13 章与文献 [218]。在算法优化的过程中，我们可能会面临死循环或不收敛的情况，相关研究请参考文献 [218, 288, 210, 118]。

7.10 基于牛顿迭代法的全耦合相场裂缝问题线性求解

本节我们将介绍基于牛顿迭代法的全耦合相场裂缝问题求解。首先我们回忆一下相关方程：

$$A'\left(U_h^{n,j}\right)\left(\delta U_h^n, \Psi\right) = -A\left(U_h^{n,j}\right)(\Psi) + F(\Psi)$$

$$U_h^{n,j+1} = U_h^{n,j} + \lambda \delta U_h^n$$

改写成矩阵形式的线性方程组为

$$M\delta U = B \tag{7.32}$$

这里 B 为离散化后的残差：

$$B \sim -A\left(U_h^{n,j}\right)(\Psi) + F(\Psi)$$

其中解向量 δU 为

$$\delta U = \left(\delta u_1, \cdots, \delta u_{M_u}, \delta \varphi_{M_u+1}, \cdots, \delta \varphi_{M_\varphi}\right)^{\mathrm{T}}$$

7.10.1 预处理矩阵 P^{-1}

接下来我们回顾 7.9.1.2 节中的块结构并简单阐述预处理器的处理方法。给定方程组：

$$\begin{pmatrix} M_{uu} & M_{u\varphi} \\ M_{\varphi u} & M_{\varphi\varphi} \end{pmatrix} \begin{pmatrix} \delta u \\ \delta \varphi \end{pmatrix} = \begin{pmatrix} B_u \\ B_\varphi \end{pmatrix}$$

其中 B_u、B_φ 分别为相场问题与位移场问题的残差部分。通常相关方程需要通过迭代处理进行求解。在这里我们可以通过对块结构进行预处理从而达到更好的求解效果。首先从式 (7.32) 开始，构造一个预处理矩阵 P^{-1} 使：

$$P^{-1}M\delta U = P^{-1}B$$

理想状态下预处理矩阵应满足 $P^{-1} := A^{-1}$。在实际操作中，人们常采用以下方式构建 P^{-1}

$$P^{-1}M = \begin{pmatrix} I & * \\ 0 & I \end{pmatrix}$$

这里 P^{-1} 是一个三角块矩阵：

$$P^{-1} = \begin{pmatrix} P_1^{-1} & 0 \\ P_3^{-1} & P_4^{-1} \end{pmatrix}$$

其详细步骤如下：

$$M = \begin{pmatrix} M_{uu} & M_{u\varphi} \\ M_{\varphi u} & M_{\varphi\varphi} \end{pmatrix} \begin{pmatrix} I & 0 \\ 0 & I \end{pmatrix} = \begin{pmatrix} I & M_{uu}^{-1}M_{u\varphi} \\ M_{\varphi u} & M_{\varphi\varphi} \end{pmatrix} \begin{pmatrix} M_{uu}^{-1} & 0 \\ 0 & I \end{pmatrix}$$

$$= \begin{pmatrix} I & M_{uu}^{-1}M_{u\varphi} \\ 0 & \underbrace{M_{\varphi\varphi} - M_{\varphi u}M_{uu}^{-1}M_{u\varphi}}_{=S} \end{pmatrix} \begin{pmatrix} M_{uu}^{-1} & 0 \\ -M_{\varphi u}M_{uu}^{-1} & I \end{pmatrix}$$

$$= \begin{pmatrix} I & M_{uu}^{-1}M_{u\varphi} \\ 0 & I \end{pmatrix} \underbrace{\begin{pmatrix} M_{uu}^{-1} & 0 \\ -S^{-1}M_{\varphi u}M_{uu}^{-1} & S^{-1} \end{pmatrix}}_{=P^{-1}}$$

其中 $S = M_{\varphi\varphi} - M_{\varphi u}M_{ul}^{-1}M_{u\varphi}$ 也被称舒尔 (Schur) 补矩阵。P^{-1} 是矩阵 M 的预处理矩阵。为此我们进行验算如下：

$$\begin{pmatrix} M_{uu}^{-1} & 0 \\ -S^{-1}M_{\varphi u}M_{uu}^{-1} & S^{-1} \end{pmatrix} \begin{pmatrix} M_{uu} & M_{u\varphi} \\ M_{\varphi u} & M_{\varphi\varphi} \end{pmatrix} = \begin{pmatrix} I & M_{uu}^{-1}M_{u\varphi} \\ 0 & I \end{pmatrix}$$

在以上过程中，我们默认矩阵 S、M_{uu} 是可逆的。接下来我们执行矩阵向量乘法可知：

$$\begin{pmatrix} X_{\text{new}} \\ Y_{\text{new}} \end{pmatrix} = \begin{pmatrix} P_1^{-1} & 0 \\ P_3^{-1} & P_4^{-1} \end{pmatrix} \begin{pmatrix} X \\ Y \end{pmatrix}$$

因此我们有

$$X_{\text{new}} = P_1^{-1}X \tag{7.33}$$

$$Y_{\text{new}} = P_3^{-1}X + P_4^{-1}Y \tag{7.34}$$

7.10.2 外推格式的块-对角预处理器

根据 7.9.1.3 节可知，准全耦合方案的块结构如下：

$$\begin{pmatrix} M^{uu} & 0 \\ M^{\varphi u} & M^{\varphi\varphi} \end{pmatrix}$$

在这里我们可以使用一个简单的对角预处理矩阵：

$$P^{-1} = \begin{pmatrix} (P^{uu})^{-1} & 0 \\ 0 & (P^{\varphi\varphi})^{-1} \end{pmatrix}$$

该预处理器的相关研究可参考文献 [217, 218]。基于 2.3 节中的原型问题，我们构造了一个椭圆型方程结构，在这种情况下构建近似逆矩阵 [220]：

$$P_{uu} \approx -\nabla \cdot \left(\varphi^2 \nabla u\right)$$

$$P_{\varphi\varphi} \approx -\varepsilon\Delta\varphi - \frac{1}{\varepsilon}(1-\varphi)$$

基于该预处理矩阵的数值模拟请读者参考文献 [391, 217, 419, 283, 317]。

7.10.3 无矩阵几何多重网格求解器

近期我们提出了一种无矩阵几何多重网格方法 (GMG)[246]。相关技术细节请读者参考文献 [245, 246]，该方案求解性能良好，在二维 L 型平板测试算例中平均每牛顿迭代步仅需 5~50 次 GMRES 迭代。在单边切口剪切实验模拟中，使用 AMG 求解器平均需要 5~20 次迭代，而 GMG 求解器则需要 5~6 次迭代。

7.10.4 自适应网格并行求解

本书大部分数值模型均基于 Deal.II 有限元库搭建。针对单边缺口剪切实验，图 7.3 提供了局部细化网格的处理示意图。即使引入自适应网格后，基于不同子域划分的并行计算求解器仍展现出良好的求解性能。

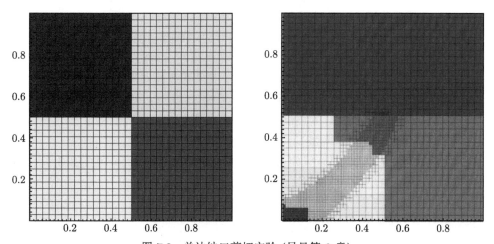

图 7.3　单边缺口剪切实验 (另见第 8 章)

在 4 个处理器上对模型进行并行计算。不同的子域采取不同的处理器进行并行计算。根据网格细化，每个处理器的工作负载在每个时间步动态调整。原始代码请参考文献 [217, 218]

7.11 增广 Lagrangian 补偿法与原始-对偶活动集法

在处理命题 19、20 时，补偿参数 γ 过小可能会导致系统求解困难。一种解决方案是采用较小的补偿参数，随着模型求解不断增大补偿值。当然另外一种更好的方案是采用随迭代步更新的增广 Lagrangian 补偿系数 Ξ。这就需要在每个时间点 t_n 求解 PDEs 时均进行更新迭代。为降低计算成本，我们提出了一个基于启发式的自适应准则增广 Lagrangian 循环模式。主要算法流程如下。

算法 12 (基于内置牛顿求解器的增广 Lagrangian 循环) 在每个时间 $t_n(n = 1, 2, 3, \cdots, N)$，给定 $\Xi_{h,0}$，例如：$\Xi_{h,0} = 0$。另外固定 $\lambda > 0$。每个时间步内的迭代顺序 $j = 0, 1, 2, \cdots$。

(1) 给定 Ξ_h^j，接下来利用牛顿迭代法求解命题 30 或 31 中的增广 Lagrangian 方程，从而得到 $U_h^{n,j+1} = (u_h^{n,j+1}, \varphi_h^{n,j+1})$。

(2) 补偿系数更新：

$$\Xi_h^{n,j+1} = \left[\Xi_h^{n,j} + \gamma\left(\varphi_h^{n,j} - \varphi_h^{n-1}\right)\right]^+$$

(3) 检查是否满足终止准则：

$$\left\{\|u_h^{n,j+1} - u_h^{n,j}\|_{L^2}, \|\Xi_h^{n,j+1} - \Xi_h^{n,j}\|_{L^2}\right\} \leqslant \text{TOL}_{\text{AL}} \tag{7.35}$$

(4) 分情况讨论：
(a) 当满足终止准则时，令 $U_h^n := U_h^{n,j^*}$，其中 j^* 为终止迭代次数；
(b) 当不满足终止准则时，令 $j \to j + 1$ 并返回步骤 (1) 重新计算。

备注 52 与 7.6.1 节相比，本模型中补偿系数随迭代次数同时更新。

为降低计算成本，我们制定了自适应的终止准则。该类方法常应用于不精确牛顿法中[127,97,128]。在这类方法中每个时间步仅求解线性方程的近似解[347]，这类自适应不精确方法中，为保证不受到外界收敛模式的干扰，需要设置内置精度，如：$\text{TOL}_{\text{New}} < \text{TOL}_{\text{AL}}$。因此，需要满足 $\alpha < 1$。在实际计算中我们常设置 $\alpha = 10^{-3}$，相关计算请读者参考文献 [438]。

命题 62 当 $j = 0$ 时，设定 Ξ_h^j，$\text{TOL}_{\text{New}} = 10^{-8}$，当 $j = 0, 1, 2, 3, \cdots$，时，根据算法 12 中的步骤 (4) 计算：

$$\Delta := \|\Xi_h^{j+1} - \Xi_h^j\|_{L^2}$$

在每个迭代步中，我们采用以下终止准则：

$$\text{TOL}_{\text{New}} := \alpha \Delta$$

其中 $\alpha = 10^{-3} < 1$。

在前几节中，为处理裂缝不可逆约束，我们主要介绍了两种方法：

(1) 增广 Lagrangian 牛顿 (AL) 迭代法；

(2) 原始-对偶活动集 (PDAS) 法。

理论上，原始-对偶活动集法应优于增广 Lagrangian 牛顿迭代法，因为这更接近半光滑的牛顿法[227]，在某些假设下具有更好的收敛性。但实际上这两种方法在模拟相场裂缝扩展问题中均表现良好。根据我们的研究结果总结可知：

(1) 原始-对偶活动集法没有补偿参数 γ，但却有一个相似的参数 c，但该参数对计算的影响较小；

(2) 原始-对偶活动集法可能会陷入死循环，当然文献 [218, 118] 中针对该问题介绍了一个解决方案；

(3) 这两种方法均需要足够的迭代计算才能收敛，这两种方法的具体表现尚待进一步研究。

相关代码请读者参考第 13 章。

接下来我们回顾命题 32，其中直接导出了两个不包括不等式约束的方程。结合本章我们介绍的方法，可以提出两种模型：

(1) 基于位移方程的全耦合处理方案；

(2) 利用外推法处理 φ。

当然这两种模型均可通过牛顿法来求解，尤其是第二种模型可以参考 7.7 节中介绍的外推算法。

算法 13 (基于应变场迭代的相场裂缝模型) 在时间 t_n 中初始化 $\mathcal{H}(x, t_0) = \mathcal{H}^0$。

(1) 给定 $\mathcal{H}^{n-1}$，求解命题 32。

(2) 对于每一个点 $x_q \in B$：

$$\mathcal{H} = \mathcal{H}^n = \begin{cases} \psi_E^+(e), & \psi_E^+(e(u(x_q, t_n))) > \mathcal{H}^{n-1} \\ \mathcal{H}^{n-1}, & \text{其他} \end{cases}$$

(3) 时间步递增 $n \to n+1$。

备注 53 应变函数需要在有限元网格中的每个点进行求值并储存，相关算法可参考 Deal.II 教程中的 step-18。若采用外推法 (7.7.3 节)，则 $\mathcal{H}^n$ 可以在每个外推迭代步中更新处理。

第 8 章 数值模拟 II：单边剪切实验

本章我们将对单边剪切实验进行数值模拟[312,308,65,17,217,438,439,86,296,152,245]。

8.1 模型框架

命题 20 与 24 给出了 CVIS/PDE 模型框架，具体细节请参考 5.2.2.1 节、5.2.3 节以及 7.9.5 节。

8.2 测试案例

接下来我们提供了几个算例并进行分析，这部分算例主要关注于：
(1) 裂缝不可逆约束方法，即增广 Lagrangian 法、原始-对偶活动集法等；
(2) 基于外推法对位移方程第一项的非线性处理；
(3) 迭代外推法 (参考 7.7.3 节)；
(4) 分部耦合方案；
(5) 涉及时间收敛性的加载步长重置。
我们基于以下几个目标参数对数值模拟进行分析：
(1) 裂缝到达下边界的时间步；
(2) 载荷-位移曲线；
(3) 牛顿迭代次数。
本节数值模拟基于第 13 章中提供的开源代码实现[203,218] (https://github.com/tjhei/cracks)。

8.3 建模方案

模型几何形状与材料属性请参考 5.1 节、5.2 节与文献 [308]，全域 $B = (10\text{mm}, 10\text{mm})^2$，几何形状与初始条件如图 8.1 所示。在初始几何中预置一条狭缝，并对初始网格进行四次均匀细化 (包括 1024 个网格单元，2210 个固体自由度，1105 个裂缝自由度，总计 3315 个自由度)。$h = 0.044\text{mm}$。

在该模型中，初始裂缝通过几何形状进行设定，同时伴随裂缝扩展，因此需要重点关注模型几何边界以及应力张量分布，相关研究请参考文献 [17]。

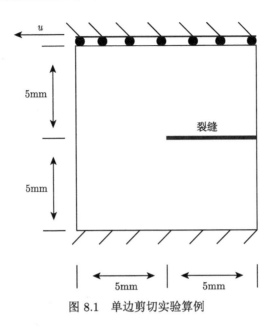

图 8.1 单边剪切实验算例

8.4 边界条件

我们将边界条件分解为

$$\partial B = \Gamma_{\text{left}} \cup \Gamma_{\text{right}} \cup \Gamma_{\text{bottom}} \cup \Gamma_{\text{top}} \cup \Gamma_{\text{slit,bottom}} \cup \Gamma_{\text{slit,top}}$$

此时我们增加顶边位移, 即在模型中引入一个时间依赖的非齐次 Dirichlet 条件

$$u_x = t\bar{u}, \quad \bar{u} = 1\text{mm/s} \tag{8.1}$$

这里 t 代表总时间。相场变量则在全局应用齐次 Neumann 条件 (无牵引), 即

$$\sigma(u)n e_x = 0, \quad 在 \Gamma_{\text{left}} \times I \text{ 上}$$
$$u_y = 0, \quad 在 \Gamma_{\text{left}} \times I \text{ 上}$$

$$\sigma(u)n e_x = 0, \quad 在 \Gamma_{\text{right}} \times I \text{ 上}$$
$$u_y = 0, \quad 在 \Gamma_{\text{right}} \times I \text{ 上}$$

$$u = 0, \quad 在 \Gamma_{\text{bottom}} \times I \text{ 上}$$

$$u_x = t\bar{u}, \quad 在 \Gamma_{\text{top}} \times I \text{ 上}$$
$$u_y = 0, \quad 在 \Gamma_{\text{top}} \times I \text{ 上}$$

$$\sigma(u)n e_x = 0, \quad 在 \Gamma_{\text{slit,bottom}} \times I \text{ 上}$$
$$u_y = 0, \quad 在 \Gamma_{\text{slit,bottom}} \times I \text{ 上}$$

$$\sigma(u)n = 0, \quad 在 \Gamma_{\text{slit,top}} \times I 上$$
$$\varepsilon\partial_n\varphi = 0, \quad 在 \partial B \times I 上$$

其中 $e_x = (1,0)^{\mathrm{T}}$ 表示笛卡儿坐标系中单位向量。

8.5 初 始 条 件

初始相场参数设定为

$$\varphi(0) = 1, \quad 在 B \times \{0\} 中$$

由于初始裂缝设置在几何图像中，因此只需将该区域初始自由度加倍，人为产生内部狭缝边界即可。

8.6 模 型 参 数

令 $\mu = 80.77\mathrm{kN/mm}^2$，$\lambda = 121.15\mathrm{kN/mm}^2$，$G_c = 2.7\mathrm{N/mm}$。同时我们令 $\kappa = 10^{-12}\mathrm{mm}$，$\varepsilon = 2h$。

时间步长设置为 $k = 10^{-4}\mathrm{s}$，同时设置加载时间总长为 $I = (0, 0.014\mathrm{s})$。由于我们处理的是主方程不含时间导数的增量问题，因此在设置加载步长 k 与 $\bar{u}$ 之间的关系时有多种方案。

方案 1：$u_x = t\bar{u} = 10^{-4}\mathrm{mm}$，时间 t 以增量 $k = 10^{-4}\mathrm{s}$ 递增，此时 $\bar{u} = 1\mathrm{mm/s}$。
方案 2：$u_x = t\bar{u} = 10^{-4}\mathrm{mm}$，时间 t 以增量 $k = 1\mathrm{s}$ 递增，此时 $\bar{u} = 10^{-4}\mathrm{mm/s}$。

为了验证结果正确与否，我们需要对裂缝扩展状态进行观察，特别是裂缝开始扩展以及扩展至边界时的状态。其次我们对 $\Gamma_{\text{top}} := \{(x,y) \in B | 0\mathrm{mm} \leqslant x \leqslant 10\mathrm{mm}, y = 10\mathrm{mm}\}$ 区域的表面载荷进行计算

$$\tau = (F_x, F_y) := \int_{\Gamma_{\text{top}}} \sigma(u)n\mathrm{d}s$$

其中 n 为法向量。

8.7 本 章 小 结

8.7.1 不同耦合方案下模拟结果研究

图 1.6 (1.3.3 节) 展示了单边缺口剪切实验不同时间下的裂缝扩展状态[217]，其结果与其他文献中的结论保持一致[308,65,217,438,439,17]。图 8.2~图 8.7 展示了不同求解方法下的载荷曲线。

8.7 本章小结

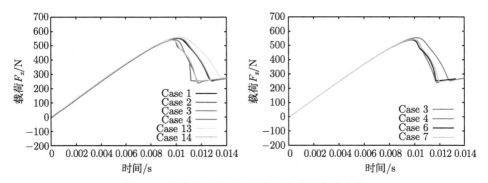

图 8.2 单边剪切实验不同求解方式下载荷曲线

其中 Case2 采用了 AL 全耦合方案，Case14 采用了迭代 PDAS 方案，Case13 中时滞误差最为明显

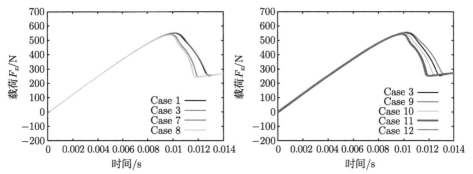

图 8.3 单边剪切实验不同求解方式下载荷曲线

我们发现细化加载步后能显著提升精确度

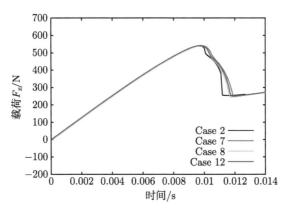

图 8.4 单边剪切实验不同求解方式下载荷曲线

基于应变场迭代方案与全耦合方案对比

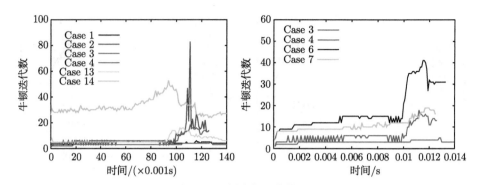

图 8.5 牛顿迭代次数统计

裂缝起裂与扩展时迭代次数均出现显著增加,其中 AL 全耦合方案最大迭代次数较多,基于应变场迭代方案迭代次数较少

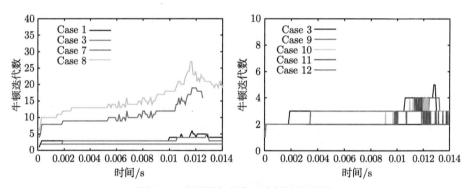

图 8.6 不同耦合方案下牛顿迭代细节

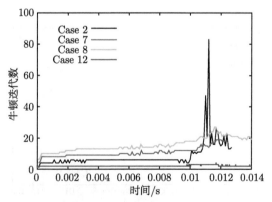

图 8.7 牛顿迭代次数统计

基于应变场迭代方案迭代次数较少,精确度较高

接下来我们展示了引入稳定项的分部耦合方案数值模拟结果 (7.6 节),不同稳定项求解过程中的数值解实际上是相同的,只是迭代次数略有差异。为便于后

续对比分析，令 $L = L_u = L_\varphi$，选取 $L = 0$ 以及 $L = 10^{-2}$ 两种方案分别进行模拟。另外我们还对比了动态稳定项分部耦合模拟结果 [152]。模拟结果如图 8.8 和图 8.9 所示。

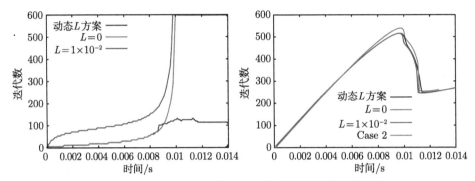

图 8.8 不同稳定项取值情况下迭代次数与载荷变化曲线

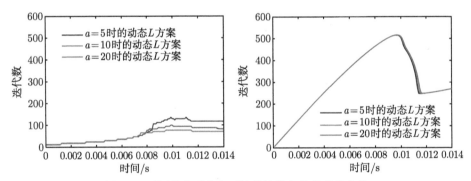

图 8.9 不同稳定项方案下迭代次数与载荷变化曲线

我们观察到当使用动态稳定项方案时，计算成本显著降低。最大迭代次数为 21，当 $a = 20$ 时降低至 12，精确度变化较小。

8.7.2 不同加载步条件下模拟结果研究

接下来我们根据算例 2 (AL 全耦合方案) 与算例 13 (准 PDAS 全耦合方案)，然后我们细化加载步探究解变量的变化。一般而言，加载步减小，收敛性会随之加强，但到目前为止我们还未得到针对该模型的严格收敛性证明。

模型网格为均匀细化网格方案，共有 3315 个自由度。模拟结束时间为 0.014s，其中 $N = 140$ 时，$k = 10^{-4}$s。模拟结果如表 8.1 所示。

根据模拟结果可知，即使加载步不断细分，但裂缝实际起裂时间没有明显变化。

表 8.1 细化加载步模拟结果表

	加载步数 N	起裂时间/s	最大载荷/N
分部耦合方案	140	0.0101	551.784
	160	0.009975	550.277
	180	0.00995556	549.2
	200	0.00994	548.138
	280	0.0098	546.365
	700	0.00974	544.732
	1400	0.00968	553.503
全耦合方案	140	0.0096	538.596

第 9 章 数值建模 Ⅲ：自适应方法

本章我们将讨论数值模型中的自适应方法及相关技术。

9.1 自适应方法简介

衡量数模好坏的标准除了准确性与收敛性之外还有计算量的大小。我们希望以最小的计算成本获得足够精确的模拟结果。在通常情况下，研究人员仅关心模型中的部分区域的解变量，如界面周边、边界区域以及某些特定位置，这类区域可以通过目标函数 $J: X \to \mathbb{R}$ 来描述。其中解变量为 $U_{hk} \to J(U_{hk})$，通常为研究人员关心的物理量 (QoI)。同时我们可以有针对性地计算目标函数的误差：

$$J(U) - J(U_{hk})$$

当 $h \to 0$, $k \to 0$ 时，我们希望：

$$J(U) - J(U_{hk}) \to 0$$

在实际数值模拟时则设定：

$$|J(U) - J(U_{hk})| < \text{TOL}$$

为达成计算目标，我们有以下几种求解思路：
(1) 在均匀网格上进行暴力求解，但随着物理场的增加，计算量与计算难度也会随之上升；
(2) 引入多个计算核心，进行并行计算处理；
(3) 开发网格自适应离散化方案；
(4) 线性和非线性求解器终止准则自适应处理；
(5) 自适应多尺度建模。

本章后续部分将着重介绍自适应方法，包括空间网格自适应方法。同时我们将简要介绍相应的误差控制方法。

9.1.1 时间/空间自适应相场裂缝模型简介

研究人员对空间网格细化方法进行了大量研究,包括各向异性模型 [328],基于残差的自适应有限元算法 [92,93],各向异性自适应网格细化技术 [30],网格预处理技术 [65],预测-校正法 [217,281],面向目标函数的自适应法 [436],多尺度模型中的网格细化技术 [344] 以及其他相关方法 [35]。

研究裂缝扩展路径时依赖于高精度的网格。为此我们展开了相关网格细化的研究 [217,281,276,424],使用对偶参数的后验误差分析 [250]。面向目标的后验误差分析 [397,362,59,60,399,398,363,385,226]。同时可参考障碍问题的相关研究 [450],包括障碍问题的先验有限元误差分析等 [82,83]。

9.1.2 相关研究进展

在第 5 章中,我们讨论了关于相场模型和离散化误差分析。接下来我们参考相关领域的误差控制和计算研究,如 Allen-Cahn 相场模型, Ginzburg-Landau 型方程、有界变差函数 [41,39,40,457,161,162],以及离散化的多尺度问题 [1]。在模型离散化误差分析 [75,338]、非线性迭代误差分析 [365,148]、线性误差分析 [50,302] 等相关方面,研究人员同样进行了大量研究。目标导向方法最近的研究重点主要在于时间步长控制 [304,157]、多尺度问题 [289]、随机估计 [300]、机器学习 [81] 和最佳收敛率 [160,49] 这几个方向。后验误差控制方向则请读者参考相关文献 [4, 37, 330, 366, 209, 422, 398]。

9.1.3 有效性,可靠性以及基本自适应算法

在相关文献中 [52,51,37,359],我们可得到关于范数与一般可微泛函 $J: X \to \mathbb{R}$ 的后验误差估计。这使得我们能够初步估计算法应用 (机械和土木工程、流体动力学、固体动力学等) 的工作量。在范数的误差估计中,$U - U_{hk}$ 可以用 $J(\cdot)$ 表示。我们令 U 为解变量,X 或 V 为函数空间。

当进行后验误差分析时,我们常关心以下关系:

$$C_1 \eta \leqslant |J(U) - J(U_{hk})| \leqslant C_2 \eta \tag{9.1}$$

其中 $\eta := \eta(U_{hk})$ 为误差估计量,C_1, C_2 为正常数,$J(U) - J(U_{hk})$ 为真实误差。

定义 60 (有效性与可靠性) 一个好的误差估计量 $\eta := \eta(U_{hk})$ 应满足以下条件:

(1) 当误差估计量满足以下条件时称其为有效的

$$C_1 \eta \leqslant |J(U) - J(U_{hk})|$$

此时误差估计量的上界即为真实误差。

(2) 当误差估计量满足以下条件时称其为可靠的

$$|J(u) - J(U_{hk})| \leqslant C_2 \eta$$

此时真实误差受误差估计量控制。

备注 54 一般情况下，确定可靠的误差估计量更为简单[52,37]。由于可能出现 $J(U) - J(U_{hk}) = 0$ 的情况，因此能否确定有效的误差估计量需要单独讨论。相关研究内容请读者参考文献 [375, 150, 149]。

定义 61 将后验误差估计与局部网格自适应应用于有限元框架中，即可获得自适应有限元模型。自适应有限元基本算法为：

(1) 求解当前时空网格 $\mathcal{T}_h$ 上的偏微分方程；
(2) 通过后验误差估计得到 η；
(3) 标记重点关注区域的相关元素；
(4) 使用特定的处理策略细化/粗化相关网格区域。

9.1.4 误差估计量 η

我们引入有效性指数 I_{eff} 作为衡量估计量与真实误差的一个指标。

定义 62 (有效性指数) 有效性指数 I_{eff} 定义为

$$I_{\text{eff}} := \left| \frac{\eta}{J(U) - J(U_{hk})} \right| \tag{9.2}$$

有效性指数 I_{eff} 最好应满足 $h \to 0, k \to 0$ 时 $I_{\text{eff}} \to 1$。

(1) $I_{\text{eff}} > 1$，则说明高估了误差值；
(2) $I_{\text{eff}} < 1$，则说明低估了误差值。

在有效性指数计算时，整体误差估计量与局部误差估计量 η_i 关系为 $\sum_i \eta_i =:$ η。由于误差存在正负，因此求和时可能存在相互抵消的情况。为此我们对上述公式进行了修正。

定义 63 考虑局部的有效性指数计算公式：

$$I_{\text{ind}} := \frac{\sum_i |\eta_i|}{|J(U) - J(U_{hk})|} \tag{9.3}$$

指数 i 为每个网格或自由度的序数。

9.2 面向目标的稳态相场裂缝后验误差分析

本节将简要介绍基于目标函数的双重加权残差 (dual-weighted residual (DWR)) 法。以及非线性求解器的自适应终止准则,相关研究可参考文献 [365, 52, 37, 144, 163, 164, 165, 182]。接下来我们将忽略时间误差,仅研究解变量 U_h。当然本章最后一节将对时间误差进行简要介绍。到目前为止,全时空自适应方法尚未成熟,因此我们需要对其分开讨论。

9.2.1 问题陈述与设置

在命题 19、20、25、27 及 28 中我们获得了相应的有限元离散公式。参考文献 [398],在相应模型中设置针对变分不等式的双重加权残差法可以进行以下处理。

公式 20 加载步 $n = 1, 2, 3, \cdots, N$,给定 φ^{n-1},令 $U := U^n = (u^n, \varphi^n) \in X$,其中 $\varphi(0) = \varphi_0$,则

$$A(U)(\Psi) := \bar{A}(U)(\Psi) + \left(\gamma \left[\varphi - \varphi^{n-1}\right]_+, \psi\right) = 0, \quad \forall \Psi = (w, \psi) \in X$$

其中 $\bar{A}(U)(\Psi)$ 未进行补偿处理 (命题 19,27)。

为构建基于伴随误差的 DWR 方法,需要对伴随进行定义。为此我们参考前文的半线性形式问题,对单个时间点 t_n 进行讨论。在处理相场裂缝扩展的过程中,我们将模型离散为不同时间点上的稳态问题,由此可将时间依赖问题在时间上离散化。在这类数值模型中,每一个时间步中的误差有可能会线性累积。具体来说,模型网格在牛顿迭代过程中保持固定,并且仅在单个时间步内进行细化处理,此时全局空间误差是一系列稳态问题误差的累积。

定义 64 (初始问题) 令 $U \in X$,则

$$A(U)(\Psi) = 0, \quad \forall \Psi \in X \tag{9.4}$$

为了方便起见,即使存在右手项 $F(\Psi)$,也将其移至左边项中。

命题 63 (伴随问题) 给定目标函数 $J(U)$,令 $Z \in X$,则

$$A'(U)(\Phi, Z) = J'(U)(\Phi), \quad \forall \Phi \in X \tag{9.5}$$

例 12 接下来我们演示一下障碍问题 (2.4.1 节) 中伴随问题的计算方法,原始问题为

$$-\Delta u - \gamma [g - u]^+ = f$$

9.2 面向目标的稳态相场裂缝后验误差分析

伴随问题的处理方式如下:
(1) $A(u)(\phi) = (\nabla u, \nabla \phi) - \gamma([g-u]^+, \phi) - (f, \phi)$ (弱形式);
(2) $A'(u)(\delta u, \phi) = (\nabla \delta u, \nabla \phi) + \gamma(\delta u, \phi)_{B(u)}$;
(3) $A'(u)(\phi, \delta u) = (\nabla \phi, \nabla \delta u) + \gamma(\phi, \delta u)_{B(u)}$;
(4) $A'(u)(\phi, z) = (\nabla \phi, \nabla z) + \gamma(\phi, z)_{B(u)}$。

9.2.2 非线性问题中的双重加权残差法

9.1 节简单介绍了非线性偏微分方程中基于伴随的误差估计。需要注意的是我们在公式 20 中仅讨论了 PDE 系统 (未讨论 CVIS 系统)。

公式 21 给定 U_h,令 $J(U)$ 与 $A(U)(\Psi)$ 可微,求解

$$\min_{U \in X} (J(U) - J(U_h)) \quad \text{s.t.} \quad A(U)(\Psi) = 0$$

命题 64 令 $U \in X$,求解式 (9.4),同时令 $Z \in X$,求解式 (9.5)。令 $\tilde{U}, \tilde{Z}$ 为近似值,则

$$J(U) - J(\tilde{U}) = \frac{1}{2}\rho(\tilde{U})(Z - \tilde{Z}) + \frac{1}{2}\rho^*(\tilde{U}, \tilde{Z})(U - \tilde{U}) + \rho(\tilde{U})(\tilde{Z}) + \mathcal{R}^{(3)}$$

式中

$$\rho(\tilde{U})(\cdot) := -A(\tilde{U})(\cdot)$$

$$\rho^*(\tilde{U}, \tilde{Z})(\cdot) := J'(\tilde{U})(\cdot) - A'(\tilde{U})(\cdot, \tilde{Z})$$

其中 $\mathcal{R}^{(3)}$ 为三阶余项。由此我们可得原始误差与伴随误差:

$$e = U - \tilde{U}, \quad e^* = Z - \tilde{Z}$$

推理 6 命题 64 中全局误差估计量 (全误差估计) 如下:

$$\eta = J(U) - J(\tilde{U}) = \frac{1}{2}\rho(\tilde{U})((Z - \tilde{Z})) + \frac{1}{2}\rho^*(\tilde{U}, \tilde{Z})((U - \tilde{U})) + \rho(\tilde{U})(\tilde{Z}) + \mathcal{R}^{(3)}$$

定义 65 全局误差估计量 (全误差估计) 可分解为以下几种误差的累加:

$$\eta = \eta_p + \eta_a^* + \eta_m + \mathcal{R}^{(3)}$$

其中

$$\eta_p := \frac{1}{2}\rho(\tilde{U})(Z - \tilde{Z}) \quad (\text{原始误差估计量})$$

$$\eta_a^* := \frac{1}{2}\rho^*(\tilde{U},\tilde{Z})(U-\tilde{U}) \quad \text{(伴随误差估计量)}$$

$$\eta_m := \rho(\tilde{U})(\tilde{Z}) \quad \text{(迭代误差估计量)}$$

9.2.3 局部误差估计

本节我们介绍针对自适应局部网格的局部误差估计量，包括根据弱形式或者补偿法构建的基于有限元的误差估计[52]以及基于自由度的误差估计法[375]。

首先，我们引入一个通过有限元函数实现的单位分解量 (PU)。

定义 66 (单元分解法) 我们可将单元分解为

$$V_{\rm PU} := \{\psi_1, \cdots, \psi_M\}$$

式中 $\dim(V_{\rm PU}) = M$。其中 PU 具有如下性质：

$$\sum_{i=1}^{M} \psi_i \equiv 1 \tag{9.6}$$

备注 55 PU 可以简单理解为低阶有限元空间中的线性或双线性元素：

$$V_{\rm PU} = V_h^{(1)}$$

在变分处理中的作用请参考相关文献 [102]。

命题 65 (局部误差估计) 给定前一个 PU，局部误差估计量为

$$J(U)-J(\tilde{U}) = \eta = \sum_{i=1}^{M} \frac{1}{2}\rho(\tilde{U})\left((Z-\tilde{Z})\psi_i\right)$$
$$+ \frac{1}{2}\rho^*(\tilde{U},\tilde{Z})\left((U-\tilde{U})\psi_i\right) + \rho(\tilde{U})(\tilde{Z}) + \mathcal{R}^{(3)}$$

不考虑余项的情况下，局部误差估计量为

$$J(U)-J(\tilde{U}) \approx \eta = \sum_{i=1}^{M} \frac{1}{2}\rho(\tilde{U})\left((Z-\tilde{Z})\psi_i\right)$$
$$+ \frac{1}{2}\rho^*(\tilde{U},\tilde{Z})\left((U-\tilde{U})\psi_i\right) + \rho(\tilde{U})(\tilde{Z})$$

定义 67 局部误差估计量由离散误差与非线性迭代误差组成：

$$\eta = \eta_{\rm dis} + \eta_m := \sum_{i=1}^{M}(\eta_i + \eta_i^*) + \eta_m$$

其中 η_{dis} 为离散误差，η_m 为非线性迭代误差

$$\eta_i := \frac{1}{2}\rho(\tilde{U})\left((Z-\tilde{Z})\psi_i\right)$$

$$\eta_i^* := \frac{1}{2}\rho^*(\tilde{U},\tilde{Z})\left((U-\tilde{U})\psi_i\right)$$

$$\eta_m := \rho(\tilde{U})(\tilde{Z})$$

9.2.4 基于伴随误差的有限元近似问题

在计算过程中，我们需要对未知数进行空间离散近似处理。

命题 66 (实际误差估计) 令 $U_h := U_h^{(1)} \in X_h^{(1)}$ 为低阶解，$U_h^{(2)} \in X_h^{(2)}$ 为高阶解，$Z_h := Z_h^{(1)} \in X_h^{(1)}, Z_h^{(2)} \in X_h^{(2)}$ 为近似解，相应地，局部误差估计量为

$$J(U) - J(U_h) \approx \eta = \sum_{i=1}^{M} \frac{1}{2}\rho(U_h)\left(\left(Z_h^{(2)} - Z_h\right)\psi_i\right)$$

$$+ \frac{1}{2}\rho^*(U_h, Z_h)\left(\left(U_h^{(2)} - U_h\right)\psi_i\right) + \rho(U_h)(Z_h)$$

误差可表示为

$$\eta = \eta_h + \eta_m := \sum_{i=1}^{M}(\eta_i + \eta_i^*) + \eta_m$$

其中

$$\eta_i := \frac{1}{2}\rho(U_h)\left(\left(Z_h^{(2)} - Z_h\right)\psi_i\right)$$

$$\eta_i^* := \frac{1}{2}\rho^*(U_h, Z_h)\left(\left(U_h^{(2)} - U_h\right)\psi_i\right)$$

$$\eta_m := \rho(U_h)(Z_h)$$

基于伴随误差的误差估计有以下特点：
(1) 可以对多目标函数进行定量处理；
(2) 可应用于局部自适应网格与自适应求解器。

现阶段，任意目标函数的自适应方案收敛性尚未完成有效论证，相关误差估计的收敛性研究请参考文献 [160, 49]。

9.2.5 PU-DWR 准静态相场裂缝误差分析

针对稳态相场裂缝问题，我们引入一个原始误差估计量 ρ。对相场裂缝模型 (第 10 章与第 12 章) 的时间步 t_0 至 t_1 进行研究，通过分析每一个时间步中的相关误差，推广到整个准静态相场模型中的总误差。相关研究请参考文献 [436]。

命题 67 (PU-DWR 相场裂缝模型原始误差估计) 令 $U = (u, \varphi) \in X^D$，$U_h = (u_h, \varphi_h) \in X_h^D$ 分别为连续解与离散解，令 $Z = (z_u, z_\varphi) \in X^0$，$Z_h \in X_h^0$ 为连续伴随解与离散伴随解。上标 (2) 表示高阶解。针对命题 26，其原始后验误差为

$$|J(U) - J(U_h)| \leq \eta := \left| \sum_{i=1}^{M} \eta_i \right| \leq \sum_{i=1}^{M} |\eta_i|$$

其中

$$\begin{aligned}
\eta_i &= -A(U_h)\left(\left(Z_h^{(2)} - i_h Z_h^{(2)}\right)\psi_i\right) \\
&= -\left(g(\varphi_h)\sigma(u_h), \nabla\left(\left(z_{u,h}^{(2)} - i_h z_{u,h}^{(2)}\right)\psi_i\right)\right) \\
&\quad - \left((1-\kappa)\varphi_h \sigma(u_h) : \nabla u_h, \left(\left(z_{\varphi,h}^{(2)} - i_h z_{\varphi,h}^{(2)}\right)\psi_i\right)\right) \\
&\quad + \left(\frac{G_c}{\varepsilon}(1-\varphi_h), \left(\left(z_{\varphi,h}^{(2)} - i_h z_{\varphi,h}^{(2)}\right)\psi_i\right)\right) \\
&\quad - \left(G_c \varepsilon \nabla \varphi_h, \nabla\left(\left(z_{\varphi,h}^{(2)} - i_h z_{\varphi,h}^{(2)}\right)\psi_i\right)\right) \\
&\quad - \gamma\left([\varphi_h - \varphi_h^0]^+, \left(\left(z_{\varphi,h}^{(2)} - i_h z_{\varphi,h}^{(2)}\right)\psi_i\right)\right)
\end{aligned}$$

命题 68 (PU-DWR 相场裂缝模型误差估计) 令 $U = (u, \varphi) \in X^D$，$U_h = (u_h, \varphi_h) \in X_h^D$ 分别为连续解与离散解，令 $Z = (z_u, z_\varphi) \in X^0$，$Z_h \in X_h^0$ 为连续伴随解与离散伴随解。上标 (2) 表示高阶解。针对命题 26，各个加载步误差满足：

$$|J(U) - J(U_h)| \leq \eta := |\eta_h| + |\eta_m| \leq \sum_{i=1}^{M} |\eta_i + \eta_i^*| + |\eta_m|$$

式中 η_h 的取值请参考命题 66。这里我们有

$$\eta_i = -\frac{1}{2} A(U_h)\left(\left(Z_h^{(2)} - i_h Z_h^{(2)}\right)\psi_i\right)$$

其中迭代误差为

$$\eta_m = -A(U_h)(Z_h)$$

伴随误差为

$$\eta_i^* = \frac{1}{2}\left(J'(U_h)\left(\left(U_h^{(2)} - i_h U_h^{(2)}\right)\psi_i\right) - A'(U_h)\left(\left(U_h^{(2)} - i_h U_h^{(2)}\right)\psi_i, Z_h\right)\right)$$

备注 56 相关准线性问题的理论与算法后续有以下几个主要的研究方向：
(1) 计算效率与可靠性研究 [150]；
(2) 考虑不等式约束的时空自适应模型；
(3) 多目标函数模型 [148]；
(4) 线性和非线性迭代误差与离散化误差研究。

9.2.6 网格细化与标记方法

现在我们对每个元素 $K_m \in \mathcal{T}_h$ 均有一个误差值，以此为标准可以建立一个计算方案，对目标元素进行改进，从而提升模型的准确性。首先我们给出一个误差指标 TOL，基于局部误差进行网格自适应处理：

$$|J(U) - J(U_h)| \leqslant \eta := \sum_{K \in \mathcal{T}_h} \eta_K$$

要获得 η_K，可通过对每个结点的 η_i 进行累加处理。相应地，可以借助 PU 对每个节点进行判断

$$|J(U) - J(U_h)| \leqslant \eta := \sum_{i=1}^{M} \eta_i$$

网格细分具体算法如下：

算法 14
(1) 计算当前网格 $\mathcal{T}_h$ 中的解 U_h 与 Z_h；
(2) 计算每个单元 K 的误差估计量 η_K 或每个节点自由度的 η_i；
(3) 计算总误差 $\eta := \sum_{K \in \mathcal{T}_h} \eta_K$ 或 $\eta := \sum_i \eta_i$；
(4) 检查是否满足终止准则 $|J(U) - J(U_h)| \leqslant \eta \leqslant \text{TOL}$，若满足，则表示符合要求，若不满足，则标记相应单元。

为了后续对网格进行细化操作，我们标记需要细化的元素。在这里我们介绍三种方法。

(1) 平均值法。
若满足

$$\eta_i > \frac{\alpha \eta}{M_{\text{el}}}$$

则标记相关元素，这里 α 常取 1。

(2) 残差法。

针对函数 X, Y，令 $1 - X > Y$。指示函数 $M_*, M^* \in \{1, \cdots, M_{el}\}$ 满足

$$\sum_{i=1}^{M^*} \eta_i \approx X\eta, \quad \sum_{i=M_*}^{M_{el}} \eta_i \approx Y\eta$$

随后对元素 $K_1, \cdots, K_{M^*}$ 细化，对元素 $K_{M_*}, \cdots, K_{M_{el}}$ 粗化 [135]。

(3) 相对残差法。

与残差法相似，但在该方案中我们将参考误差的最大值与最小值确定细化和粗化的元素。

备注 57 TOL 的值正常情况下远大于实际数值求解的误差值。

算法 15 (拟稳态问题的网格自适应方案) 针对加载步 $n = 1, 2, 3 \cdots$，执行以下操作：

(1) 计算网格 $\mathcal{T}_h$ 中的原始解 U_h；

(2) 计算当前网格上的高阶伴随解 Z_h；

(3) 计算误差估计量并确定每一个 PU 节点上的局部估计量；

(4) 计算局部估计量之和 $\eta := \sum_i \eta_i$；

(5) 检查误差值是否满足终止标准 $|J(U) - J(U_h)| \leqslant \eta \leqslant \text{TOL}$，若满足，则接受当前解，否则进行以下处理：

(6) 标记所有需要细分的 PU 节点；

(7) 进行网格细分处理；

(8) 回到步骤 (1) 重新检验是否满足终止标准。

具体实施案例请参考文献 [362, 364]。

9.3 静态二维 Navier-Stokes 模型误差分析

本节将介绍流体力学中的一个著名算例，其基本方程为不可压缩流体的 Navier-Stokes(N-S) 方程 [405,357,198]。主要分析流动阻力，障碍物周边压差等相关参数。需要注意的是我们仅研究了误差估计量 ρ，忽略伴随部分 ρ^*。

9.3.1 基本方程

公式 22 (不可压缩流体 N-S 方程) 令 $\Omega \subset \mathbb{R}^2$ 且 $\partial\Omega = \Gamma_{\text{wall}} \cup \Gamma_{\text{in}} \cup \Gamma_{\text{out}}$，此时流速 $v: \Omega \to \mathbb{R}^2$ 为向量值，压力 $p: \Omega \to \mathbb{R}s$ 为标量值：

$$\rho(v \cdot \nabla)v - \text{div}(\sigma) = \rho f$$
$$\text{div}(v) = 0, \quad \text{在 } \Omega \text{ 中}$$

9.3 静态二维 Navier-Stokes 模型误差分析

$$v = 0, \quad 在 \ \Gamma_{\text{wall}} \ 上$$
$$v = g, \quad 在 \ \Gamma_{\text{in}} \ 上$$
$$\nu \partial_n v - p n = 0, \quad 在 \ \Gamma_{\text{out}} \ 上$$

式中 n 为法向量；ρ 为密度；$\sigma = \sigma(v, p) = -pI + \rho\nu \left(\nabla v + \nabla v^{\mathrm{T}}\right)$，$\sigma$ 为牛顿流体的柯西应力张量，单位为 Pa；雷诺数为 $\mathrm{Re} = \dfrac{LU}{\nu}$，其中 L 为特征长度，U 为特征速度。

备注 58 若忽略黏性流体对流项，则可得到线性 Stokes 方程。

在变分形式中，我们有

$$V_v := \{v \in H^1(\Omega)^2 | v = 0, 在 \Gamma_{\text{in}} \cup \Gamma_{\text{wall}} \cup S \ 上\}$$

$$V_p = L^2(\Omega)/\mathbb{R}$$

在流出边界上我们有边界条件：

$$\rho\nu \partial_n v - p n = 0$$

备注 59 在应用 $\rho\nu \partial_n v - pn = 0$ 的流出边界条件时，与总应力方程 $\sigma \cdot n = -pI + \rho\nu \left(\nabla v + \nabla v^{\mathrm{T}}\right) \cdot n$ 存在出入，相关处理细节请参考文献 [225, 434]。

命题 69 令 $v \in \{g + V_v\}$，$p \in V_p$ 此时我们有

$$(\rho(v \cdot \nabla)v, \phi) + (\sigma, \nabla \phi) = (\rho f, \phi), \quad \forall \phi \in V_v$$

$$(\mathrm{div}(v), \psi) = 0, \quad \forall \psi \in V_p$$

选取的有限元空间需要满足相应的离散条件[198]。这里我们选取四边形网格进行划分，为此我们令 $(v_h, p_h) \in V_h^2 \times V_h^1$ 且

$$V_h^s := \left\{u_h \in C(\bar\Omega) | u_{h|K} \in Q_s(K), \forall K \in \mathcal{T}_h\right\}$$

这样我们就可获得一个协调的有限元空间，即 $V_h^s \subset H^1(\Omega)$。

命题 70 令 $v_h \in \{g + V_h^2\}$，$p_h \in V_h^1$，则

$$(\rho(v_h \cdot \nabla)v_h, \phi_h) + (\sigma, \nabla \phi_h) = (\rho f, \phi_h), \quad \forall \phi_h \in V_h^2$$

$$(\mathrm{div}(v_h), \psi_h) = 0, \quad \forall \psi_h \in V_h^1$$

半线性形式如下所示，令 $U_h := (v_h, p_h) \in \{g + V_h^2\} \times V_h^1$，$\forall \Psi_h = (\phi_h, \psi_h) \in V_h^2 \times V_h^1$，有

$$a(U_h)(\Psi_h) = (\rho(v_h \cdot \nabla)v_h, \phi_h) + (\sigma, \nabla \phi_h) - (\rho f, \phi_h) + (\mathrm{div}(v_h), \psi_h) = 0$$

9.3.2 目标函数

在该模型中我们主要研究阻力、浮力以及液压压差，其目标函数如下：

$$J_1(U) = \int_S \sigma \cdot ne_1 \mathrm{d}s, \quad J_2(U) = \int_S \sigma \cdot ne_2 \mathrm{d}s$$

$$J_3(U) = p(x_0, y_0) - p(x_1, y_1)$$

其中 $x_0, x_1, y_0, y_1 \in \Omega$，$e_1, e_2$ 分别为 x，y 方向的单位向量，阻力系数与浮力系数分别为

$$c_D = \frac{2J_1}{\rho V^2 D}, \quad c_L = \frac{2J_2}{\rho V^2 D}$$

模型管径 $D = 0.1\mathrm{m}$，平均速度为 $V = V(t) = \frac{2}{3} v_x(0, H/2)$，其中 $H = 0.41\mathrm{m}$。

9.3.3 基于对偶的后验误差估计

命题 71 伴随残差的计算如下：

$$a'_U(U_h)(\Psi_h, Z_h) = (\rho(\phi_h \cdot \nabla)v_h + \rho(v_h \cdot \nabla)\phi_h, z_h^v)$$

$$+ (-\psi_h I + v\rho(\nabla \phi_h + \nabla \phi_h^\mathrm{T}), \nabla z_h^v) + (\mathrm{div}(\phi_h), z_h^p)$$

命题 72 (伴随问题) 伴随问题与特定的目标函数 $J_i(U), i = 1, 2, 3$ 相关，以阻力为例，令 $Z_h = (z_h^v, z_h^p) \in V_h^4 \times V_h^2$，有

$$a'_U(U_h)(\Psi_h, Z) = J_1(\Psi_h), \quad \forall \Psi_h = (\phi_h, \psi_h) \in V_h^4 \times V_h^2$$

其中 $\Psi_h = (\phi_h, \psi_h)$：

$$J_1(\Psi_h) = \int_S \left(-\psi_h I + \rho v (\nabla \phi_h + \nabla \phi_h^\mathrm{T})\right) \cdot ne_1 \mathrm{d}s$$

原始 PU-DWR 误差估计量如下。

命题 73 我们有

$$|J(U) - J(U_h)| \leqslant \eta := \left|\sum_i \eta_i\right| \leqslant \sum_i |\eta_i|$$

$$\eta_i = -a(U_h)\left(\left(Z_h^{(2)} - i_h Z_h^{(2)}\right)\chi_i\right), \quad Z_h^{(2)} = (z_h^v, z_h^p), \quad \chi_i \in V_\mathrm{PU}$$

其中 $Z_h^{(2)}$ 为高阶离散解。

备注 60 在误差估计量计算中，假设非齐次 Dirichlet 边界 Γ_{in} 数据上的 g 无限小并忽略。

9.3.4 二维几何模型

模型形状请参考图 9.1。

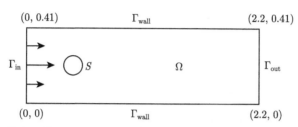

图 9.1 静态二维 Navier-Stokes 算例几何模型，圆柱障碍位于 (0.2, 0.2)

9.3.5 模型参数

我们设定所有边界均符合以下条件：

$$g = (g_1, g_2)^{\text{T}} = (v_x(0,y), v_y)^{\text{T}} = \left(4v_m y \frac{H-y}{H^2}, 0\right)^{\text{T}}$$

其中 $v_m = 0.3 \text{m/s}$，雷诺数为 $\text{Re} = 20$。

同时我们令 $\rho = 1 \text{kg/m}^3$，$\nu = 10^{-3} \text{m}^2/\text{s}$，$f = 0$。

9.3.6 算例 1

在文献 [382] 中，我们给出以下取值范围：
(1) 拖曳力 $c_D: 5.57 \sim 5.59$；
(2) 升力 $c_L: 0.0104 \sim 0.0110$；
(3) 压力差 $\Delta p: 0.1172 \sim 0.1176$。

根据有限元计算可得，相关参数如表 9.1 所示。

表 9.1 数值计算参考结果

	取值
拖曳力	$5.5754236905876873 \times 10^0$
升力	$1.0970590684824442 \times 10^{-2}$
压力差	$1.1745258793930979 \times 10^{-1}$

9.3.7 算例 2

阻力的目标函数计算如下：

$$J_1(U) := 5.5787294556197073$$

在不同网格划分情况下计算结果如表 9.2 所示。

表 9.2 PU-DWR 的 2D 数值模拟结果：阻力误差评估和有效性指数分析

网格细分次数	$J_1(U) - J_1(U_h)$	η	I_{eff}
0	3.51×10^{-1}	1.08×10^{-1}	3.07×10^{-1}
1	9.25×10^{-2}	2.08×10^{-2}	2.25×10^{-1}
2	1.94×10^{-2}	6.07×10^{-3}	3.12×10^{-1}
3	3.31×10^{-3}	2.45×10^{-3}	7.40×10^{-1}

根据模拟结果可知，随着网格细化，真实误差与误差估计量 η 均在减少。

图 9.2 中比较了局部细分与均匀细分情况下的真实误差变化情况。同样的网格数下，自适应网格误差降低更明显，但收敛速度并无显著提升。相关研究请读者参考文献 [326, 47, 52, 76, 382, 150, 146, 236]。仅在极端细化的网格中，局部自适应网格的收敛速度有可能超过粗网格的收敛速度。

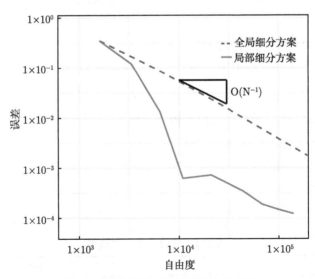

图 9.2 细分网格下真实误差变化曲线

9.3.8 模拟云图

网格细分情况与变量云图如图 9.3 和图 9.4 所示。

9.3 静态二维 Navier-Stokes 模型误差分析

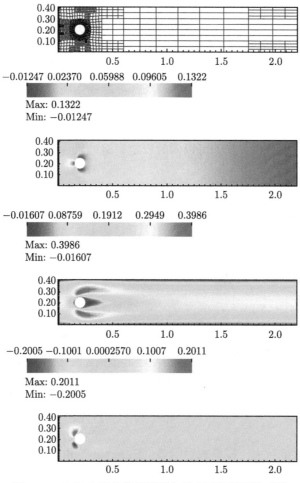

图 9.3 2D N-S 流体数值模型自适应网格下模拟云图
$J_1(U)$、压力解、x 速度、y 速度

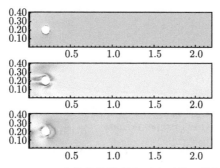

图 9.4 2D N-S 流体数值模型自适应网格下模拟云图
压力伴随解、x 速度、y 速度

9.4 面向多目标的后验误差估计

我们现在将单目标误差分析推广到多目标函数误差分析，这在多物理场问题中有着广泛的应用空间，其中每个物理场都可能成为研究人员的关注对象。

9.4.1 多目标函数介绍

在相场裂缝模型中，我们可以同时设定裂缝能量、裂缝体积以及位移值为目标函数，如：

$$J_1(\varphi) = E_s(\varphi) = \frac{1}{2} \int_B G_c \left(\frac{1}{\varepsilon}(1-\varphi)^2 + \varepsilon |\nabla \varphi|^2 \right) dx$$

$$J_2(u,\varphi) = \int_B u \cdot \nabla \varphi dx$$

$$J_3(u) = u(x), \quad x \in B$$

其中系统能量、应力等也可做为目标函数使用。在其他涉及流体的多物理场问题中，目标函数还可能包括压力、浮力、雷诺数等其他值。在多目标函数问题中，比较原始的控制方法是罗列出 $S(S \in N)$ 个目标函数，参考命题 64，$i = 1, \cdots, S$，令

$$|J_i(U) - J_i(U_h)| \leqslant \text{TOL}_i$$

但这种方式计算量巨大，模型求解困难。Hartmann 和 Houston 提出综合考虑相对误差，对多目标函数进行综合判断[212]：

$$\sum_{i=1}^{S} \frac{|J_i(U) - J_i(U_h)|}{|J_i(U_h)|} \tag{9.7}$$

这就需要我们首先假定 $J_i(U_h) \neq 0$。

9.4.2 多目标函数组合方法

在文献 [212, 211] 的基础上，作者在过去四年中在面向多目标的误差估计法方面进行了大量研究[151,148,150,145,147,146]。其他研究人员的研究成果请参考文献 [342, 15, 253, 417]。我们常将 p-Laplace 问题作为研究对象[351,113,3,286,129,409]。其与相场裂缝模型一样均为准线性模型。同时我们还对以 Navier-Stokes 方程为代表的准静态模型进行了相关研究[150,146]。

9.4 面向多目标的后验误差估计

接下来,我们给定 S 个目标函数 $J_i(\cdot), i = 1, \cdots, S$。根据式 (9.7),可以将多个单目标函数整合为一个总的目标函数

$$J_c(U) := \sum_{i=1}^{S} w_i J_i(U) \tag{9.8}$$

式中权重 $w_i \in \mathbb{R}$,令泛函 $\Phi \in X$,同时有

$$J_c(\Phi) := \sum_{i=1}^{S} w_i J_i(\Phi) \tag{9.9}$$

其中权重参数是总目标函数的一个重要部分,这里我们首先假定对于 $i = 1, \cdots, S$, $w_i = 1$。若奇数 i 均满足 $J_i = -1$,偶数 i 均满足 $J_i = 1$,则 $J_c \equiv 0$。因此,我们在计算权重时需要考虑到符号的问题:

$$w_i := \frac{\text{sign}(J_i(U) - J_i(U_h))\,\omega_i}{|J_i(U_h)|} \tag{9.10}$$

其中 $\omega_i > 0$,则

$$J_c(\Phi) = \sum_{i=1}^{S} \omega_i \frac{\text{sign}(J_i(U) - J_i(U_h))}{|J_i(U_h)|} J_i(\Phi) \tag{9.11}$$

若权重 ω_i 取值相似,则所有单个泛函的相对误差均为同阶的。令 $\Phi = U - U_h$,则线性情况下误差表示为

$$J_c(U - U_h) = \sum_{i=1}^{S} \omega_i \frac{\text{sign}(J_i(U) - J_i(U_h))}{|J_i(U_h)|} J_i(U - U_h) \tag{9.12}$$

在自适应算法中,我们需要解决初始问题(可能是非线性的)、伴随问题(线性)以及离散误差问题。由于式 (9.11) 中的 $\Phi := U$,即

$$J_c(U) = \sum_{i=1}^{S} \omega_i \frac{\text{sign}(J_i(e_h))}{|J_i(U_h)|} J_i(U) \tag{9.13}$$

此时的总目标函数是可计算的,可用于伴随问题的右手项。

命题 74 (基于组合目标函数的伴随问题) 根据式 (9.13),$J_c'(U)(\Phi)$ 为 $U \in X$ 在 $\Phi \in X$ 方向上的 Fréchet 导数,随后我们给定 $Z \in X$:

$$A'(U)(\Phi, Z) = J_c'(U)(\Phi), \quad \forall \Phi \in X \tag{9.14}$$

关于 $J'_c(U)(\Phi)$ 的细节请参考文献 [148]。

证明：相关证明请参考命题 65，将右手项替换为式 (9.13) 即可。

命题 75 (高阶解与离散误差问题) 对于线性问题，我们可以将三个问题的解联系起来，令 $U_h \in X_h$，$U_h^{(2)} \in X_h^{(2)}$，$e_h^{(2)} \in X_h^{(2)}$，有

$$A(U_h)(\Psi_h) = F(\Psi_h) \quad \text{（低阶初始问题）}$$

$$A\left(U_h^{(2)}\right)\left(\Psi_h^{(2)}\right) = F\left(\Psi_h^{(2)}\right) \quad \text{（高阶初始问题）}$$

$$A\left(e_h^{(2)}\right)\left(\Psi_h^{(2)}\right) = F\left(\Psi_h^{(2)}\right) - A(U_h)\left(\Psi_h^{(2)}\right) \quad \text{（离散误差问题）}$$

对于 $X_h \subset X_h^{(2)}$，求解离散误差问题获得的解 $e_h^{(2)}$ 可等价为求解：

$$e_h^{(2)} = U_h^{(2)} - U_h$$

根据命题 75，我们有

$$J_c(U) = \sum_{i=1}^{S} \omega_i \frac{\text{sign}\left(J_i\left(U_h^{(2)} - U_h\right)\right)}{|J_i(U_h)|} J_i(U) \tag{9.15}$$

在文献 [148] 中，我们介绍了非线性问题的相关程序，在文献 [149] 中，我们尝试了计算低阶有限元原始解并扩展至高阶空间的方案，相关研究请读者参考文献 [52, 37, 375, 149]。

备注 61 虽然在处理多目标函数的问题中采取全局细化的方案更有助于简化程序，但是考虑到计算成本与内存需求，实际情况下我们很少采用全局细化的方案。

9.4.3 面向多目标函数的稳态相场模型后验误差估计

参考准静态相场裂缝模型，其在 $n=1$ 时，我们可得以下命题：

命题 76 $i = 1, \cdots, S$ 时，给定 $J_i(U)$，$J_c(U)$ 为组合目标函数，令 $U = (u, \varphi) \in X^D$，$U_h = (u_h, \varphi_h) \in X_h^D$ 分别为初始问题连续解与离散解。相应地 $U_h^{(2)} = \left(u_h^{(2)}, \varphi_h^{(2)}\right) \in V_{u,h}^{(2)} \times V_{\varphi,h}^{(2)}$ 为初始问题的高阶解，$Z = (z_u, z_\varphi) \in X^0$ 为组合目标函数伴随问题的连续解。此时我们有

$$|J_c(U) - J_c(U_h)| \leqslant \eta := \left|\sum_{i=1}^{M} \eta_i\right| \leqslant \sum_{i=1}^{M} |\eta_i|$$

其中

$$\eta_i = -A(U_h)\left(\left(Z_h^{(2)} - i_h Z_h^{(2)}\right)\psi_i\right)$$
$$= -\left(g(\varphi_h)\sigma(u_h), \nabla\left(\left(z_u^{(2)} - i_h z_u^{(2)}\right)\psi_i\right)\right)$$
$$- \left((1-\kappa)\varphi_h\sigma(u_h) : \nabla u_h, \left(\left(z_\varphi^{(2)} - i_h z_\varphi^{(2)}\right)\psi_i\right)\right)$$
$$+ \left(\frac{G_c}{\varepsilon}(1-\varphi_h), \left(\left(z_\varphi^{(2)} - i_h z_\varphi^{(2)}\right)\psi_i\right)\right) - \left(G_c\varepsilon\nabla\varphi_h, \nabla\left(\left(z_\varphi^{(2)} - i_h z_\varphi^{(2)}\right)\psi_i\right)\right)$$
$$- \gamma\left(\left[\varphi_h - \varphi_h^0\right]^+, \left(\left(z_\varphi^{(2)} - i_h z_\varphi^{(2)}\right)\psi_i\right)\right)$$

其全误差估计量为

$$|J_c(U) - J_c(U_h)| \leqslant \eta := |\eta_h| + |\eta_m| \leqslant \sum_{i=1}^M |\eta_i + \eta_i^*| + |\eta_m|$$

其中

$$\eta_h = \sum_{i=1}^M (\eta_i + \eta_i^*)$$

需要注意的是，此时我们研究的仍是稳态相场裂缝问题。

9.4.4 面向多目标函数的自适应算法

算法 16 (基于离散化误差的自适应算法) 如下：
(1) 求解原始问题低阶解 U_h 与高阶解 $U_h^{(2)}$。
(2) 构造组合目标函数 J_c。
(3) 求解伴随问题：计算伴随 $A'(U_h)\left(\Phi_h, Z_h^{(2)}\right) = J_c'(U_h)(\Phi_h)$ 获得 $Z_h^{(2)}$。
(4) 基于命题 76 确定每个节点的 η_i，并计算累加和 $\eta := \sum_i \eta_i$，检验是否满足终止准则：$\eta < \text{TOL}_c$。这里 $\text{TOL}_c := \inf_{1 \leqslant i \leqslant S}\left\{\frac{\omega_i \text{TOL}_i}{|J_i(U_h)|}\right\}$，若满足终止准则，则停止计算，否则进行下一步。
(5) 标记需要细化的节点并进行细化处理。
(6) 返回步骤 (1)。

备注 62 假设已知伴随解，为确保所有泛函均能有 $|J_i(U) - J_i(U_h)| < \text{TOL}_i$，我们令 $\text{TOL}_c := \inf_{1 \leqslant i \leqslant S}\left\{\frac{\omega_i \text{TOL}_i}{|J_i(U_h)|}\right\}$。

接下来我们考虑离散化与非线性迭代误差。以下算法的结构与文献 [365, 154, 148] 中的结构相似，其中 i 为细分次数，j 为牛顿迭代次数。在 l 级网格中的误差估计量 η_h, η_m 分别为 η_h^l, η_m^l。

算法 17 (l 级多目标泛函的自适应牛顿算法) 如下：

(1) 给定初始解 $U_h^{l,0} \in X_h^l$，$j = 0$。

(2) 引入 $Z_h^{l,0}$，则

$$A'\left(U_h^{l,0}\right)\left(\Phi_h, Z_h^{l,0}\right) = \left(J_c^{(0)}\right)'\left(U_h^{l,0}\right)(\Psi_h), \quad \forall \Phi_h \in X_h^l$$

(3) 当 $\left|A\left(U_h^{l,j}\right)\left(Z_h^{l,j}\right)\right| =: \eta_m^l > 10^{-2}\eta_h^{l-1}$ 时，进行以下步骤。

(4) 求解：

$$A'\left(U_h^{l,j}\right)\left(\delta U_h^{l,j}, \Psi_h\right) = -A\left(U_h^{l,j}\right)(\Psi_h), \quad \forall \Psi_h \in X_h^l$$

(5) 令 $U_h^{l,j+1} = U_h^{l,j} + \alpha \delta U_h^{l,j}$，其中 $\alpha \in (0, 1]$。

(6) 令 $j = j + 1$。

(7) 针对 $Z_h^{l,j}$ 求解：

$$A'\left(U_h^{l,j}\right)\left(\Phi_h, Z_h^{l,j}\right) = \left(J_c^{(j)}\right)'\left(U_h^{l,j}\right)(\Phi_h), \quad \forall \Phi_h \in X_h^l$$

(8) 结束并回到步骤 (3)。

备注 63 关于该算法终止准则的相关研究请参考文献 [365, 148]。

算法 18 (面向多目标函数误差估计的自适应算法) 如下：

(1) 给定初始解 $U_h^{0,(2)}$，U_h^0，令 $l = 1, \text{TOL}_{\text{dis}} > 0$。

(2) 根据初始解 $U_h^{l-1,(2)}$，利用牛顿法求解 (非线性) 高阶初始问题。

(3) 根据初始解 U_h^{l-1}，利用牛顿法求解低阶初始问题与 (线性) 伴随问题。

(4) 构造组合目标函数 J_c 并求解伴随问题。

(5) 计算误差估计量 η。

(6) 检验是否满足 $|\eta| < \text{TOL}_{\text{dis}}$，若满足，则意味着满足求解精度，终止计算。

(7) 标记所有不满足误差要求的有限元元素。

(8) 细分网格，并令 $l = l + 1$。

(9) 返回步骤 (2) 重新计算。

9.5 预测-矫正-自适应方法

本节我们将利用两种手段相结合的方式实现预测-矫正的自适应方法。
(1) 裂缝周边的网格自适应细化处理；
(2) 自适应尺度选择方法。

在裂缝扩展过程中，首先预测未知裂缝路径，随后计算当前裂缝路径，从而判断是否激活预测矫正自适应方法。

9.5.1 扩展裂缝周边局部网格细化方案

为获取高精度裂缝形态、相场裂缝周边，我们需要对网格进行细化处理，这样才能在模拟裂缝扩展的过程中保证足够的精度。因此，我们希望能够以一个足够小的正则化参数 $\varepsilon(h \leqslant \varepsilon)$ 完全细化裂缝区域。在整个模拟过程中 ε 保持不变，如 $\varepsilon = 2h$。

由于模拟前无法预知裂缝最终的扩展路径，我们首先预测一个可能的路径，随后在细分网格中重新计算数值解。该方案首次在文献 [217] 中提出，并在后续研究中成功应用于三维模型 [442,281]。该方案可分为四个步骤 (图 9.5)。

算法 19 (预测-矫正自适应方法) 在时间 t_n 时：
(1) 计算出当前时间 t_n 时的解 (u_n, φ_n)；
(2) 预测步：求解 $(\tilde{u}, \tilde{\varphi})$，同时预测 t_{n+1} 时的裂缝扩展路径；
(3) 根据 $(\tilde{u}, \tilde{\varphi})$ 完成网格细化，并回到时间 t_n 时的解 (u_n, φ_n)；
(4) 矫正步：根据新的网格分布计算时间 t_{n+1} 时的解 (u_{n+1}, φ_{n+1})。

随后令 $n \to n+1$，同时重新按照以上流程进行网格自适应处理。

根据相应的研究结果，这个方法能够有效处理相场裂缝问题 [217,442,281]。在细化处理过程中，我们简单设定一个相场变量的阈值 η_C。例如 $\eta_C = 0.6$，则当 $\varphi < \eta_C$ 时，对网格进行细化处理。

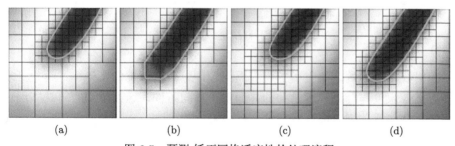

图 9.5 预测-矫正网格适应性的处理流程
(a) t_n 时的旧状态；(b) 预测裂缝扩展路径，并对裂缝扩展路径细化处理；(c) 网格细化并重置解变量 (以旧时间步的解为基准进行插值处理)；(d) 在细化网格上重新计算 t_{n+1} 时间步的解

接下来，我们在图 9.6 中展示了我们基于文献 [217] 的数值模型 (图 9.6 单边切口剪切实验) 进行的相关研究。根据计算结果可知，预测-矫正自适应方法计算更快，优势更为明显。

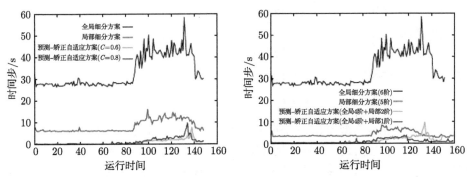

图 9.6 单边切口剪切实验的不同网格细化方案计算时间对比

9.5.2 非侵入性式自适应方法

在本节中，我们提出了一个基于预测-矫正自适应方案的非侵入式方法。在该方法中，我们根据裂缝路径动态选择局部细化尺度。相关概念在文献 [188] 中首次提出，并在文献 [191] 中成功应用于相场裂缝模型。

公式 23 (宏观问题/全局问题) 令 $u \in V_u^D$, $\forall w \in V_u^0$ 有

$$A_M(u)(w) = l(w)$$

其中 $A_M(u)(w) := (\sigma(u), \nabla w)$, $l(w) := (f, w)$, 例如线性化应力张量 $\sigma(u) = 2\mu e(u) + \lambda \mathrm{tr}(e) I$，需要注意的是在这里我们不讨论相场参数 φ。

在微观尺度上，参考命题 20，考虑非线性本构的全耦合相场模型。

公式 24 (微观问题/局部问题) 令 $U = (u, \varphi) \in X^D$, $\Phi \in X^0$ 有

$$A_m(U)(\Psi - \Phi) \geqslant 0$$

此时我们研究的是第三类扩展界面问题，这就要求我们在计算过程中必须时刻调整局部尺度与全局尺度之间的界面以及界面条件，这给我们的算法实现造成了很大挑战。相关细节请读者参考文献 [191] 与图 9.7。

算法 20 (自适应全局/局部方法) 如下：

(1) 求解时间 t_n 时的全局-局部耦合问题

$$\text{当 } \tilde{u} \in V_u \text{ 时}, \quad w \in V_u : A_M(\tilde{u})(w) = 0$$

$$当\tilde{U} \in V_u \times V_\varphi 时, \quad \Psi \in V_u \times V_\varphi : A_m(\tilde{U})(\Psi) = 0$$

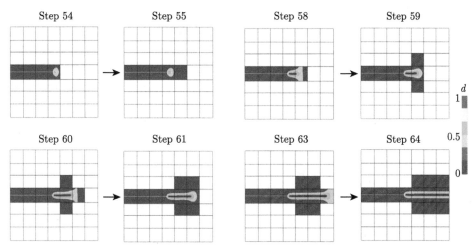

图 9.7　8 种不同时间点的预测-矫正自适应全局/局部方法的处理方式[336]

(2) 预测步：

(a) 若在全局单元 K_M 的内边缘处和局部单元 K_m 处 $\tilde{\varphi} < \eta_C$，则 $K_M \to K_m$，并重新设定新的全局-局部离散网格。

(b) 否则令 $n \to n+1$，并回到步骤 (1)。

(3) 将全局解变量 u 通过插值分配到新的局部网格 K_m 上，此时我们需要注意，在新的局部网格上 u 是已知的，但 φ 不是。

(4) 矫正步：求解新的全局-局部解

$$当 u_{n+1} \in V_u 时, \quad w \in V_u : A_M(u_{n+1})(w) = 0$$

$$当 U_{n+1} \in V_u \times V_\varphi 时, \quad \Psi \in V_u \times V_\varphi : A_m(U_{n+1})(\Psi) = 0$$

(5) 令 $n \to n+1$，并回到步骤 (1)。

9.6　自适应时间步长控制算法

本节我们将介绍自适应时间步长控制算法，由于相场裂缝模型多为准静态模型，所以该算法并非在所有情况下均能正常使用。在裂缝扩展问题中，主要有三种情况：

(1) $G < G_c$，裂缝不扩展 (此时仅需较大的时间步即可满足计算要求)。

(2) $G = G_c$，裂缝扩展阶段 (此时需要较小的时间步才能满足计算要求)。

(3) $G > G_c$，裂缝扩展结束或材料已完全断裂 (此时仅需较大的时间步即可满足计算要求)。

在本节中，我们的目标是对目标函数 $J(U)$ 的精度进行控制。主要思路是基于 ODE 理论，利用局部截断误差实现时间步自适应处理 [358]。因此，我们需要假设离散时间导数差分形式的截断误差具有正则性和稳定性。具体而言，在每个时间步中，我们首先计算相场裂缝模型的解，以及目标函数 $J(U)$，并计算两个连续的步长 k 下的解，获得目标函数的绝对值的差：$\text{abs}_{\text{err}} := |J(U_{2k}) - J(U_k)|$。

命题 77 给定 TOL_{TS}，此时我们计算

$$\text{abs}_{\text{err}} := |J(U_{2k}) - J(U_k)| \tag{9.16}$$

$$\theta = \gamma * \left(\frac{\text{TOL}_{\text{TS}}}{\text{abs}_{\text{err}}}\right)^K \tag{9.17}$$

其中 $\gamma \approx 1$ 为一个常系数 $\left(\text{在文献 [209] 中}, \gamma = 0.9, K = \frac{1}{15}\right)$ 随后我们更新时间步长

$$k_{\text{new}} = \begin{cases} k_{\text{old}}, & 1 \leqslant \theta \leqslant 1.2 \\ k_{\text{old}} * \theta, & \text{其他} \end{cases} \tag{9.18}$$

此外我们需要检验 $k_{\min} \leqslant k_{\text{new}} \leqslant k_{\max}$，若 $k_{\text{new}} < 0.5 k_{\text{old}}$，则说明新的时间步长比旧的要小得多，此时需要重新设定当前时间步长。

备注 64 在自适应时间步长控制算法中，需防止出现时间步长振荡现象，若时间步长前后差别不是很大，则保留原来的时间步长即可。另外尽管在 $k_{\text{new}} < 0.5 k_{\text{old}}$ 时需要重新设定当前时间步长，这一操作看来十分浪费，但确是有必要的，因为时间步长的大幅度减少意味着之前的参数设定不够精确，继续计算会影响后续结果的准确性。

算法 21 在时间步 t_n 内：

(1) 令 $k_{\text{old}} := k_{n-1}$；

(2) 首先计算时间步长 $2k_{\text{old}}$ 下一个时间步后的解，即计算 $t_n + 2k_{\text{old}}$ 时刻的 $J(U_{2k})$。随后计算时间步长 k_{old} 下两个时间步后的解，并计算 $t_n + k_{\text{old}} + k_{\text{old}}$ 时的 $J(U_k)$；

(3) 计算式 (9.16) 与 (9.17)；

(4) 根据式 (9.18) 计算新的时间步长 k_{new}；

(5) 若 $k_{\text{new}} < 0.5k_{\text{old}}$ 则令 $k_{\text{new}} = k_{\text{old}}$ 并回到步骤 (2) 重新计算，否则令 $k_n = k_{\text{new}}$；

(6) 令 $n \to n+1$，并回到步骤 (1) 继续计算。

该时间步长控制方法已被初步应用，并获得较好的模拟效果，但尚无严格的计算分析。

第 10 章 数值模拟 III：面向目标的相场狭缝误差控制和螺杆时间控制

10.1 基于双加权残差法的相场裂缝模拟

本节我们将基于第 6 章的狭缝模型继续进行讨论，其中模型参数、边界条件、初始条件等相关信息请参考第 6 章。

10.1.1 目标函数

我们选取上半域的位移值作为目标泛函：

$$J^D(u) = \int_{\{x \geqslant 0\}} u \mathrm{d}x \quad (例\ 1)$$

目标点的值为

$$J^D(u) = u(0.75, 0.75) \quad (例\ 2)$$

例 1 和例 2 模拟结果如图 10.1 与图 10.2 所示。首先我们在一个 10 阶细分的网格中计算出参考值：

$$J^D\left(u_{\mathrm{ref}}\right) := 0.345876861918397 84$$

$$J^D\left(u_{\mathrm{ref}}\right) := 0.401367604270908 23$$

我们采取了两种不同的标记方案：第一种，固定残差法中 $X = 0.1, Y = 0.0$。第二种令 $\alpha = 10$。对比两种方案可知，细化方案不同，计算效率也有所不同。为此，我们进行以下研究。

10.1.2 面向目标的后验误差分析

我们采用命题 67 中的误差估计法，需要注意的是，我们在处理的是一个经典泊松问题。其中 $U = (u, \varphi), u : B \to \mathbb{R}, \varphi : B \to \mathbb{R}$ 且 $Z = (z_u, z_\varphi), z_u : B \to \mathbb{R}, z_\varphi : B \to \mathbb{R}$。

10.1 基于双加权残差法的相场裂缝模拟

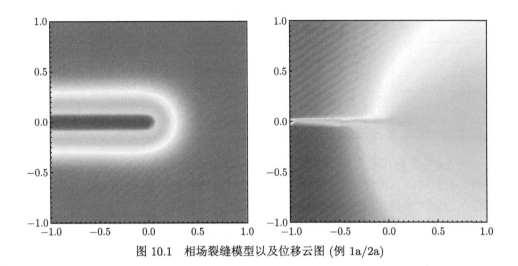

图 10.1 相场裂缝模型以及位移云图 (例 1a/2a)

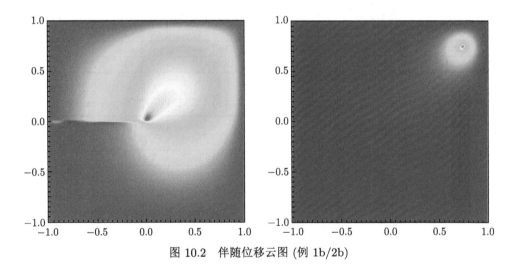

图 10.2 伴随位移云图 (例 1b/2b)

例 1a/b 上半域分析结果如表 10.1 和表 10.2 所示, 在例 1a 中总误差约减少了一个半数量级, 有效性指数 I_{eff} 指示性较好。在例 1b 中误差估计量比真实误差低了一个数量级。

表 10.1 例 1a: 基于相场裂缝上半域的 PU-DWR 方法, $X = 0.1$

自由度	$\left\vert J^D\left(u_h\right)-J^D\left(u_{\text{ref}}\right)\right\vert$	η	η_{ind}	I_{eff}	I_{ind}
2178	1.02×10^{-2}	4.59×10^{-4}	1.20×10^{-3}	4.50×10^{-2}	1.18×10^{-1}
3370	7.12×10^{-6}	1.01×10^{-3}	1.18×10^{-3}	1.42×10^{2}	1.66×10^{2}
5442	7.24×10^{-4}	1.03×10^{-3}	1.98×10^{-3}	1.42×10^{0}	2.73×10^{0}
8862	5.14×10^{-4}	5.65×10^{-4}	1.24×10^{-3}	1.10×10^{0}	2.42×10^{0}

表 10.2 例 1b：基于相场裂缝上半域的 PU-DWR 方法，$\alpha = 10$

Dofs	$\|J^D(u_h) - J^D(u_{\text{ref}})\|$	η	η_{ind}	I_{eff}	I_{ind}
2178	1.02×10^{-2}	4.59×10^{-4}	1.20×10^{-3}	4.50×10^{-2}	1.18×10^{-1}
2834	4.84×10^{-4}	2.69×10^{-4}	4.27×10^{-4}	5.56×10^{-1}	8.82×10^{-1}
4470	1.38×10^{-3}	1.53×10^{-4}	2.53×10^{-4}	1.10×10^{-1}	1.83×10^{-1}
6938	9.81×10^{-4}	9.64×10^{-5}	1.86×10^{-4}	9.83×10^{-2}	1.89×10^{-1}
10094	4.62×10^{-4}	6.44×10^{-5}	1.44×10^{-4}	1.39×10^{-1}	3.11×10^{-1}

例 2a/b 目标函数分析结果如表 10.3 和表 10.4 所示。在例 2a 中真实误差低了两个数量级，这可能是由于固定点选取位置远离裂缝区域 (图 10.3)，因此受影响较小。总体来说，固定点目标函数误差较小，表现良好。

表 10.3 例 2a：基于相场裂缝区域固定点的 PU-DWR 方法，$X = 0.1$

Dofs	$\|J^D(u_h) - J^D(u_{\text{ref}})\|$	η	η_{ind}	I_{eff}	I_{ind}
2178	3.87×10^{-3}	3.36×10^{-5}	6.63×10^{-4}	8.66×10^{-3}	1.71×10^{-1}
3286	2.44×10^{-4}	1.69×10^{-4}	7.19×10^{-4}	6.94×10^{-1}	2.95
5470	1.22×10^{-4}	1.12×10^{-4}	7.51×10^{-4}	9.13×10^{-1}	6.14
9622	1.28×10^{-4}	8.84×10^{-5}	5.97×10^{-4}	6.90×10^{-1}	4.66
16010	5.02×10^{-5}	4.30×10^{-5}	2.55×10^{-4}	8.55×10^{-1}	5.08

表 10.4 例 2b：基于相场裂缝区域固定点的 PU-DWR 方法，$\alpha = 10$

Dofs	$\|J^D(u_h) - J^D(u_{\text{ref}})\|$	η	η_{ind}	I_{eff}	I_{ind}
2178	3.87×10^{-3}	3.36×10^{-5}	6.63×10^{-4}	8.66×10^{-3}	1.71×10^{-1}
3062	1.89×10^{-4}	4.42×10^{-5}	2.74×10^{-4}	2.34×10^{-1}	1.45
4794	2.22×10^{-4}	3.16×10^{-5}	1.72×10^{-4}	1.42×10^{-1}	7.74×10^{-1}
7046	2.00×10^{-4}	3.31×10^{-5}	1.04×10^{-4}	1.66×10^{-1}	5.19×10^{-1}
10778	1.07×10^{-4}	9.26×10^{-5}	2.79×10^{-4}	8.62×10^{-1}	2.60

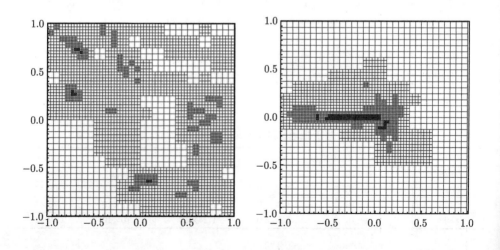

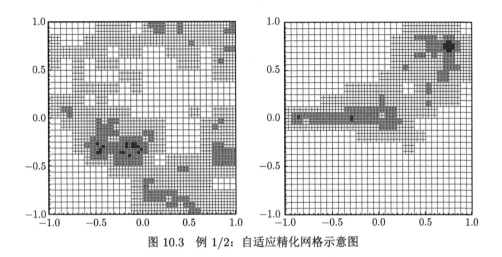

图 10.3 例 1/2: 自适应精化网格示意图

10.2 非相场法多目标误差分析

本节我们不考虑变分不等式的因素，仅在几何体内部建立狭缝模型。此时问题转化为一个具有非均匀边界条件的简单泊松问题。分别采用单目标函数与多目标函数对模型进行分析。原始模型来自于文献 [151]。

10.2.1 模型设置

令
$$B = (-1,1) \times (-1,1) \setminus \{(x,0) | -1 \leqslant x \leqslant 0\}$$

Γ_D, Γ_N 为外部边界条件，Γ_i 为内部狭缝边界条件。令 $u: B \to \mathbb{R}$，我们有

$$-\Delta u = 0, \quad \text{在 } B \text{ 中}$$

$$u = g, \quad \text{在 } \Gamma_D \text{ 上}$$

$$\partial_n u = 0, \quad \text{在 } \Gamma_N \text{ 上}$$

图 10.4 分别展示了相场法与几何法处理的狭缝模型。

10.2.2 研究对象

该模型中我们有以下几个目标函数 (需要注意的是这些目标函数并不需要同时考虑)

$$J_0(u) := u(0.75, 0.75)$$

$$J_1(u) := u(-0.5, -0.25)$$

$$J_2(u) := \int_{\Gamma_1} \nabla u \cdot n \mathrm{d}x$$

$$J_3(u) := \int_{\Omega_1} u \, \mathrm{d}x$$

其中 $\Gamma_1 = \{-1\} \times (-1, -0.25)$，$\Omega_1 = (0, 1) \times (-1, 0)$。通过细化网格从而获得泛函的近似值：

$$J_0^{\mathrm{ref}}(u) \approx +0.18949212064$$
$$J_1^{\mathrm{ref}}(u) \approx -0.66061009755$$
$$J_2^{\mathrm{ref}}(u) \approx -0.54411579542$$
$$J_3^{\mathrm{ref}}(u) \approx -0.18268521784$$

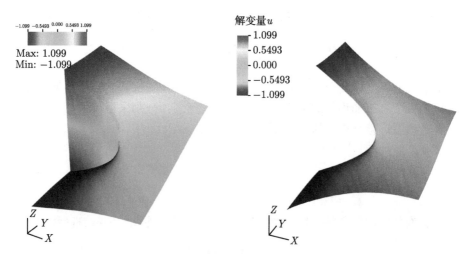

图 10.4 相场裂缝与几何裂缝中位移参数分部对比图

10.2.3 单目标函数误差分析

首先我们考虑单目标函数相关的研究，相关结果如表 10.5 所示，这里我们观察到有效性指数虽然不是很好，但相对稳定，同时误差收敛性好。

表 10.5 单目标函数狭缝模型研究结果

细分等级	元素个数	$J(u_h)$	$\left|J_0^{\mathrm{ref}}(u) - J(u_h)\right|$	η	I_{eff}
0	64	$1.7935965773 \times 10^{-1}$	5.65×10^{-2}	4.21×10^{-2}	7.45×10^{-1}
1	208	$1.8491217214 \times 10^{-1}$	2.48×10^{-2}	1.84×10^{-2}	7.42×10^{-1}
2	652	$1.8835424968 \times 10^{-1}$	6.04×10^{-3}	4.59×10^{-3}	7.60×10^{-1}
3	4720	$1.8891444682 \times 10^{-1}$	3.06×10^{-3}	2.19×10^{-3}	7.16×10^{-1}
4	11608	$1.8920500524 \times 10^{-1}$	1.52×10^{-3}	1.25×10^{-3}	7.05×10^{-1}
5	24124	$1.8935107801 \times 10^{-1}$	7.45×10^{-4}	5.40×10^{-4}	7.25×10^{-1}
6	54160	$1.8942207581 \times 10^{-1}$	3.70×10^{-4}	2.65×10^{-4}	7.17×10^{-1}

10.2.4 多目标函数误差分析

在多目标误差分析过程中，目标函数选取如下：

$$J_c(\Phi) = \sum_{i=1}^{4} J_i(\Phi) \frac{\text{sign}\left(J_i\left(e_h^{(2)}\right)\right)}{|J_i(u_h)|} \omega_i$$

其中 $\omega_i=1$，表 10.6、图 10.5 显示了相关模拟结果。更多研究请参考文献 [151]。

表 10.6 多目标函数狭缝模型研究结果

细分等级	元素个数	$|J_c^{\text{ref}}(u_h) - J_c(u_h)|$	η	I_{eff}
0	64	1.65×10^{-1}	1.16×10^{-1}	7.02×10^{-1}
1	184	8.16×10^{-2}	5.92×10^{-2}	7.25×10^{-1}
2	580	3.89×10^{-2}	2.76×10^{-2}	7.09×10^{-1}
3	1420	1.92×10^{-2}	1.36×10^{-2}	7.08×10^{-1}
4	3088	9.71×10^{-3}	6.84×10^{-3}	7.04×10^{-1}
5	6784	4.88×10^{-3}	3.47×10^{-3}	7.13×10^{-1}
6	13648	2.44×10^{-3}	1.73×10^{-3}	7.08×10^{-1}
7	27472	1.23×10^{-3}	8.81×10^{-4}	7.14×10^{-1}
8	54712	6.17×10^{-4}	4.39×10^{-4}	7.12×10^{-1}

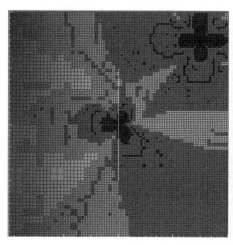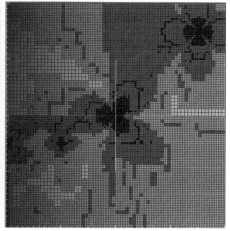

图 10.5 单目标函数狭缝模型网格细分

左图模型网格细分等级为 6 级，共有 54160 个元素，右图网格细分等级为 7 级，共有 54712 个元素

10.3 螺钉断裂实验中的误差控制

本节我们主要研究自适应加载步模型。所选取的螺纹断裂实验模拟也很有代表性，在模拟过程中并未预置初始裂缝。随着不断加载，高应力区域首先发生起裂，随后裂缝逐渐扩展，具体模拟结果请参考文献 [427, 426, 438]。

10.3.1 模型配置

本模型基于命题 32，采用了准全耦合方案。其几何设置如图 10.6 所示，其总长度为 17.20mm，初始网格包括 3440 个节点，10800 个自由度[192]。

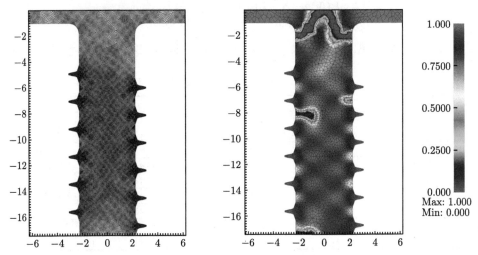

图 10.6 螺钉断裂实验数值模拟

10.3.2 边界条件与初始条件

我们在螺钉顶部施加一个非齐次 Dirichlet 边界条件，螺钉顶部 $y = 0.0$。在每个时间步中螺钉顶部均存在一个拉伸载荷。

$$u_y = t \times \bar{u}, \quad \bar{u} = 1.0 \text{mm}$$

t 为当前时间，其他边界的边界条件为

$$u_x = 0, \quad 在 \Gamma_{\text{top}} \times I 上$$

$$u = 0, \quad 在 \Gamma_{\text{bottom}} \times I 上$$

$$\sigma(u)n = 0, \quad 在 \Gamma_{\text{sides}} \times I 上$$

在初始时刻，全局相场变量均满足 $\varphi(0) = 1$，确保无初始裂缝，便于我们后续研究裂缝的起裂与扩展。

10.3.3 模型参数

在模拟过程中，令 $\gamma = 1$，$\kappa = 10^{-10}h$，$\varepsilon = 2h$，$\text{TOL}_{\text{New}} = 10^{-8}$，$\mu = 80.77\text{kN/mm}^2$，$\lambda = 121.15\text{kN/mm}^2$，$G_c = 2.7\text{N/mm}$，初始时间步长 $k = 0.01\text{s}$，自适应时间步控制参数设置如下：

10.3 螺钉断裂实验中的误差控制

$\text{TOL}_{\text{TS}} = 10^{-2}, \text{TOL}_{\text{TS}} = 10^{-3}, k_{\max} = 10^{-2}, k_{\min} = 10^{-4}, \gamma = 1, K = 1/15$。

10.3.4 模拟结果

我们设置两个目标函数：
(1) 整体能量函数：

$$E_b = \int_B g(\varphi)\psi(e)\mathrm{d}x \qquad (10.1)$$

(2) 裂缝表面能函数：

$$E_s(\varphi) = \frac{G_c}{2}\int_B \left(\frac{(1-\varphi)^2}{\varepsilon} + \varepsilon|\nabla\varphi|^2\right)\mathrm{d}x \qquad (10.2)$$

以及断裂能函数：

$$\psi(e) := \mu\mathrm{tr}\left(e(u)^2\right) + \frac{1}{2}\lambda\mathrm{tr}(e(u))^2$$

其中 $|\nabla\varphi|^2 := \nabla\varphi \cdot \nabla\varphi$。

为了控制时间步长，我们引入目标函数：

$$J(\varphi) := E_s(\varphi)$$

模拟结果如图 10.7～图 10.9 所示。

随着裂缝不断扩展，裂缝能量逐渐上升，随后螺丝断裂，裂缝能量保持不变，相对误差趋向于 0。此时增大模型时间步长，相对误差变化不大。

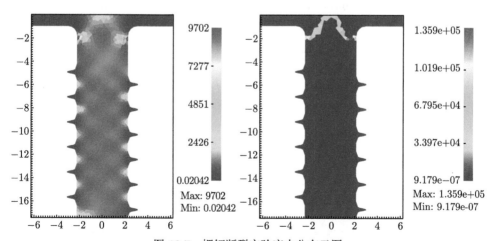

图 10.7 螺钉断裂实验应力分布云图
$t = 9.92\times 10^{-2}$s (左); $t = 1.03\times 10^{-1}$s (右)

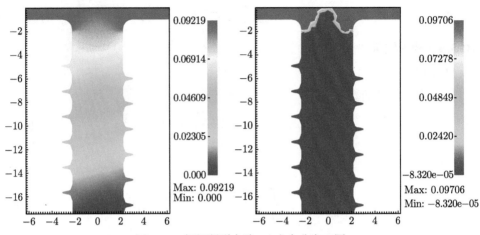

图 10.8 螺钉断裂实验 u_y 方向分布云图
$t = 9.92 \times 10^{-2}$ s (左); $t = 1.03 \times 10^{-1}$ s (右)

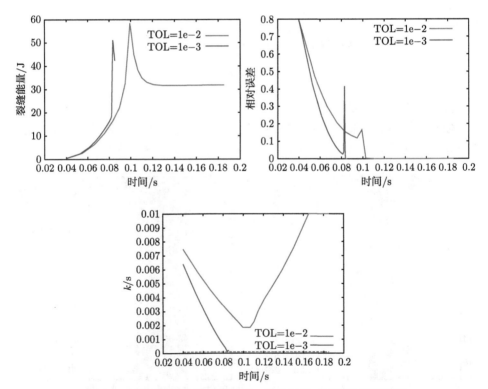

图 10.9 螺钉断裂实验裂缝能量、相对误差、时间步长变化图

第 11 章 多物理场相场裂缝模型

本章我们将集中讨论多物理场相场裂缝模型，相关概念已在第 5、7、9 章中进行了介绍。本章主要讨论裂缝的界面条件 $\partial\mathcal{C}$、相场参数的相关特征、裂缝宽度等参数计算、欧拉-拉格朗日坐标系中的固体大变形问题以及流固耦合相场裂缝模型与非等温裂缝模型。

11.1 界面定义与界面条件

界面 $\Gamma := \Gamma(t) := \bar{\Omega}_1(t) \cap \bar{\Omega}_2(t)$ 的定义请参考 1.3 节。在处理多种界面时同时会采用 $\partial\mathcal{C}$ 与 Γ_i 等符号来表示相关界面。根据 3.3.2 节中相关定义可知，在界面处解变量 u 的耦合条件为

$$\begin{aligned} u_1 &= u_2, \\ \sigma_1 n &= \sigma_2 n, \end{aligned} \quad \text{在 } \Gamma \text{ 上}$$

其中法向量 n 满足 $n_1 = -n_2$。

11.2 水平集方法

在 $\varepsilon > 0$ 的情况下 (参考图 2.1)，裂缝将结构域 B 分为两个部分。由于相场裂缝模型基于欧拉法建立坐标系，因此可能会出现裂缝边界位于网格中间的情况 (图 1.7 与图 1.8 所示)。因此，我们在建模时面临界面准确位置无法判断的问题，需要耦合指示函数与位移函数并捕获裂缝界面，建立时空离散的有限元模型。在此基础上，我们采用了水平集函数，将整个系统划分为裂缝区域 Ω_F 与非裂缝区域 Ω_R。

定义 68(指示函数) 给定域 B，裂缝 C，边界 $\partial\mathcal{C}$。我们定义以下函数为指示函数

$$\chi_R := \chi_R(x,t) = 1, \quad \text{在 } \Omega_R(t) \text{ 中}$$
$$\chi_F := \chi_F(x,t) = 1, \quad \text{在 } \Omega_F(t) \text{ 中}$$

指示函数满足 $\chi_F = 1 - \chi_R$。

接下来我们使用 χ_R 与 χ_F 划分非裂缝域 Ω_R 与裂缝域 Ω_F。

命题 78 采用 φ 来构造指示函数, 方程如下所示:

$$\chi_F(x,t,\varphi) := \begin{cases} 1, & \varphi(x,t) \leqslant x_1 \\ 0, & \varphi(x,t) \geqslant x_2 \\ -\dfrac{\varphi - x_2}{x_2 - x_1}, & \text{其他} \end{cases}$$

$$\chi_R(x,t,\varphi) := \begin{cases} 1, & \varphi(x,t) \geqslant x_2 \\ 0, & \varphi(x,t) \leqslant x_1 \\ \dfrac{\varphi - x_1}{x_2 - x_1}, & \text{其他} \end{cases}$$

根据相场参数定义, $0 \leqslant \varphi \leqslant 1$, 当极端情况, 即 $x_1 = 1, x_2 = 0$ 或 $x_1 = 0, x_2 = 1$ 时, 显然符合以上设置。在过渡区中, 我们则采用线性插值的方式保证指示函数的连续性。

由于相场参数本身是一个未知变量, 这意味着当我们求解非线性全耦合模型时, 通常需要采用牛顿迭代法。

算法 22(引入指示函数的迭代耦合法) 令 $U = (u, \varphi) \in X$ 为相场裂缝问题 $A_U(U,p)(\Psi) = 0$ 的解, 其中 $p \in V_p$ 为 $A_p(U,p)(\psi_p) = 0$ 的解, 迭代顺序为 $j = 1, 2, 3, \cdots$。

(1) 给定 $p^{j-1} \in V_p$, 令 $U^j \in X$, 有

$$A_U\left(U^j, p^{j-1}\right)(\Psi) = 0, \quad \forall \Psi \in X$$

(2) 给定 $U^j = (u^j, \varphi^j) \in X$, 令 $p^j \in V_p$, 有

$$A_p\left(U^j, p^j\right)(\psi_p) = 0, \quad \forall \psi_p \in V_p$$

在步骤 (2) 中, $p^j = p_F^j \chi_F\left(\varphi^j\right) + p_R^j \chi_R\left(\varphi^j\right)$, 其中 φ^j 可通过步骤 (1) 求解得到。

接下来我们将基于指示函数构造水平集函数 (如图 11.1 所示), 完成界面的相关定义。

命题 79(基于相场参数构造一个水平集变量) 令 $0 < c_{ls} < 1$ 为裂缝边界等值面, 此时水平集变量定义为

$$\varphi_{ls} := \varphi - c_{ls}$$

我们常令 $c_{ls} := x_1$(如 $c_{ls} := x_1 = 0.3$)。

定义 69(裂缝边界 $\partial \mathcal{C}$) 根据水平集变量, 我们可以定义裂缝边界为

$$\partial \mathcal{C} := \{x \in B \mid \varphi_{ls}(x) = 0\}$$

其中 $\varphi = c_{\mathrm{ls}}$，$\varphi_{\mathrm{ls}} = 0$。

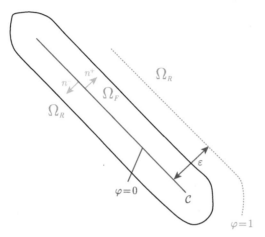

图 11.1　根据相场变量构造水平集，将系统分解为裂缝域与非裂缝域

在泊松问题处理中，我们仅利用指示函数定义了界面 Γ，此时并未明显区分域 Ω_1 与 Ω_2。在引入水平集函数后我们可获得如下命题。

命题 80　给定指示函数 χ_1 与 χ_2，令 $u = (u_1, u_2) \in X$，其中 $u_1 := \chi_1 u$，$u_2 := \chi_2 u$，我们有
$$A(u, \psi) = F(\psi), \quad \forall \psi \in X$$
其中
$$A(u, \psi) := (\chi_1 \nabla u, \nabla \psi) + (\chi_2 \nabla u, \nabla \psi)$$

11.3　缝宽与裂缝总体积计算

缝宽以及裂缝开度等参数是相场裂缝模拟中的重要参数。

定义 70　在数值模型中，裂缝宽度定义为
$$w = [u] \cdot n = u_1 n - u_2 n \tag{11.1}$$

其中 u_1，u_2 代表裂缝两侧的位移，即 $u_1 := u|_{\mathcal{C}^+}$，$u_2 := u|_{\mathcal{C}^-}$。

在相场裂缝模型中，裂缝边界 $\mathcal{C}$ 通常位置相对模糊，因此我们可以根据 11.2 节的思路构建裂缝边界。在研究裂缝问题时，我们主要分析以下物理量：
(1) 界面 $\partial \mathcal{C}$ 的定义（11.2 节）；
(2) 界面 $\partial \mathcal{C}$ 的法向量 n；
(3) 边界位移 u。

命题 81(直角坐标系下相场裂缝开度计算) 令 $B=(a,b)\times(c,d)$，相场裂缝与笛卡儿坐标系中 X 轴平行，此时法向量 n 即为 Y 轴方向，裂缝在 $x_0 \in B$ 点的开度计算公式为

$$\text{COD}(x_0) := w := w(x_0, y) = \int_c^d u(x_0, y) \cdot \nabla\varphi(x_0, y)\,\mathrm{d}y \tag{11.2}$$

同时我们有

$$\int_\Gamma [u]\cdot n \mathrm{d}s \approx \int_B u \cdot \nabla\varphi \mathrm{d}x \tag{11.3}$$

相关证明请参考文献 [110]。

接下来我们推导出与坐标轴不平行的裂缝开度计算公式，令 $x_c \in \Gamma$ 为裂缝中的一点，我们可得 $x_\varepsilon^+ \in B$，$x_\varepsilon^- \in B$ 在法向量 n 方向的距离 ε，则

$$\text{COD}(x_c) = \int_{x_E^+}^{x_\varepsilon^-} u(x,y) \cdot \nabla\varphi(x,y)\mathrm{d}n \tag{11.4}$$

在法向量与坐标轴平行的情况下，可得式 (11.2)。

在实际算例中，我们经常会遇到裂缝与坐标轴不平行的情况，此时就需要用到式 (11.4)。这类情况下需要设法获取裂缝面法向量。

定义 71(法向量) 基于相场参数的法向量计算公式如下：

$$n := -\frac{\nabla\varphi}{\|\nabla\varphi\|}$$

当然在实际操作中我们还面临以下两个难题：

(1) 模型中网格位置固定，但法向量求解过程中需要任意方向的积分，因此我们仅能计算特定点位置的裂缝宽度。

(2) 位移量有正负之分，裂缝中间等值面点上的位移值为 0(图 11.2)，这会对我们求解裂缝宽度造成影响。

接下来，以 I 型裂缝 (拉伸裂缝) 为例，介绍裂缝宽度计算流程。

(1) 计算位移值 $u: B \to \mathbb{R}^d$。

(2) 取绝对值 $\tilde{w} := \|u\|_{L^\infty}$。

(3) 假设裂缝左右对称 $u^+ = -u^-$(参考图 11.2)，令 $w_\varepsilon := 2\tilde{w}$，证明如下：

$$w = [u]\cdot n = u^+ n - u^- n = 2u^+ n \approx 2\tilde{w} =: w_\varepsilon$$

(4) 在裂缝边界 $\partial\mathcal{C}$ 处，令 $w_D := w_\varepsilon$。

11.3 缝宽与裂缝总体积计算

(5) 在裂缝区域对 w_D 插值。

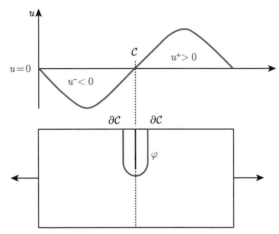

图 11.2 拉伸裂缝模型中位移 u 与相场变量 φ 的分布情况

接下来我们详细讲解步骤 (5) 的处理方法。

(1) 若 $u^+ = -u^-$，可采用常数插值，即：给定 $\partial\mathcal{C}$ 上 w_D，在 Ω_F 中，$w_\varepsilon := w_D$，如图 11.3 所示。

(2) Laplace 插值，给定 $w_\varepsilon : B \to \mathbb{R}$，则

$$-\Delta w_\varepsilon = g, \quad \text{在 } B \text{ 中}$$

$$w_\varepsilon = w_D, \quad \text{在 } \partial\mathcal{C} \text{ 上}$$

$$w_\varepsilon = 0, \quad \text{在 } \partial B \text{ 上}$$

其中，$g := \beta \|w_D\|_{L^\infty}$，$\beta > 0$，此时该问题仅需在裂缝域 Ω_F 中求解，但为了计算方便，我们在整个空间域 B 中进行计算。

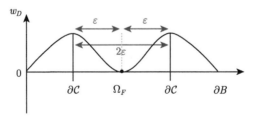

图 11.3 裂缝域 Ω_F 中 w_D 分部剖面图

定义 72(裂缝体积 TCV) 令 B 为整个系统空间域，裂缝体积计算公式为

$$\mathrm{TCV}_\varepsilon = \int_B u \cdot \nabla\varphi \mathrm{d}x \tag{11.5}$$

11.4 相场裂缝模型建模

11.4.1 弹性固体压裂模型

首先我们在纯固体模型进行分析,施加裂缝内部压力,以此作为多物理场裂缝模型的基础。该模型可应用于多孔介质裂缝模拟、地下建模、地热能回收、核废料、地下水流或主动脉夹层的生物医学等多方面的应用中。

11.4.1.1 裂缝界面弹性耦合处理

首先利用水平集函数将系统域 B 划分为 Ω_F, Ω_R。在 Ω_F 中,基于 Stokes 方程建立流动方程 [405,198,360]。在 Ω_R 中,假定模型基质满足弹性固体力学方程。

接下来我们首先分析裂缝域的不可压缩流体部分。

公式 25(不可压缩流体 Stokes 方程) 令 $v: \Omega_F \to \mathbb{R}^d$, $p: \Omega_F \to \mathbb{R}$,我们有

$$-\nabla \cdot \sigma_f(v, p) = 0 \tag{11.6}$$

$$\nabla \cdot v = 0 \tag{11.7}$$

流体应力张量为

$$\sigma_f = -pI + \nu\rho\left(\nabla v + \nabla v^{\mathrm{T}}\right)$$

其中 ρ 为流体密度,ν 为运动黏度。

备注 65(符号标注问题) f 表示 "流体",F 表示 "裂缝"。

接下来我们研究加压裂缝与压裂裂缝。

定义 73 为便于后续区分,我们定义以下术语:
(1) 加压裂缝:仅在裂缝域内施加压力 p_F,无液体充填;
(2) 压裂裂缝:裂缝域内充填液体,并对裂缝边界有压力 p_F。

命题 82(加压裂缝弹性界面耦合条件) 准弹性模型加压裂缝界面耦合条件为

$$\sigma_R n = \sigma_F n, \quad 在 \Gamma 上 \tag{11.8}$$

由于裂缝域在模拟过程中未充填流体,因此在这种情况下我们仅需设定界面条件。

命题 83(加压裂缝孔隙弹性界面耦合条件) 多孔介质中孔隙弹性模型加压裂缝界面条件为

$$p_R = p_F, \quad 在 \Gamma 上 \tag{11.9}$$

$$\sigma_R n = \sigma_F n, \quad 在 \Gamma 上 \tag{11.10}$$

11.4 相场裂缝模型建模

命题 84(压裂裂缝孔隙弹性界面耦合条件) 多孔介质模型中裂缝域与非裂缝域的界面耦合条件如下：

$$p_R = p_F, \quad 在 \Gamma 上 \tag{11.11}$$

$$\sigma_R n = \sigma_F n, \quad 在 \Gamma 上 \tag{11.12}$$

备注 66 多孔介质中的界面条件非常复杂，在模拟过程中存在各种相关问题，详细研究请读者参考相关文献 [379, 242, 315, 100, 101, 89, 316]。

由于本构方程与介质相关，因此我们在搭建模型时需要首先确定介质特性：

(1) 弹性介质：在非裂缝域中，我们有

$$\sigma_R := \sigma_s(u) = 2\mu e(u) + \lambda \mathrm{tr}(e) I \tag{11.13}$$

(2) 孔隙弹性介质：在非裂缝域中，我们有

$$\sigma_R := \sigma_s(u) - \alpha p I$$

其中 $\alpha \in [0, 1]$ 为 Biot 系数，关于多孔介质中有效应力的研究可参考相关文献 [117]。在裂缝域中，我们对流体有以下几个设定：

(1) 通常而言，缝高远小于缝长，可基于雷诺润滑理论[401]，令摩擦力 $\nabla v \approx 0$，主要研究静水压力[313,319]：

$$\sigma_F = -pI \tag{11.14}$$

(2) 裂缝内 Stokes 流的黏性力方程为

$$\sigma_F = -pI + \rho_f \nu \left(\nabla v + \nabla v^\mathrm{T} \right)$$

11.4.1.2 考虑界面条件的能量泛函

界面条件引入能量泛函方程中。动态界面条件可视为一个非齐次 Neumann 条件，即裂缝界面的边界积分。另外考虑到 Neumann 条件会同时作用于两个域，最终我们可得

$$\int_{\partial B_N \cup \Gamma} \tau \cdot u \, ds$$

修改后的能量泛函可写为

$$E_b(u, \varphi) = \frac{1}{2} \int_B g(\varphi) \sigma_R(u) : e(u) \mathrm{d}x - \int_{\partial B_N \cup \Gamma} \tau \cdot u \, ds$$
$$+ \frac{1}{2} \int_B G_c \left(\frac{1}{\varepsilon} (1-\varphi)^2 + \varepsilon |\nabla \varphi|^2 \right) \mathrm{d}x \tag{11.15}$$

其中 τ 为 Neumann 边界上的表面力。边界位置可以通过水平集函数确定，但需要切割单元或者进行相应处理。在此我们通过散度 (2.2 节) 定义界面矢量力：

$$\tau = \sigma_R n, \quad \text{在 } \Gamma \text{ 上}$$

假定在外边界 ∂B_N 上存在其他牵引力 $\tau : \partial B_N \to \mathbb{R}^d$。

$$\sigma_R n = \sigma_F n, \quad \text{在 } \Gamma \text{ 上}$$

基于式 (11.13) 与式 (11.14)，在裂缝边界处

$$\begin{aligned}
-\int_\Gamma \tau \cdot u \mathrm{d}s &= -\int_\Gamma \sigma_F n \cdot u \mathrm{d}s \\
&= \int_\Gamma p n \cdot u \mathrm{d}s \\
&= \int_{\Omega_R} \nabla \cdot (pu) \mathrm{d}x - \int_{\partial B} pn \cdot u \mathrm{d}s \\
&= \int_{\Omega_R} (u \nabla p + p \nabla \cdot u) \mathrm{d}x - \int_{\partial B} pn \cdot u \mathrm{d}s
\end{aligned} \quad (11.16)$$

接下来我们暂时忽略 ∂B_N，研究 ∂B 上的齐次 Dirichlet 位移条件，此时有

$$-\int_{\partial B} pn \cdot u = 0 \quad (11.17)$$

假定裂缝内部压力无变化：

$$\nabla p \approx 0 \quad \Rightarrow \quad \int_{\Omega_R} u \nabla p \mathrm{d}x \approx 0$$

最终式 (11.16) 可简化为

$$-\int_\Gamma \tau \cdot u \mathrm{d}s = \int_{\Omega_R} p \nabla \cdot u \mathrm{d}x \quad (11.18)$$

最终我们将积分扩展到全域 [313,318,319]：

$$\int_{\Omega_R} p \nabla \cdot u \mathrm{d}x \to \int_B g(\varphi) p \nabla \cdot u \mathrm{d}x \quad (11.19)$$

由此我们可得如下命题。

11.4 相场裂缝模型建模

命题 85 给定 $p \in L^{\infty}(B)$，假定 $\nabla p \approx 0$，令 $u \in H_0^1(B)$，$\varphi \in K$，有以下能量方程：

$$E_T(u,\varphi) = \int_B g(\varphi)\sigma_R(u):e(u) + \int_B g(\varphi)p\nabla \cdot u \mathrm{d}x$$

$$+ \int_B G_c\left(\frac{1}{2\varepsilon}(1-\varphi)^2 + \frac{\varepsilon}{2}|\nabla\varphi|^2\right)\mathrm{d}x \qquad (11.20)$$

命题 86(弹性加压裂缝模型) 令 $\sigma := \sigma_R$，$p \in L^{\infty}(B)$，给定 φ^{n-1}，令 $(u,\varphi) \in H_0^1(B) \times K$，此时有

$$(g(\varphi)\sigma(u),e(w)) + (g(\varphi)p, \nabla \cdot w) = 0, \quad \forall w \in H_0^1(B) \qquad (11.21)$$

$$(1-\kappa)(\varphi\sigma(u):e(u),\psi-\varphi) + 2(1-\kappa)(\varphi p\nabla \cdot u,\psi-\varphi)$$

$$+ \left(-\frac{1}{\varepsilon}\left(G_c(1-\varphi),\psi-\varphi\right) + \varepsilon\left(G_c\nabla\varphi,\nabla(\psi-\varphi)\right)\right) \geqslant 0, \quad \forall \psi \in K \qquad (11.22)$$

接下来推导命题 86 的强形式。利用分部积分，可得考虑位移和相场变量的准稳态椭圆型方程系统。

命题 87 给定 $p: B \to \mathbb{R}$，令 $u: B \to \mathbb{R}^d$

$$-\nabla \cdot (g(\varphi)\sigma(u)) - \nabla \cdot (\varphi^2 p) = 0, \quad 在 B 中$$

$$u = 0, \quad 在 \partial B 上$$

令 $\varphi: B \to [0,1]$，有

$$(1-\kappa)\sigma(u):e(u)\varphi - G_c\varepsilon\Delta\varphi - \frac{G_c}{\varepsilon}(1-\varphi) + 2\varphi p\nabla \cdot u \leqslant 0, \quad 在 B 中$$

$$\partial_t\varphi \leqslant 0, \quad 在 B 中$$

$$\left[(1-\kappa)\sigma(u):e(u)\varphi - G_c\varepsilon\Delta\varphi - \frac{G_c}{\varepsilon}(1-\varphi) + 2\varphi p\nabla \cdot u\right] \cdot \partial_t\varphi = 0, \quad 在 B 中$$

$$\partial_n\varphi = 0, \quad 在 \partial B 上$$

11.4.1.3 考虑压力梯度与边界压力的相场裂缝模型

接下来我们考虑式 (11.16) 中的剩余项，并将边界条件分解为 Dirichlet 边界条件和 Neumann 边界条件，即 $\partial B = \partial B_D \cup \partial B_N$。

命题 88 令 $\sigma := \sigma R$, Neumann 边界 ∂B_N 上的牵引力为 τ, 令 $p \in W^{1,\infty}(B)$ 且给定 φ^{n-1}, 则当 $u \in V_u^D$ 且 $\varphi \in K$ 时, 我们可获得能量泛函:

$$E_T(u,\varphi) = \frac{1}{2}\int_B g(\varphi)\sigma(u):e(u)\mathrm{d}x + \int_B g(\varphi)(p\nabla \cdot u + \nabla p \cdot u)\mathrm{d}x$$

$$- \int_{\partial B_N} (\tau + pn) \cdot u \mathrm{d}s + \int_B G_c\left(\frac{1}{2\varepsilon}(1-\varphi)^2 + \frac{\varepsilon}{2}|\nabla\varphi|^2\right)\mathrm{d}x \quad (11.23)$$

相应地 Euler-Lagrange 耦合系统如下。

命题 89 令 $p \in W^{1,\infty}(B)$, 给定 φ^{n-1}, 当 $(u,\varphi) \in V_u^D \times K$, $\forall w \in V_u^0$ 时, 有

$$(g(\varphi)\sigma(u), e(w)) + (g(\varphi)p, \nabla \cdot w) + (g(\varphi)\nabla p, w) - \int_{\partial B_N}(\tau + pn)\cdot w \mathrm{d}s = 0 \quad (11.24)$$

且对于 $\forall \psi \in K$

$$(1-\kappa)(\varphi\sigma(u):e(u), \psi - \varphi) + 2(1-\kappa)(\varphi(p\nabla \cdot u + \nabla p \cdot u), \psi - \varphi)$$

$$+ G_c\left(-\frac{1}{\varepsilon}(1-\varphi, \psi - \varphi) + \varepsilon(\nabla\varphi, \nabla(\psi - \varphi))\right) \geqslant 0 \quad (11.25)$$

11.4.1.4 孔隙弹性模型中的加压裂缝模型

孔隙弹性模型中, 我们有

$$\sigma := \sigma_R = \sigma_s - \alpha p I$$

基于命题 83 中介绍的界面条件, 可得如下命题。

命题 90 给定 ∂B_N 边界上的应力 τ, 令 $p \in W^{1,\infty}(B)$, 给定 φ^{n-1}, 当 $u \in V_u^D$, $\varphi \in K$ 时, 能量方程为

$$(g(\varphi)\sigma(u), e(w)) + (1-\alpha)(g(\varphi)p, \nabla \cdot w) + (g(\varphi)\nabla p, w)$$

$$- \int_{\partial B_N}(\tau + pn)\cdot w \mathrm{d}s = 0, \quad \forall w \in V_u^0 \quad (11.26)$$

在中间积分 $\int_B g(\varphi)((1-\alpha)p\nabla \cdot u + \nabla p \cdot u)\mathrm{d}x$ 中, 有运动条件 $p := p_R = p_F$。因此可得相应的 Euler-Lagrange 如下。

命题 91 令 $p \in W^{1,\infty}(B)$, 给定 φ^{n-1}, 令 $(u,\varphi) \in V_u^D \times K$, 则对于 $\forall w \in V_u^0$:

$$(g(\varphi)\sigma(u), e(w)) + (1-\alpha)(g(\varphi)p, \nabla \cdot w) + (g(\varphi)\nabla p, w)$$

11.4 相场裂缝模型建模

$$-\int_{\partial B_N} (\tau + pn) \cdot w \mathrm{d}s = 0 \tag{11.27}$$

且对于 $\forall \psi \in K$

$$(1-\kappa)(\varphi\sigma(u):e(u), \psi-\varphi) + 2(1-\kappa)(\varphi((1-\alpha)p\nabla \cdot u + \nabla p \cdot u), \psi-\varphi)$$
$$+ G_c \left(-\frac{1}{\varepsilon}(1-\varphi, \psi-\varphi) + \varepsilon(\nabla\varphi, \nabla(\psi-\varphi))\right) \geqslant 0 \tag{11.28}$$

11.4.2 非等温裂缝模型

本节将介绍非等温裂缝模型[337,410]，该模型中引入了导热系数以及 11.4.1 节的界面条件。此时我们在裂缝边界上需要增加温度参数 ϑ 并修改热孔弹性介质中的本构方程。

定义 74 在各向同性热孔弹性介质模型中，应力张量的本构表达式为[117]

$$\sigma := \sigma_R := \sigma(u, p, \vartheta) = \sigma_0 + \sigma_s(u) - \alpha_B(p - p_0)I - 3\alpha_T K(\vartheta - \vartheta_0)I$$

其中 σ_0 为初始应力，α_B 为 Biot 系数，p 为孔隙压力，p_0 为初始孔隙压力，$3\alpha_T$ 为骨架体积热膨胀系数，$K := \frac{2}{3}\mu + \lambda$，$\vartheta$ 为温度，ϑ_0 为初始温度。

由此我们可得系统能量方程:

$$\begin{aligned} E_b(u) &= \frac{1}{2}(\sigma_s(u), e(u))_\Omega - \langle \tau, u \rangle_{\partial B_N} \\ &= \frac{1}{2}(\sigma_s(u), e(u))_\Omega - \langle \tau, u \rangle_{\partial B_N} - (\alpha_B(p-p_0), \nabla \cdot u) \\ &\quad - (3\alpha_T K(\vartheta - \vartheta_0), \nabla \cdot u) \end{aligned} \tag{11.29}$$

存在裂缝 C 时，我们可获得以下正则化公式。

命题 92 令 $U \in V_u^D \times K$，我们有

$$\begin{aligned} E_\varepsilon(u, \varphi) &= \frac{1}{2}(g(\varphi)\sigma_s(u), e(u))_\Omega - \langle \tau, u \rangle_{\Gamma_n \cup \partial C} \\ &\quad - (\alpha_B(p-p0)\varphi^2, \nabla \cdot u) - (3\alpha_T K(\vartheta - \vartheta_0)\varphi^2, \nabla \cdot u) \\ &\quad + \int_B G_c \left(\frac{1}{2\varepsilon}(1-\varphi)^2 + \frac{\varepsilon}{2}|\nabla\varphi|^2\right) \mathrm{d}x \end{aligned} \tag{11.30}$$

该方程可引入增广的 Lagrange 补偿项形成 Euler-Lagrange 耦合系统。

命题 93 令 $p, p_0 \in L^\infty(B)$，$\vartheta, \vartheta_0 \in L^\infty(B)$，给定 φ^{n-1}，$u \in V_u^D$，加载步 $n = 1, 2, 3, \cdots, N$，有

$$A_1(U)(w) := (g(\varphi)\sigma_s(u), e(w)) - \langle \tau, w \rangle$$

$$- \left(\alpha_B (p - p_0)\varphi^2, \nabla \cdot w\right) - \left(3\alpha_T K (\vartheta - \vartheta_0)\varphi^2, \nabla \cdot w\right) = 0, \quad \forall w \in V_u^0$$

令 $\varphi \in V_\varphi$，对于 $\forall \psi \in V_\varphi$，有

$$A_2(U)(\psi) := (1 - \kappa)\left(\varphi \sigma_s(u) : e(u), \psi\right)$$

$$- 2(\alpha_B - 1)\left((p - p_0)\nabla \cdot u\varphi, \psi\right) - 2(3\alpha_T K)\left((\vartheta - \vartheta_0)\nabla \cdot u\varphi, \psi\right)$$

$$- G_c \left(\frac{1}{\varepsilon}(1 - \varphi, \psi) + \varepsilon(\nabla\varphi, \nabla\psi)\right)$$

$$+ \left(\left[\Xi + \gamma(\varphi - \varphi^{n-1})\right]^+, \psi\right) = 0$$

接下来我们考虑裂缝中压力与温度的变化，并将其作为多孔介质与裂缝界面上的额外作用量。根据文献 [313]，我们可以得到一个包含传热方程的表达式，将裂缝视为一个高渗透区，且该区域的宽度远小于其长度，则在 C 中压力应力为

$$-(p - p_0)I \tag{11.31}$$

另外我们假定裂缝边界的热应力为

$$C_\vartheta(x, t)(\vartheta_F - \vartheta_0)I \tag{11.32}$$

其中常数 $C_\vartheta(x, t)$ 通过裂缝附近温度分布获得[410]。在这里我们忽略对流项，热传导方程可写为

$$\partial_t \vartheta - \kappa \Delta \vartheta = 0 \tag{11.33}$$

其中传热系数 $\kappa = \dfrac{K_r}{\rho_r C_r}$。$\rho_r$ 为岩石密度，C_r 为岩石比热，K_r 为导热率。通过热传导方程，可以确定裂缝区域界面温度梯度，此时我们有

$$C_\vartheta = A_\vartheta \left(\frac{\lambda_\vartheta}{2\lambda_\vartheta + 1}\right), \quad A_\vartheta = \frac{E\beta}{1 - \nu_s}$$

其中 β 为线性热膨胀系数，E 为杨氏模量，ν_s 为泊松比，λ_ϑ 由 Hagoort 基于热传导方程得到[208]。

$$\lambda_\vartheta = \sinh^{-1}\left(\frac{\gamma_\vartheta}{0.5 l_0}\sqrt{\pi \kappa t}\right)$$

11.4 相场裂缝模型建模

这里 $1 \leqslant \gamma_\vartheta \leqslant \dfrac{4}{\pi}$，且 l_0 为半缝长。

备注 67 我们正在尝试基于相场法将裂缝与多孔介质的传热方程相结合 [267,193]，一个代表性尝试方程如下：

$$\rho_r C_r \partial_t \vartheta + \alpha_p \vartheta \partial_t p - \nabla \cdot (K_r \nabla \vartheta) + \nabla \cdot (\alpha_c \vartheta K_p \nabla p) = q_\vartheta$$

其中 p 是压强，α_c 是对流热容，α_p 是孔隙压力和温度之间的关系系数，K_p 为流体的渗透张量，q_ϑ 是热源注入项。令 $\alpha_p = 0$ 并忽略对流项，则可得到以上方程。关于非断裂热孔弹性模型的最新研究成果，请读者参考相关文献 [418,420]。

通过推导多孔介质域 Ω 与裂缝边界 $\partial \mathcal{C}$ 之间的耦合关系可知，裂缝边界上的法向应力满足：

$$\sigma_R(u, p, \vartheta) n = \sigma_F(p, \vartheta) n$$

结合运动学条件可得以下命题。

命题 94 加压非等温裂缝模型界面耦合关系如下：

$$\begin{aligned} p_R &= p_F, \quad \text{在 } \Gamma \text{ 上} \\ \vartheta_R &= \vartheta_F, \quad \text{在 } \Gamma \text{ 上} \\ \sigma_R(u, p, \vartheta) n &= \sigma_F(p, \vartheta) n, \quad \text{在 } \Gamma \text{ 上} \end{aligned}$$

其中

$$\begin{aligned} \sigma_R &= \sigma_0 + \sigma_s - \alpha_B (p - p_0) I - 3\alpha_\vartheta K (\vartheta - \vartheta_0) I \\ \sigma_F &= -(p - p_0) + C_\vartheta (\vartheta_F - \vartheta_0) \end{aligned}$$

在这里，我们应用散度定理，将面积分改写为区域积分，以便将其与多孔介质相结合。对于式 (11.30) 中的牵引项，我们可得

$$\begin{aligned} \langle \sigma n, u \rangle_{\partial \mathcal{C}} &= -\langle (p - p_0) n, u \rangle_{\partial \mathcal{C}} + \langle C_\vartheta (\vartheta - \vartheta_0) n, u \rangle_{\partial \mathcal{C}} - (\nabla \cdot (p - p_0) u) \\ &\quad + \langle (p - p_0) n, u \rangle_{\Gamma_N} - (C_\vartheta \nabla \cdot (\vartheta - \vartheta_0) u) + \langle C_\vartheta (\vartheta - \vartheta_0) n, u \rangle_{\Gamma_N} \\ &= -((p - p_0), \nabla \cdot u) - (\nabla (p - p_0), u) + \langle (p - p_0) n, w \rangle_{\Gamma_N} \\ &\quad - (C_\vartheta (\vartheta - \vartheta_0), \nabla \cdot u) - (C_\vartheta \nabla (\vartheta - \vartheta_0), u) + \langle (\vartheta - \vartheta_0) n, u \rangle_{\Gamma_N} \end{aligned}$$

相关参数已在全域 B 中实现了定义，在多孔介质中：

$$\begin{aligned} \langle \sigma n, u \rangle_{\partial \mathcal{C}} &+ (\alpha_B (p - p_0), \nabla \cdot u) + (3\alpha_\vartheta K (\vartheta - \vartheta_0), \nabla \cdot u) \\ &= (\alpha_B (p - p_0), \nabla \cdot u) \end{aligned}$$

$$- ((p - p_0), \nabla \cdot u) - (\nabla (p - p_0), u) + \langle (p - p_0) n, u \rangle_{\Gamma_N}$$
$$+ (3\alpha_\vartheta K (\vartheta - \vartheta_0), \nabla \cdot u)$$
$$- (C_\vartheta (\vartheta - \vartheta_0), \nabla \cdot u) - (C_\vartheta \nabla (\vartheta - \vartheta_0), u) + \langle (\vartheta - \vartheta_0) n, u \rangle_{\Gamma_N}$$

最终我们有

$$- \langle \tau, u \rangle_{\Gamma_N \cup \partial \mathcal{C}} - (\alpha_B (p - p_0), \nabla \cdot u) - (3\alpha_\vartheta K (\vartheta - \vartheta_0), \nabla \cdot u)$$
$$= - \langle \tau, u \rangle_{\Gamma_N} - \langle \sigma n, u \rangle_{\mathcal{C}} - (\alpha_B (p - p_0), \nabla \cdot u) - (3\alpha_\vartheta K (\vartheta - \vartheta_0), \nabla \cdot u)$$
$$= - \langle \tau, u \rangle_{\Gamma_N} - (\alpha_B (p - p_0), \nabla \cdot u)$$
$$+ ((p - p_0), \nabla \cdot u) + (\nabla (p - p_0), u) - \langle (p - p_0) n, u \rangle_{\Gamma_N}$$
$$- (3\alpha_\vartheta K (\vartheta - \vartheta_0), \nabla \cdot u) - (C_\vartheta (\vartheta - \vartheta_0), \nabla \cdot u)$$
$$- (C_\vartheta \nabla (\vartheta - \vartheta_0), u) + \langle C_\vartheta (\vartheta - \vartheta_0) n, u \rangle_{\Gamma_N}$$

随后将这些表达式插入式 (11.30)

$$E_\varepsilon(u, \varphi) = \frac{1}{2} \left(((1 - \kappa)\varphi^2 + \kappa) \sigma_s(u), e(u) \right)_\Omega - \langle \tilde{\tau}, u \rangle_{\Gamma_N}$$
$$- ((\alpha_B - 1) (p - p_0) \varphi^2, \nabla \cdot w) + (\nabla (p - p_0) \varphi^2, w)$$
$$- ((3\alpha_\vartheta K + C_\vartheta) (\vartheta_F - \vartheta_0) \varphi^2, \nabla \cdot w) - (C_\vartheta \nabla (\vartheta - \vartheta_0) \varphi^2, w)$$
$$+ G_c \left(\frac{1}{2\varepsilon} \int_B (1 - \varphi)^2 \mathrm{d}x + \frac{\varepsilon}{2} \int_B |\nabla \varphi|^2 \mathrm{d}x \right)$$

其中 $\tilde{\tau} = \tau + (p - p_0) n - (\vartheta - \vartheta_0) n$, 随后我们可以推导出 Euler-Lagrange 耦合方程 (增广 Lagrange 法)。

命题 95 令 $p \in W^{1,\infty}(B)$, $\vartheta \in W^{1,\infty}(B)$, 给定 φ^{n-1}, $u \in V_u^D$, 在加载步 $n = 1, 2, 3, \cdots, N$ 中, 对于 $\forall w \in V_u^0$, 有

$$A_1(U)(w) = \left(((1-\kappa)\varphi^2 + \kappa) \sigma(u), e(w) \right) - \langle \tilde{\tau}, w \rangle_{\Gamma_N}$$
$$+ (1 - \alpha_B) ((p - p_0) \varphi^2, \nabla \cdot w) + (\nabla (p - p_0) \varphi^2, w)$$
$$- (3\alpha_\vartheta K + C_\vartheta) ((\vartheta - \vartheta_0) \varphi^2, \nabla \cdot w) - (C_\vartheta \nabla (\vartheta_F - \vartheta_0) \varphi^2, w)$$
$$= 0$$

令 $\varphi \in V_\varphi$, 对于 $\forall \psi \in V_\varphi$:

$$A_2(U)(\psi) = (1 - \kappa)(\varphi \sigma(u) : e(u), \psi)$$

11.4 相场裂缝模型建模

$$+ 2(1-\alpha_B)((p-p_0)\nabla \cdot u\varphi, \psi) + 2(\nabla(p-p_0)u\varphi, \psi)$$

$$- 2(3\alpha_\vartheta K + C_\vartheta)((\vartheta-\vartheta_0)\nabla \cdot u\varphi, \psi) - 2(C_\vartheta \nabla(\vartheta-\vartheta_0)u\varphi, \psi)$$

$$- G_c\left(\frac{1}{\varepsilon}(1-\varphi, \psi) + \varepsilon(\nabla\varphi, \nabla\psi)\right)$$

$$+ \left([\Xi + \gamma(\varphi - \varphi^{n-1})]^+, \psi\right)$$

$$= 0$$

备注 68 若 $\vartheta < \vartheta_0$,即注入温度低于初始温度时,热应力会促使裂缝开启,从物理角度分析,这是由于温度降低,岩石体积收缩。从数学角度分析,则是由于与压力项符号不同造成的结果。

11.4.3 面向目标的误差分析

本节,我们基于命题 95,提出了一个针对准稳态加压非等温相场裂缝模型的面向目标的误差分析法。

命题 96(加压非等温相场裂缝模型) 给定 $p, p_0, \vartheta, \vartheta_0$,令 $U = (u, \varphi) \in X^D$, $U_h = (u_h, \varphi_h) \in X_h^D$ 分别为连续以及离散的原始解。令 $Z = (z_u, z_\varphi) \in X^0$, $Z_h \in X_h^0$ 分别为连续以及离散的伴随解。上标 (2) 代表高阶解。由此可得误差估计如下:

$$|J(U) - J(U_h)| \leqslant \eta := \left|\sum_{i=1}^M \eta_i\right| \leqslant \sum_{i=1}^M |\eta_i|$$

其中

$$\eta_i = -A(U_h)\left(\left(Z_h^{(2)} - i_h Z_h^{(2)}\right)\psi_i\right)$$

$$A(U)\left((Z - i_h Z)\psi_i\right)$$

$$= A_1(U)\left((z_u - i_h z_u)\psi_i\right) + A_2(U)\left((z_\varphi - i_h z_\varphi)\psi_i\right)$$

$$= (g(\varphi)\sigma(u), \nabla(z_u - i_h z_u)\psi_i) - \left\langle\tilde{\mathcal{T}}, (z_u - i_h z_u)\psi_i\right\rangle_{\Gamma_N}$$

$$+ (1-\alpha_B)((p-p_0)\varphi^2, \nabla \cdot (z_u - i_h z_u)\psi_i) + (\nabla(p-p_0)\varphi^2, (z_u - i_h z_u)\psi_i)$$

$$- (3\alpha_\vartheta K + C_\vartheta)((\vartheta - \vartheta_0)\varphi^2, \nabla \cdot (z_u - i_h z_u)\psi_i)$$

$$- (C_\vartheta \nabla(\vartheta_F - \vartheta_0)\varphi^2, (z_u - i_h z_u)\psi_i) + (1-\kappa)(\varphi\sigma(u) : e(u), (z_\varphi - i_h z_\varphi)\psi_i)$$

$$+ 2(1-\alpha_B)((p-p_0)\nabla \cdot u\varphi, (z_\varphi - i_h z_\varphi)\psi_i) + 2(\nabla(p-p_0)u\varphi(z_\varphi - i_h z_\varphi)\psi_i)$$

$$-2\left(3\alpha_\vartheta K + C_\vartheta\right)\left((\vartheta - \vartheta_0)\nabla \cdot u\varphi, (z_\varphi - i_h z_\varphi)\psi_i\right)$$

$$-2\left(C_\vartheta \nabla(\vartheta - \vartheta_0)u\varphi, (z_\varphi - i_h z_\varphi)\psi_i\right)$$

$$-G_c \frac{1}{\varepsilon}\left(1 - \varphi, (z_\varphi - i_h z_\varphi)\psi_i\right) + \varepsilon G_c\left(\nabla\varphi, \nabla(z_\varphi - i_h z_\varphi)\psi_i\right)$$

$$+\left(\left[\Xi + \gamma(\varphi - \varphi^{n-1})\right]^+, (z_\varphi - i_h z_\varphi)\psi_i\right).$$

在实际建模过程中,考虑到不同研究人员的关注点不同,我们还需要考虑是否有必要通过如此多的项来实现误差估计。

11.5 多孔介质中的压力传播方程

前面章节中我们介绍了相场裂缝中孔隙压力场与温度场的相关公式。这些公式常应用于多孔介质模型中,在岩石力学等相关领域有广泛应用。本节我们将介绍多孔介质中的压力传播方程。

首先我们确定了两个流动区域,在相场参数 φ 的辅助下可将系统分为裂缝域 Ω_F 与非裂缝域 Ω_R,并利用 $\chi_F(\varphi)$ 与 $\chi_R(\varphi)$ 对区域属性进行标定。在不同区域需要用到不同的流动方程,因此 $\chi_F(\varphi)$ 与 $\chi_R(\varphi)$ 将作为全局流体压力方程中的时空相关系数。如 3.4 节所示,多孔介质中的压力方程是抛物线型的。在文献 [316] 中,研究人员在裂缝域与非裂缝域均使用抛物线型压力方程,构成了一个全局抛物线型 PDE 系统。这些问题被统一称为广义抛物方程 (generalized parabolic equations)[284] 或者衍射方程 (diffraction equations)[268]。

算法 23 计算多孔介质中的 (u, φ, p):
(1) 给定压力 p,求解 (u, φ)(参考 11.4.1 节命题 91);
(2) 确定 $\chi_F(\varphi)$ 与 $\chi_R(\varphi)$;
(3) 根据 (u, φ),求解广义抛物型方程 $p = \chi_R p + \chi_F p$。

命题 97(压力方程强形式) 给定 $U = (u, \varphi)$,令 $p: B \times I \to \mathbb{R}$ 且 $p = \chi_R p + \chi_F p$,初始压力 $p(0) = p_0$,则:

$$\theta \partial_t p - \nabla \cdot K_{\text{eff}}\left(\nabla p - \rho^0 g\right) = \tilde{q}, \quad 在 B \times I 中 \tag{11.34}$$

其中

$$\theta := \theta(x, t) := \chi_R \theta_R + \chi_F \theta_F = \chi_R\left(\left(\frac{3\alpha^2}{3\lambda + 2\mu} + \frac{1}{M}\right)\right) + \chi_F c_F$$

$$\tilde{q} := \tilde{q}(x, t) := \chi_R q_R - \chi_{\Omega_R}\partial_t\left(\frac{3\alpha}{3\lambda + 2\mu}\bar{\sigma}\right) + \chi_F q_F$$

其中 $M = c_B^{-1}$ 为多孔介质中的压缩系数，c_F 为裂缝中流体压缩系数，K_{eff} 为有效压缩系数，$\tilde{q}$ 为源项，$\bar{\sigma}$ 为应力项。

界面条件隐藏在弱形式中，由于我们研究全局压力空间，因此有

$$p_R = p_F \tag{11.35}$$

$$\left[K_{\text{eff}}\left(\nabla p - \rho^0 g\right)\right] \cdot n = 0 \tag{11.36}$$

式中

$$K_{\text{eff}} = \chi_F \frac{K_F}{\eta_F} + \chi_R \frac{K_R}{\eta_R}$$

其中 K_F, K_R 为各区域渗透率，η_F, η_R 为对应的运动黏度。

命题 98 令 $p \in V_p := H^1(B)$，$p(0) := p_0$，则方程弱形式为

$$(\theta \partial_t p, \phi) + \left(K_{\text{eff}}\left(\nabla p - \rho^0 g\right), \nabla \phi\right) - (q, \phi) = 0, \quad \forall \phi \in V_p$$

11.6 连续介质力学简介

在前面的章节中，我们主要研究无穷小位移问题。但当涉及固体大变形问题时，我们则需要用到不同的坐标系统，即 Lagrangian 系统与 Eulerian 系统。其中 Lagrangian 域表达符号为 $\hat{\Omega}$ 或 $\hat{B}$，Eulerian 域表达符号为 Ω 或 B。

定义 75 关于域 (见图 11.4)，我们有如下定义：
(1) $\hat{\Omega}$：参考域或未变形域；
(2) $\Omega(t)$：当前域或变形域；

$\Omega(t=0)$ 即所谓的初始状态，通常情况下 $\hat{\Omega} := \Omega(t=0)$，后续为了方便起见，我们常表达为 $\Omega := \Omega(t)$。

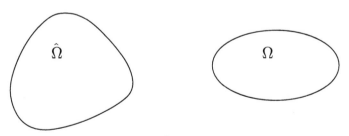

图 11.4 域 $\hat{\Omega}$ 与 Ω 变形情况

备注 69 在 Lagrangian 系统中，我们主要研究特定的物质点及其变形情况，而在 Eulerian 系统中，我们主要观察空间点上的情况，随着时间的推移，该空间点可能被不同的物质占据。

物质点 $\hat{x} \in \hat{\Omega}$ 的时间演化规律可以用以下映射描述：

$$t \mapsto x(t, \hat{x})$$

接下来我们介绍不同系统中所使用的符号：
(1) Lagrange 系统：$\hat{f}(t, \hat{x}) : \hat{\Omega} \times I \to \mathbb{R}$；
(2) Euler 系统：$f(t, x) : \Omega \times I \to \mathbb{R}$。

随后我们处理两种类型的领域中的点以及函数方程，为便于表述，后面可能会出现缩写情况。

定义 76 Lagrange 坐标系中的点表示为

$$\hat{x} := \hat{x}(t, x) \in \hat{\Omega}$$

Euler 坐标系中的点表示为

$$x := x(t, \hat{x}) \in \Omega$$

其中 $\hat{x}$ 与 x 之间的关系参考定义 77。

假设 3 我们假定：
(1) $(t, \hat{x}) \mapsto x(t, \hat{x})$ 是连续可微的；
(2) 在所有 $t \geqslant t_0$ 时刻中，映射 $\hat{x} \mapsto x(t, \hat{x})$ 均是可逆的；
(3) Jacobi 矩阵恒为正。

定义 77 假设域 $\hat{\Omega}$ 的变形过程是光滑的一对一的保定向映射

$$\hat{T} : \hat{\Omega} \to \Omega, \quad 在 \ (t, \hat{x}) \mapsto (t, x) = (t, \hat{T}(t, \hat{x})) \ 内$$

该映射将参考域的每个点 $\hat{x} \in \hat{\Omega}$ 与物理域的一个新位置 $x \in \Omega$ 联系起来。由此，我们可将变形后的配置表示为 $\Omega = \hat{T}(\hat{\Omega})$，即

$$\hat{x}(t, x) = \hat{T}^{-1}(t, x)$$
$$x(t, \hat{x}) = \hat{T}(t, \hat{x})$$

定义 78(位移场 Lagrange 描述) 对于 $\hat{x} \in \hat{\Omega}$，$x \in \Omega$，存在以下位移关系：

$$\hat{u} : (t, \hat{x}) \to \hat{u}(t, \hat{x}) = x(t, \hat{x}) - \hat{x}$$

它将物质点在参考配置 $\hat{x}$ 中的位置与当前时间 t 时的配置 x 关联起来。

定义 79(位移场 Euler 描述) 位移场的 Eulerian 描述如下：

$$u(t, x) = x - \hat{x}(t, x)$$

同时，我们有 $x = \hat{T}(t, \hat{x})$。

定义 80 为了保证一个系统中的函数在其他系统中也具有相同的值，当 $\hat{u} := \hat{u}(t, \hat{x}) : \hat{\Omega} \times I \to \mathbb{R}^d, u := u(t, x) : \Omega \times I \to \mathbb{R}^d$ 时，我们有

$$\hat{u}(t, \hat{x}) = u(t, x)$$

11.6.1 Euler-Lagrange 坐标系

在之前的章节中，我们有

$$\hat{T}(\hat{x}, t) : \hat{\Omega}_s \times I \to \Omega_s$$

在流固耦合模型中，我们常使用一种称为 ALE (arbitrary Lagrangian-Eulerian) 的方法实现公式变形 [131,230,239,132,166]，耦合固体域与流体域：

$$\widehat{\mathcal{A}}(\hat{x}, t) : \hat{\Omega}_f \times I \to \Omega_f$$

接下来，我们将介绍连续介质力学中的相关重要符号。

命题 99 变形、变形梯度以及变形梯度的行列式分别定义为

$$\widehat{\mathcal{A}}(\hat{x}, t) = x = \hat{x} + \hat{u}(\hat{x}, t), \quad \hat{F} := \hat{\nabla}\widehat{\mathcal{A}} = \hat{I} + \hat{\nabla}\hat{u}, \quad \hat{J} := \det(\hat{F})$$

这里 $\hat{x} \in \hat{\Omega}$ 为参考域中的某点，$\hat{u}$ 为 Lagrange 坐标表示的位移。

命题 100(变形梯度 $\hat{F}$) 当 $x := x(t, \hat{x}) = \hat{T}(t, \hat{x}) = \hat{x} + \hat{u}$, 时，针对无穷小变形，我们有

$$\mathrm{d}x = \hat{F} \cdot \mathrm{d}\hat{x}$$

其中 $\hat{F} = \hat{\nabla}x$，即 $\hat{F}_{ij} = \dfrac{\partial x_i}{\partial \hat{x}_j}$。

证明：根据 $x = \hat{x} + \hat{u}$，我们有

$$\hat{F} = \frac{\partial x}{\partial \hat{x}} = \frac{\partial}{\partial \hat{x}}(\hat{x} + \hat{u}) = \frac{\partial \hat{x}}{\partial \hat{x}} + \frac{\partial \hat{u}}{\partial \hat{x}} = \hat{I} + \hat{\nabla}\hat{u}$$

即

$$\hat{F} = \hat{\nabla}\hat{T} = \hat{I} + \hat{\nabla}\hat{u}$$

定义 81 我们令：

$$\hat{J} := \det(\hat{F}) > 0$$

并将其用于关联无穷小参考域和当前域之间的体积变化：

$$\mathrm{d}\Omega = \hat{J}\mathrm{d}\hat{\Omega} \tag{11.37}$$

命题 101 Euler 展开式可写为 [232]

$$\partial_t \hat{J}(t,\hat{x}) = \nabla \cdot v(t,x) \hat{J}(t,\hat{x})$$

将式 (11.37) 扩展到整个计算域，我们可以计算域的体积变化：

$$|\Omega| = \int_\Omega 1 \mathrm{d}x = \int_{\hat{\Omega}} \hat{J}(t,\hat{x}) \mathrm{d}\hat{x} \tag{11.38}$$

进一步可得

$$\frac{\mathrm{d}}{\mathrm{d}t}|\Omega| = \int_\Omega \nabla \cdot v(t,x) \mathrm{d}x$$

定义 82 两种系统的逆变换定义为

$$\mathcal{A}(x,t) : \Omega \times I \to \hat{\Omega}$$

其中 $\hat{x} = \mathcal{A}(x,t) = x - u(x,t)$。

可参考图 11.5 与图 11.6。

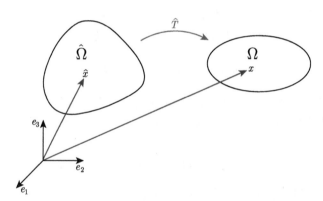

图 11.5 变形映射情况 $\hat{T} : \hat{\Omega} \to \Omega$

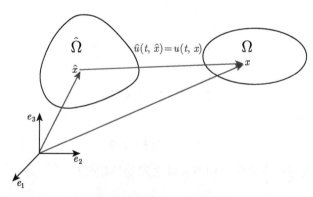

图 11.6 位移描述关系图，其中满足 $\hat{x} + \hat{u} = x$

11.6 连续介质力学简介

由此可得以下恒等式。

命题 102 针对两种变换的复合情况，我们有

$$\mathcal{A}(\hat{\mathcal{A}}(\hat{x},t),t) = \hat{x} \tag{11.39}$$

证明 显然

$$\mathcal{A}(\hat{\mathcal{A}}(\hat{x},t),t) = \mathcal{A}(x,t) = \hat{x}$$

命题 103 当 I、$\hat{I}$ 分别代表单位矩阵时，针对 Euler 与 Lagrange 系统中的变形梯度，我们有

$$\underbrace{(I - \nabla u)}_{=F}\underbrace{(\hat{I} + \hat{\nabla}\hat{u})}_{=\hat{F}} = \hat{I}$$

证明 基于前文的描述，我们有

$$\nabla \mathcal{A}(\hat{\mathcal{A}}, t) = \nabla \hat{x}$$
$$\Leftrightarrow \nabla \mathcal{A} \cdot \hat{\nabla}\hat{\mathcal{A}} = \hat{I}$$
$$\Leftrightarrow \nabla(x - u)\hat{\nabla}(\hat{x} + \hat{u}) = \hat{I}$$
$$\Leftrightarrow (I - \nabla u)(\hat{I} + \hat{u}) = \hat{I}$$

由此我们可得

$$F = (I - \nabla u)$$
$$J = \det(F)$$

命题 104 基于以上表述，我们有

$$\hat{F} = (\hat{I} + \hat{\nabla}u) = (I - \nabla u)^{-1} =: F^{-1}$$
$$\hat{J} = \det(\hat{F}) = \det(F^{-1}) =: J^{-1}$$

其中，若 $\hat{J} > 0$，则我们可得 $J > 0$。

接下来我们可以给出应变张量的形式。

命题 105(Green-Lagrange 应变张量) 在 Euler 和 Lagrange 系统中，我们分别有

$$E := \frac{1}{2}\left(F^{-T}F^{-1} - I\right), \quad \hat{E} := \frac{1}{2}\left(\hat{F}^{T}\hat{F} - \hat{I}\right)$$

命题 106 针对无穷小变形，我们有

$$E \approx \hat{E} \approx e(u) = \frac{1}{2}\left(\nabla u + \nabla u^{T}\right)$$

更进一步可得
$$\hat{J} \approx J \approx 1 + \nabla \cdot u$$

证明 在处理无穷小变换的条件下，我们有一阶近似：
$$\|\nabla u\| \ll 1$$

由此我们可令：
$$\Omega \approx \hat{\Omega}$$
$$\hat{\nabla} \approx \nabla$$

针对应变 $\hat{E}$，我们在忽略二阶项 $\nabla u^{\mathrm{T}} \cdot \nabla u$ 时可得 $e(u)$，其与 ∇u 同阶，即 $\|\nabla u\| \ll 1$ 可得 $\|E\| \ll 1$，但反之并不成立。同时我们可得

$$\hat{J} \approx J \approx 1 + \partial_z u_z + \partial_y u_y + \partial_x u_x = 1 + \nabla \cdot u = 1 + (e(u))_{ii=1,2,3} \tag{11.40}$$

根据前文可知，当 $x \in \Omega$，$\hat{x} \in \hat{\Omega}$ 时，我们有 $\hat{f}(t,\hat{x}) = f(t,x)$，其中 $\hat{f}: \hat{\Omega} \times I \to \mathbb{R}$，$f: \Omega \times I \to \mathbb{R}$。

定义 83(Lagrangian 场全局时间导数) 我们有

$$\frac{\mathrm{d}}{\mathrm{d}t}\hat{f}(t,\hat{x}) = \hat{\partial}_t \hat{f}(t,\hat{x})$$

该式表示跟随粒子的路径线观察下，物质随时间变化的速率，即同一个例子在任意时刻的速率变化情况。

定义 84(Eulerian 场局部时间导数) Eulerian 场时间导数定义为

$$\partial_t f(t,x)$$

其表示测量的固定空间点处的时间变化率，这意味着每个时刻 f 处都可能是一个新粒子。

命题 107(Eulerian 场全局时间导数) 令 $f(t,x): \Omega \times I \to \mathbb{R}$，其中 $t \in I \subset \mathbb{R}$，$x := (x_1, \cdots, x_d) \in B \subset \mathbb{R}^d$，有

$$\frac{\mathrm{d}}{\mathrm{d}t} f(t,x) = \frac{\mathrm{d}}{\mathrm{d}t} f(t, \hat{\mathcal{A}}(t,\hat{x}))$$
$$= \partial_t f(t,x) + \nabla f(t,x) \cdot \partial_t \hat{\mathcal{A}}(t,\hat{x})$$
$$= \partial_t f(t,x) + \nabla f(t,x) \cdot \partial_t x(t,\hat{x})$$
$$= \partial_t f(t,x) + \nabla f(t,x) \cdot v(t,x)$$

11.6 连续介质力学简介

其描述了一个欧拉量 f 在时间 t 时,在点 x 处随流函数 $v(t,x)$ 的变化率。该式中第一项描述了局部变化率,第二项描述了粒子的运动情况。

命题 108 令 $\hat{u}:\hat{\Omega}\times I \to R^d$,$u:\Omega\times I \to \mathbb{R}^d$,我们可得 Lagrangian 坐标系内的位移时间导数:
$$d_t\hat{u} = \partial_t\hat{u}$$

在 Eulerian 坐标系中,总时间导数为
$$d_t u = \partial_t u + (v\cdot\nabla)u$$

其中速度向量满足 $v:\Omega\times I \to \mathbb{R}^d$
$$(v\cdot\nabla)u = \nabla u\cdot v$$

命题 109 给定 Eulerian 函数 $f(t,x):\Omega\to\mathbb{R}^d$ 以及相对应的 Lagrangian 函数 $\hat{f}(t,\hat{x}):\hat{\Omega}\to\mathbb{R}^d$。为计算 Eulerian 空间导数,我们有
$$\nabla f = \hat{\nabla}\hat{f}\hat{F}^{-1}$$

11.6.2 Nanson 公式和 Piola 变换

本节我们将介绍连续介质力学中最重要的变换之一。令 $\hat{T}$ 为 $\hat{\Omega}$ 上的一组双射,则 $\hat{F} = \hat{\nabla}\hat{T}$ 在 $\hat{\Omega}$ 上的每一点均为可逆的。$\sigma := (\sigma_{ij})_{ij=1}^3$ 为 Ω 域定义在点 $x = \hat{T}(t,\hat{x}) \in \Omega$ 处的张量,借助 Piola 变换,可获得一个在 $\hat{\Omega}$ 域中的对应张量。

定理 9(Piola 变换) Piola 变换定义为
$$\hat{\sigma}(\hat{x}) := \hat{J}(\hat{x})\sigma(x)\hat{F}^{-\mathrm{T}}(\hat{x}), \quad x = \hat{T}(t,\hat{x})$$

即
$$\hat{\sigma} := \hat{J}\sigma\hat{F}^{-\mathrm{T}}, \quad x = \hat{T}(t,\hat{x})$$

证明 首先我们确定两个坐标系中法向量之间的关系,对无穷小单元而言,我们有
$$\mathrm{d}s\cdot\mathrm{d}x = \hat{J}\mathrm{d}\hat{s}\cdot\mathrm{d}\hat{x}$$

这里我们定义 $\mathrm{d}s := \mathrm{d}s\cdot n$,$\mathrm{d}\hat{s} := \mathrm{d}\hat{s}\cdot\hat{h}$,随后参考 $\mathrm{d}x = \hat{F}\mathrm{d}\hat{x}$,可得
$$\left(\hat{F}^{\mathrm{T}}\mathrm{d}s - \hat{J}\mathrm{d}\hat{s}\right)\mathrm{d}\hat{x} = 0$$

即
$$\left(\hat{F}^{\mathrm{T}}\mathrm{d}s - \hat{J}\mathrm{d}\hat{s}\right) = 0$$
$$\hat{F}^{\mathrm{T}}\mathrm{d}s = \hat{J}\mathrm{d}\hat{s}$$

由此我们可得 Nanson 公式
$$\mathrm{d}s = \hat{J}\hat{F}^{-\mathrm{T}}\mathrm{d}s$$

接下来我们应用 Cauchy 定理 [232,77,111]：
$$\sigma n \mathrm{d}s = \hat{\sigma}\hat{n}\mathrm{d}s$$

由此可得
$$\sigma n \mathrm{d}s = \sigma\hat{n} \cdot \hat{F}^{-\mathrm{T}}\mathrm{d}\hat{s} = \hat{\sigma}\hat{n}\mathrm{d}\hat{s}$$

最终可得
$$\hat{J}\sigma\hat{F}^{-\mathrm{T}} = \hat{\sigma}$$

11.6.3 Reynolds 传输定理及动量守恒

命题 110(Reynolds 传输定理) 令 $\Omega := \Omega(t)$ 为一个时空域并给定 $(t,x) \mapsto x(t,\hat{x})$。此外，令函数 $(t,x) \mapsto \partial_t x(t,\hat{x})$ 与 $(t,x) \to f(t,x)$ 连续可微，则我们有

$$\frac{\mathrm{d}}{\mathrm{d}t}\int_\Omega f(t,x)\mathrm{d}x = \int_\Omega \left[\frac{\mathrm{d}}{\mathrm{d}t}f(t,x) + f(t,x)\nabla \cdot v(t,x)\right]\mathrm{d}x$$
$$= \int_\Omega \left[\partial_t f(t,x) + \nabla \cdot (f(t,x) \cdot v(t,x))\right]\mathrm{d}x$$

右边第二项表示向外法向通量，若域 Ω 不移动，则右边第二项可省略。

接下来我们介绍动量守恒，给定体积力 $f: \Omega \to \mathbb{R}^d$ 以及表面力 $\tau := \tau(x,t,n) : \partial\Omega \to \mathbb{R}^d$，根据牛顿第二定律可知：

$$\frac{\mathrm{d}}{\mathrm{d}t}(mv) = F$$

其中 F 为体力，在连续介质力学中，我们可将牛顿第二定律表述为

$$\frac{\mathrm{d}}{\mathrm{d}t}\int_\Omega \rho v \mathrm{d}x = \int_\Omega \rho f \mathrm{d}x + \int_{\partial\Omega} \tau \mathrm{d}s$$

基于 Reynolds 传输定理，可将左手项写为

$$\int_\Omega [\partial_t(\rho v) + \nabla \cdot (\rho v v)]\mathrm{d}x = \int_\Omega \rho f \mathrm{d}x + \int_{\partial\Omega} \tau \mathrm{d}s$$

基于柯西应力张量计算可知

$$\tau = \sigma \cdot n$$

11.6 连续介质力学简介

进一步应用散度定理

$$\int_{\partial\Omega} \tau \mathrm{d}s = \int_{\partial\Omega} \sigma \cdot n \mathrm{d}s = \int_{\Omega} \nabla \cdot \sigma \mathrm{d}x$$

其中

$$\nabla \cdot \sigma = \begin{pmatrix} \sum_{j=1}^{d} \partial_{x_j} \sigma_{1j} \\ \sum_{j=1}^{d} \partial_{x_j} \sigma_{2j} \\ \vdots \\ \sum_{j=1}^{d} \partial_{x_j} \sigma_{dj} \end{pmatrix}$$

随后我们可得

$$\int_{\Omega} \left(\partial_t (\rho v_j) + \nabla \cdot (\rho v_j v) - \rho f_j - (\nabla \cdot \sigma)_j \right) \mathrm{d}x = 0 \tag{11.41}$$

根据质量守恒 (连续性方程) 可得

$$\partial_t \rho = -\nabla \cdot (\rho v_j)$$

其中式 (11.41) 前两项可写为

$$\begin{aligned}
\partial_t (\rho v_j) + \nabla \cdot (\rho v_j v) &= \rho \partial_t v_j + v_j \partial_t \rho + \nabla \cdot (\rho v_j v) \\
&= \rho \partial_t v_j - v_j \nabla \cdot (\rho v_j) + \nabla \cdot (\rho v_j v) \\
&= \rho \partial_t v_j + \rho v \cdot \nabla v_j.
\end{aligned}$$

因此 $j = 1, \cdots, d$ 时

$$\int_{\Omega} \left(\rho \partial_t v_j + \rho v \cdot \nabla v_j - (\nabla \cdot \sigma)_j - \rho f_j \right) \mathrm{d}x = 0$$

命题 111 Eulerian 坐标系中介质的动量守恒方程如下:

$$\int_{\Omega} \left(\rho \partial_t v + \rho v \cdot \nabla v - \nabla \cdot \sigma - \rho f \right) \mathrm{d}x = 0$$

针对光滑函数 ρ, v, σ 以及任意 Ω，我们有以下微分形式：

$$\rho \partial_t v + \rho(v - \nabla)v - \nabla \cdot \sigma = \rho f$$

命题 112 Lagrangian 坐标系中介质的动量守恒方程如下：

$$\int_{\hat{\Omega}} \left(\hat{\rho} \partial_t^2 \hat{u} - \hat{\nabla} \cdot \hat{\Sigma} - \hat{\rho} \hat{f} \right) d\hat{x} = 0$$

针对光滑函数 $\hat{\rho}, \hat{u}, \hat{\sigma}$ 以及任意 $\hat{\Omega}$，我们有以下微分形式：

$$\hat{\rho} \partial_t^2 \hat{u} - \hat{\nabla} \cdot \hat{\Sigma} = \hat{\rho} \hat{f}$$

11.6.4 本构关系

我们首先归纳各个坐标系下应力张量表达式：

$$\sigma_f := \sigma_f(v_f, p_f) = -p_f I + 2\rho_f \nu_f \left(\nabla v_f + \nabla v_f^{\mathrm{T}} \right) \tag{11.42}$$

$$\hat{\sigma}_f := \hat{\sigma}_f(\hat{v}_f, \hat{p}_f) = -\hat{p}_f \hat{I} + 2\hat{\rho}_f \nu_f \left(\hat{\nabla} \hat{v}_f \hat{F}^{-1} + \hat{F}^{-\mathrm{T}} \hat{\nabla} \hat{v}_f^{\mathrm{T}} \right) \tag{11.43}$$

其中 v_f 为速度，p_f 为压力，I 为单位矩阵，ρ_f 为密度。由此我们可根据应力张量的 ALE 公式：

$$\nabla v_f \to \hat{\nabla} \hat{v}_f \hat{F}^{-1}$$

在固体材料中，我们基于 Saint Venant-Kirchhoff(StVK) 模型 [112] 有

$$\sigma_s := \sigma_s(u_s) = JF^{-1} \left(\lambda_s (\mathrm{tr} E) I + 2\mu_s E \right) F^{-\mathrm{T}} \tag{11.44}$$

$$\hat{\sigma}_s := \hat{\sigma}_s(\hat{u}_s) = \hat{J}^{-1} \hat{F} \left(\lambda_s (\mathrm{tr} \hat{E}) \hat{I} + 2\mu_s \hat{E} \right) \tag{11.45}$$

11.7 固体变形中的裂缝问题

11.7.1 动态裂缝的 Lagrangian 描述

接下来我们回顾 5.6 节提出的动态裂缝模型，并应用这一章的概念进行重构。由于位移场天然使用拉格朗日法进行描述，但相场系统仍需基于欧拉法进行描述，因此我们首先得到以下 CVIS 耦合系统。

命题 113 给定初始值 $\hat{u}(0) = \hat{u}^0$，$\partial_t \hat{u}(0) = \hat{v}^0$，$\hat{\varphi}(0) = \hat{\varphi}^0$，令 $(\hat{u}, \hat{\varphi}) \in \hat{V}_u^D \times \hat{K}$，则我们有

$$\left(\hat{\rho}_s \partial_t^2 \hat{u}, \hat{\psi}^u \right) + \left(g(\hat{\varphi}) \hat{\Sigma}(\hat{u}), \hat{\nabla} \hat{\psi}^u \right) = 0, \quad \forall \hat{\psi}^u \in \hat{V}_u^0$$

11.7　固体变形中的裂缝问题

$$(1-\kappa)\left(\hat{J}\hat{\varphi}(\hat{\Sigma}(\hat{u})):e(\hat{u}),\hat{\psi}\hat{\varphi}^u\hat{\varphi}\right)+G_c\left(-\frac{1}{\varepsilon}(\hat{J}(1-\hat{\varphi}),\hat{\psi}^{\hat{\varphi}}-\hat{\varphi})\right.$$
$$\left.+\varepsilon\left(\hat{J}\left(\hat{\nabla}\hat{\varphi}\hat{F}^{-1}\right)\hat{F}^{-T},\hat{\nabla}\left(\hat{\psi}^{\varphi}-\hat{\varphi}\right)\right)\right)\geqslant 0,\quad\forall\hat{\psi}\in\hat{K}$$

证明其坐标转换关系如下，首先

$$(\varphi,\psi)_\Omega \to (\hat{J}\hat{\varphi},\hat{\psi})_{\hat{\Omega}}$$

接下来根据 2.2.3 节、命题 109，我们可以完成以下坐标转换：

$$(\nabla\varphi,\nabla\psi)_\Omega \to \left(\hat{J}\hat{\nabla}\hat{\varphi}\hat{F}^{-1},\hat{\nabla}\hat{\psi}\hat{F}^{-1}\right)_{\hat{\Omega}} \Leftrightarrow \left(\hat{J}\left(\hat{\nabla}\hat{\varphi}\hat{F}^{-1}\right)\hat{F}^{-T},\hat{\nabla}\hat{\psi}\right)_{\hat{\Omega}}$$

接下来我们将二阶弹性动力学方程分解为一个一阶系统。

命题 114　给定初始值 $\hat{u}(0)=\hat{u}^0$，$\hat{v}(0)=\hat{v}^0$，$\hat{\varphi}(0)=\hat{\varphi}^0$，令 $(\hat{v},\hat{u},\hat{\varphi})\in L^2(\hat{B})\times\hat{V}_u^D\times\hat{K}$，则有

$$\left(\hat{\rho}_s\partial_t\hat{v},\hat{\psi}^v\right)+\left(g(\hat{\varphi})\hat{\Sigma}(\hat{u}),\hat{\nabla}\hat{\psi}^v\right)=0,\quad\forall\hat{\psi}^v\in\hat{V}_u^0$$
$$\hat{\rho}_s\left(\partial_t\hat{u}-\hat{v},\hat{\psi}^u\right)=0,\quad\forall\hat{\psi}^u\in L^2(\hat{B})$$
$$(1-\kappa)(\hat{J}\hat{\varphi}\hat{\Sigma}(\hat{u}):e(\hat{u}),\hat{\psi}^{\varphi}-\hat{\varphi})+G_c\left(-\frac{1}{\varepsilon}(\hat{J}(1-\hat{\varphi}),\hat{\psi}^{\varphi}-\hat{\varphi})\right.$$
$$\left.+\varepsilon\left(\hat{J}\left(\hat{\nabla}\hat{\varphi}\hat{F}^{-1}\right)\hat{F}^{-T},\hat{\nabla}\hat{\psi}^{\varphi}-\hat{\nabla}\hat{\varphi}\right)\right)\geqslant 0,\quad\forall\hat{\psi}^{\varphi}\in\hat{K}$$

11.7.2　动态压裂裂缝的 Lagrangian 描述

我们将命题 86 (11.4.1.2 节) 扩展至域 $\hat{B}$ 中。压力项从欧拉坐标系到拉格朗日坐标系的转换公式为

$$\left(\varphi^2 p,\nabla\cdot w\right)_\Omega \to \left(\hat{\varphi}^2\hat{p},\nabla\cdot\left(\hat{J}\hat{F}^{-1}\hat{w}\right)\right)_{\hat{\Omega}}$$

命题 115　给定 $\hat{p}_F\in L^\infty(\hat{B})$，其中初始值 $\hat{u}(0)=\hat{u}^0,\partial_t\hat{u}(0)=\hat{v}^0,\hat{\varphi}(0)=\hat{\varphi}^0$，令 $(\hat{u},\hat{\varphi})\in\hat{V}_u^D\times\hat{K}$，则有

$$\left(\hat{\rho}_s\partial_t^2\hat{u},\hat{\psi}^u\right)+\left(g(\hat{\varphi})\hat{\Sigma}(\hat{u}),\hat{\nabla}\hat{\psi}^u\right)+\left(\hat{\varphi}^2\hat{p}_F,\hat{\nabla}\cdot\left(\hat{J}\hat{F}^{-1}\hat{\psi}^v\right)\right)=0,\quad\forall\hat{\psi}^u\in\hat{V}_u^0$$
$$(1-\kappa)\left(\hat{J}\hat{\varphi}(\hat{\Sigma}(\hat{u})):e(\hat{u}),\hat{\psi}^{\hat{\varphi}}-\hat{\varphi}\right)+2\left(\hat{\varphi}\hat{p}_F\hat{\nabla}\cdot\left(\hat{J}\hat{F}^{-1}\hat{u}\right),\hat{\psi}^{\hat{\varphi}}-\hat{\varphi}\right)$$
$$+G_c\left(-\frac{1}{\varepsilon}\left(\hat{J}(1-\hat{\varphi}),\hat{\psi}^{\hat{\varphi}}-\hat{\varphi}\right)+\varepsilon\left(\hat{J}\left(\hat{\nabla}\hat{\varphi}\hat{F}^{-1}\right)\hat{F}^{-T},\hat{\nabla}\left(\hat{\psi}^{\hat{\varphi}}-\hat{\varphi}\right)\right)\right)$$

$\geqslant 0, \quad \forall \hat{\psi}^{\varphi} \in \hat{K}$

与前一节相似，我们继续将二阶方程分解为一阶。

命题 116 给定 $\hat{p}_F \in L^{\infty}(\hat{B})$ 以及初始值 $\hat{u}(0) = \hat{u}^0$, $\hat{v}(0) = \hat{v}^0$, $\hat{\varphi}(0) = \hat{\varphi}^0$, 令 $(\hat{v}, \hat{u}, \hat{\varphi}) \in L^2(\hat{B}) \times \hat{V}_u^D \times K$, 则:

$$\left(\hat{\rho}_s \partial_t^2 \hat{v}, \hat{\psi}^u\right) + \left(g(\hat{\varphi})\hat{\Sigma}(\hat{v}), \hat{\nabla}\hat{\psi}^v\right) + \left(\hat{\varphi}^2 \hat{p}_F, \hat{\nabla} \cdot \left(\hat{J}\hat{F}^{-1}\hat{\psi}^v\right)\right) = 0, \quad \forall \hat{\psi}^u \in \hat{V}_u^0$$

$$\hat{\rho}_s \left(\partial_t \hat{u} - \hat{v}, \hat{\psi}^u\right) = 0, \quad \forall \hat{\psi}^u \in L^2(\hat{B})$$

$$(1-\kappa)\left(\hat{J}\hat{\varphi}\hat{\Sigma}(\hat{u}) : e(\hat{u}), \hat{\psi}^{\varphi} - \hat{\varphi}\right) + 2\left(\hat{\varphi}\hat{p}_F\hat{\nabla} \cdot \left(\hat{J}\hat{F}^{-1}\hat{u}\right), \hat{\psi}^{\varphi} - \hat{\varphi}\right)$$

$$+ G_c \left(-\frac{1}{\varepsilon}\left(\hat{J}(1-\hat{\varphi}), \hat{\psi}^{\varphi} - \hat{\varphi}\right) + \varepsilon \left(\hat{J}\left(\hat{\nabla}\hat{\varphi}^{-1}\right)\hat{F}^{-T}, \hat{\nabla}\hat{\psi}^{\varphi} - \hat{\nabla}\hat{\varphi}\right)\right)$$

$\geqslant 0, \quad \forall \hat{\psi}^{\varphi} \in \hat{K}$

11.7.3 固定网格下的欧拉框架

固体研究中的欧拉方法是一种固定网格方法。根据相场法的定义可知，在相场裂缝模型中，我们不需要根据裂缝扩展情况调整网格。此时我们回归固定网格系统，这意味着我们需要对方程做出一定的改变，将前文的位移方程从拉格朗日域 $\hat{B}$ 转移到欧拉域 B 中。作者的研究团队已经对流固耦合欧拉方法进行了大量的研究[137,372,368,431,174]，此外还开发了一个针对流固耦合裂缝扩展的数值模型[179]。

基于以上概念，我们首先推导出固定网格法的大位移相场裂缝模型框架。

命题 117(欧拉坐标系下固体方程) 令 $(v_s, u_s) \in L_s \times \{u_s^D + V_s^0\}$, $v_s(0) = v_s^0, u_s(0) = u_s^0$, 则

$$(\rho_s J \partial_t v_s, \psi^v)_{\Omega_s} + (\rho_s J (v_s \cdot \nabla) v_s, \psi^v)_{\Omega_s}$$

$$+ (\sigma_s, \nabla \psi^v)_{\Omega_s} - (\rho_s J f_s, \psi^v)_{\Omega_s} = 0, \quad \forall \psi^v \in V_s^0$$

$$\rho_s (\partial_t u_s + (v_s \cdot \nabla) u_s - v_s, \psi^u)_{\Omega_s} = 0, \quad \forall \psi^u \in L_s$$

由于我们在这里使用固定网格法，固体颗粒并不固定在初始位置上，这也就意味着我们需要在指示函数的帮助下捕获变形边界。

接下来我们结合命题 19，可获得以下命题。

命题 118(欧拉坐标系下相场模型) 令 $(v_s, u_s, \varphi) \in L_s \times \{u_s^D + V_s^0\} \times K$, 则 $v_s(0) = v_s^0$, $u_s(0) = u_s^0$, $\varphi(0) = \varphi_0$, 我们有

$$(\rho_s J \partial_t v_s, \psi^v)_{\Omega_s} + (\rho_s J (v_s \cdot \nabla) v_s, \psi^v)_{\Omega_s}$$

$$+ (g(\varphi)\sigma_s, \nabla\psi^v)_{\Omega_s} - (\rho_s J f_s, \psi^v)_{\Omega_s} = 0, \quad \forall \psi^v \in V_s^0$$

$$\rho_s \left(\partial_t u_s + (v_s \cdot \nabla) u_s - v_s, \psi^u\right)_{\Omega_s} = 0, \quad \forall \psi^u \in L_s$$

$$((1-\kappa)\varphi\sigma(u) : e(u), \psi - \varphi)_{\Omega_s}$$

$$+ \left(-\frac{G_c}{\varepsilon}(1-\varphi), \psi - \varphi\right)_{\Omega_s} + (G_c \varepsilon \nabla\varphi, \nabla(\psi - \varphi))_{\Omega_s} \geqslant 0, \quad \forall \psi \in K$$

备注 70 固定网格下的相场模型有着独到的优势,它允许较大的位移变化,且完美符合相场法的核心思路。但该系统伴随着部分技术问题,如不对齐的边界条件等相关因素会对模型求解造成影响。

11.8 流固耦合相场裂缝问题

本节我们将介绍流固耦合相场裂缝问题,数值建模与模拟详情请参考文献 [435, 437]。我们首先回顾流固耦合模型面临的几大挑战 [216,271,270,185,371,370,371,244,247]:

(1) 不同类型 (椭圆型、抛物型、双曲型) 的偏微分方程的处理以及耦合;
(2) 不同坐标系之间的耦合问题;
(3) 各个方程的非线性项以及线性化处理;
(4) 界面耦合条件;
(5) 处理移动界面时的离散化处理。

11.8.1 Lagrangian-Eulerian 耦合框架

11.8.1.1 符号与函数空间

接下来我们介绍将要额外使用到的一些符号 (图 11.7)。首先我们定义 $\Omega := \Omega(t) \subset \mathbb{R}^d$ 为一个时空域。其中包括三个子域 $\Omega_f(t), \Omega_s(t), \mathcal{C}(t)$,且 $\mathcal{C}(t) \subset \Omega_s(t)$。$\Omega_f(t)$ 与 $\Omega_s(t)$ 之间的界面定义为 $\Gamma_i(t) = \overline{\partial \Omega_f(t)} \cap \overline{\partial \Omega_s(t)}$。整个模型的初始域定义为 $\hat{\Omega}, \hat{\Omega}_f, \hat{\Omega}_s$,相应地,初始裂缝为 $\hat{\Gamma}_i = \overline{\partial \hat{\Omega}_f} \cap \overline{\partial \hat{\Omega}_s}$。外部边界定义为 $\partial \hat{\Omega} = \hat{\Gamma} = \hat{\Gamma}_{\text{in}} \cup \hat{\Gamma}_D \cup \hat{\Gamma}_{\text{out}}$,其中 $\hat{\Gamma}_D, \hat{\Gamma}_{\text{in}}$ 为 Dirichlet 边界,$\hat{\Gamma}_{\text{out}}$ 为流体 Neumann 边界。

对于固定参考域的函数空间 $\hat{\Omega}, \hat{\Omega}_f, \hat{\Omega}_s, \hat{\mathcal{C}}$,我们在流体域中有以下设置:

$$\hat{L}_f := L^2\left(\hat{\Omega}_f\right)$$

$$\hat{L}_f^0 := L^2\left(\hat{\Omega}_f\right)/\mathbb{R}$$

$$\hat{V}_f^0 := \left\{\hat{v}_f \in H^1\left(\hat{\Omega}_f\right)^d : \hat{v}_f = 0, \quad 在 \hat{\Gamma}_{\text{in}} \cup \hat{\Gamma}_D 上\right\}$$

$$\hat{V}_{f,u}^0 := \left\{ \hat{u}_f \in H^1\left(\hat{\Omega}_f\right)^d : \hat{u}_f = \hat{u}_s, \text{ 在} \hat{\Gamma}_i \text{上}, \hat{u}_f = 0, \quad \text{在} \hat{\Gamma}_{\text{in}} \cup \hat{\Gamma}_D \cup \hat{\Gamma}_{\text{out}} \text{上} \right\}$$

$$\hat{V}_{f,\hat{u},\hat{\Gamma}_i}^0 := \left\{ \hat{\psi}_f \in H^1\left(\hat{\Omega}_f\right)^d : \hat{\psi}_f = 0, \quad \text{在} \hat{\Gamma}_i \cup \hat{\Gamma}_{\text{in}} \cup \hat{\Gamma}_D \cup \hat{\Gamma}_{\text{out}} \text{上} \right\}$$

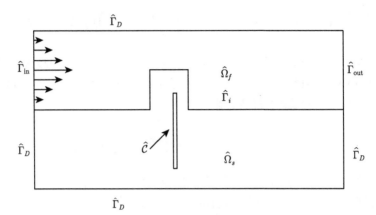

图 11.7 模型原型配置以及符号设置

在固体域中，我们有

$$\hat{L}_s := L^2\left(\hat{\Omega}_s\right)^d$$

$$\hat{V}_s^0 := \left\{ \hat{u}_s \in H^1\left(\hat{\Omega}_s\right)^d : \hat{u}_s = 0, \quad \text{在} \hat{\Gamma}_D \text{上} \right\}$$

$$\hat{W} := \left\{ \hat{\varphi} \in H^1\left(\hat{\Omega}_s \cup \hat{C}\right) : \hat{\varphi} \leqslant \hat{\varphi}^{n-1}, \quad \text{在} \hat{\Omega}_s \cup \hat{C} \text{上} \right\}$$

我们采用变分全耦合界面处理方法，为此我们定义速度空间域：

$$\hat{V}^0 := \left\{ \hat{v} \in H^1(\hat{\Omega})^d : \hat{v} = 0, \quad \text{在} \hat{\Gamma}_{\text{in}} \cup \hat{\Gamma}_D \text{上} \right\} \tag{11.46}$$

需要注意的是当前域 $\Omega, \Omega_f, \Omega_s, C$ 并无上标符号。

11.8.1.2 控制方程

针对固体域，我们采用命题 116 中相关方程。针对不可压缩流体，我们采用 Navier-Stokes 方程。

公式 26 给定 $(v_f, p_f) \in \{v_f^D + V_f^0\} \times L_f^0$，初始速度满足 $v_f(0) = v_f^0$，则

$$\rho_f \left(\partial_t v_f, \psi^v\right)_{\Omega_f} + \rho_f \left(v_f \cdot \nabla v_f, \psi^v\right)_{\Omega_f}$$

11.8 流固耦合相场裂缝问题

$$+ (\sigma_f, \nabla \psi^v)_{\Omega_f} - \langle \sigma_f n_f, \nabla \psi^v \rangle_{\Gamma_{f,N}, \Gamma_i} + \rho_f (f, \psi^v)_{\Omega_f} = 0, \quad \forall \psi^v \in V_f^0 \quad (11.47)$$

$$(\text{div} v_f, \psi^p)_{\Omega_f} = 0, \quad \forall \psi^p \in L_f^0$$

其中应力张量满足

$$\sigma_f = -p_f I + \rho_f \nu_f \left(\nabla v_f + \nabla v_f^{\mathrm{T}} \right) \quad (11.48)$$

且边界项满足:

$$g := -\rho_f \nu_f \nabla v_f^{\mathrm{T}}, \quad \text{在 } \Gamma_{f,N} = \Gamma_{\text{out}} \text{ 上}$$

11.8.1.3 Lagrange-Euler 耦合系统

本节我们将耦合 Lagrange(固体) 与 Euler(流体) 坐标系。这里我们采用 ALE (arbitrary Lagrange-Euler) 方法[131,230,239,132,166]。ALE 方法引入了一个附加的偏微分方程, 从而解决网格运动问题。相关研究请参考文献 [219,392,429,133]。在这里, 采用了二阶模型 (Poisson 型):

$$-\hat{\nabla} \cdot (\hat{\sigma}_{\text{mesh}}) = 0, \quad \hat{u}_f = \hat{u}_s, \quad \text{在} \hat{\Gamma}_i \text{上}, \quad \hat{u}_f = 0, \quad \text{在} \partial \hat{\Omega} \backslash \hat{\Gamma}_i \text{上} \quad (11.49)$$

其中 $\hat{\sigma}_{\text{mesh}} = \hat{\alpha}_u \hat{\nabla} \hat{u}_f$。

同时我们还需引入 ALE 时间导数:

$$\partial_t |_{\mathcal{A}} v_f(x,t) = w \cdot \nabla v_f + \partial_t v_f(x,t) \quad (11.50)$$

具体而言:

(1) 在 Lagrange 场中, 我们有 $w = v_f$, 即域 $\Omega(t)$ 按流体速度 v_f 移动, 且 ALE 时间导数对应于 Lagrangian 意义上的总时间导数 $\partial_t |_{\hat{\mathcal{A}}} v_f(x,t) = \dfrac{\mathrm{d}}{\mathrm{d}t} v_f(t, \hat{x})$。

(2) 在 Euler 场中, 我们有 $w = 0$, 即 $\Omega(t) =: \Omega$ 是固定的。这里我们采用经典 Euler 描述, 此时 ALE 时间导数对应于 Euler 意义上的局部时间导数 $\partial_t |_{\hat{\mathcal{A}}} v_f(x,t) = \partial_t v_f(x,t)$。

(3) 在 ALE 方法中, $0 \leqslant w \leqslant v_f$, 在界面周边, 我们有 $0 < w < v_f$, 在界面较远处, 则有 $w = 0$(Euler 场), 但全局网格并不发生移动。

11.8.1.4 界面耦合条件

在流固耦合相场裂缝模型中, 我们需要处理两种不同类型的界面:

(1) 流固界面 (FSI 界面);

(2) 固体域与裂缝域之间的界面。

在流固耦合过程中存在两个关键物理方程：速度连续性方程，法向应力平衡方程。在使用 ALE 方法时，还存在位移连续性方程，它实现了固体运动 $\hat{u}_s$ 和流体运动 $\hat{u}_f$ 耦合。

流体问题的速度连续性方程如下：

$$\hat{v}_f = \hat{v}_s, \quad 在 \hat{\Gamma}_i 上 \tag{11.51}$$

同时界面上存在法向应力平衡条件：

$$\hat{J}\hat{\sigma}_f \hat{F}^{-\mathrm{T}} \hat{n}_f + \hat{F}\hat{\Sigma}\hat{n}_s = 0, \quad 在 \hat{\Gamma}_i 上 \tag{11.52}$$

ALE 方法中存在位移连续性方程：

$$\hat{u}_f = \hat{u}_s, \quad 在 \hat{\Gamma}_i 上 \tag{11.53}$$

通过时间微分，在 $\hat{\Gamma}_i$ 上，$\partial_t \hat{u}_s = \hat{v}_s = \hat{w} = \hat{v}_f$。我们需要注意避免流体位移与固体位移的反向耦合，即 $\hat{u}_s \to \hat{u}_f$ 是一个单向条件。

根据 11.4 节，相场裂缝界面耦合条件如下：

$$\hat{F}\hat{\Sigma}\hat{n}_s = \hat{\sigma}_F \hat{n}_s, \quad 在 \partial \hat{\mathcal{C}} 上$$

其中 $\hat{\sigma}_F := -\hat{p}_F \hat{I}$，$\hat{p}_F$ 为裂缝压力，$\hat{p}_f$ 为流体压力，二者并不直接相等。

11.8.1.5 基于增广 Lagrange 补偿法的数值模型

命题 119(变分全耦合 ALE-FSI 相场裂缝模型) 给定 $\hat{p}_F \in H^1\left(\hat{\Omega}_s \cup \hat{\mathcal{C}}\right)$，并令 $(\hat{v}, \hat{u}_f, \hat{u}_s, \hat{p}_f, \hat{\varphi}) \in \left\{\hat{v}^D + \hat{V}^0\right\} \times \left\{\hat{u}_f^D + \hat{V}_{f,\hat{u}}^0\right\} \times \hat{V}_s^0 \times \hat{L}_f^0 \times \hat{V}_\varphi$，且初始状态满足 $\hat{v}(0) = \hat{v}^0, \hat{u}_f(0) = \hat{u}_f^0, \hat{u}_s(0) = \hat{u}_s^0, \hat{\varphi}(0) = \hat{\varphi}^0$，则在 $t \in I$ 时间内，我们有

$$\left(\hat{J}\hat{\rho}_f \partial_t \hat{v}, \hat{\psi}^v\right)_{\hat{\Omega}_f} + \left(\hat{\rho}_f \hat{J}\left(\hat{F}^{-1}(\hat{v} - \hat{w}) \cdot \hat{\nabla}\right)\hat{v}, \hat{\psi}^v\right)_{\hat{\Omega}_f} + \left(\hat{J}\hat{\sigma}_f \hat{F}^{-\mathrm{T}}, \hat{\nabla}\hat{\psi}^v\right)_{\hat{\Omega}_f}$$

$$+ \left\langle \rho_f \nu_f \hat{J}\left(\hat{F}^{-\mathrm{T}}\hat{\nabla}^{\mathrm{T}}\hat{n}_f\right)\hat{F}^{-\mathrm{T}}, \hat{\psi}^v \right\rangle_{\hat{\Gamma}_\mathrm{out}} + \left(\hat{\rho}_s \partial_t \hat{v}, \hat{\psi}^v\right)_{\hat{\Omega}_s} + \left(g(\hat{\varphi})\hat{F}\hat{\Sigma}, \hat{\nabla}\hat{\psi}^v\right)_{\hat{\Omega}_s}$$

$$+ \left(\hat{\varphi}_s^2 \hat{p}_F, \hat{\nabla} \cdot \left(\hat{J}\hat{F}^{-1}\hat{\psi}^v\right)\right) = 0, \quad \forall \hat{\psi}^v \in \hat{V}^0$$

$$\left(\hat{\sigma}_\mathrm{mesh}, \hat{\nabla}\hat{\psi}_f^u\right)_{\hat{\Omega}_f} = 0, \quad \forall \hat{\psi}_f^u \in \hat{V}_{f,\hat{u},\hat{\Gamma}_i}^0$$

$$\hat{\rho}_s \left(\partial_t \hat{u}_s - \hat{v}_s, \hat{\psi}_s^u\right)_{\hat{\Omega}_s} = 0, \quad \forall \hat{\psi}_s^u \in \hat{L}_s$$

$$\left(\widehat{\mathrm{div}}\left(\hat{J}\hat{F}^{-1}\hat{v}\right), \hat{\psi}_f^p\right)_{\hat{\Omega}_f} = 0, \quad \forall \hat{\psi}_f^p \in \hat{L}_f^0$$

$$(1-\kappa)\left(\hat{J}\hat{\varphi}\hat{\Sigma}:\hat{E},\hat{\psi}_s^\varphi\right)_{\hat{\Omega}_s} + 2\left(\hat{\varphi}\hat{p}_F\hat{\nabla}\cdot\left(\hat{J}\hat{F}^{-1}\hat{u}_s\right),\hat{\psi}_s^\varphi\right)_{\hat{\Omega}_s}$$

$$-\left\langle 2\hat{j}\hat{\varphi}\hat{p}_F\hat{F}^{-T}\hat{n}_s\hat{u}_s,\hat{\psi}_s^\varphi\right\rangle_{\hat{\Gamma}_i}$$

$$+G_c\left(-\frac{1}{\varepsilon}\left(\hat{J}(1-\hat{\varphi}),\hat{\psi}_s^\varphi\right)+\varepsilon\left(\hat{J}\left(\hat{\nabla}\hat{\varphi}\hat{F}^{-1}\right)\hat{F}^{-T},\hat{\nabla}\hat{\psi}_s^\varphi\right)\right)_{\hat{\Omega}_s}$$

$$+\left(\hat{J}\left[\hat{\Xi}+\gamma\left(\hat{\varphi}-\hat{\varphi}^{n-1}\right)\right]^+,\hat{\psi}_s^\varphi\right)_{\hat{\Omega}_a}=0,\quad \forall \hat{\psi}_s^\varphi\in\hat{V}_\varphi$$

备注 71 如 3.3 节所述，Neumann 边界条件是以变分全耦合的方式实现的，即

$$\left\langle\hat{J}\hat{\sigma}_f\hat{F}^{-T}\hat{n}_f,\hat{\psi}^v\right\rangle_{\hat{\Gamma}_i}+\left\langle g(\hat{\varphi})\left[\hat{F}\hat{\Sigma}-\hat{J}\hat{p}_F\hat{F}^{-T}\right]\hat{n}_s,\hat{\psi}^v\right\rangle_{\hat{\Gamma}_i}=0,\quad \forall\hat{\psi}^v\in\hat{V}^0 \quad (11.54)$$

11.8.1.6 时间离散化

首先我们介绍单次 θ 分解法。

公式 27 给定 $\hat{U}^{n-1}$，令 $\hat{U}^n=\{\hat{v}_f^n,\hat{v}_s^n,\hat{u}_f^n,\hat{u}_s^n,\hat{p}_f^n,\hat{\varphi}_s^n\}\in\hat{X}_D^0$，其中 $\hat{X}_D^0:=\left\{\hat{v}_f^D+\hat{V}_{f,\hat{v}}^0\right\}\times\hat{L}_s\times\left\{\hat{u}_f^D+\hat{V}_{f,\hat{u}}^0\right\}\times\left\{\hat{u}_s^D+\hat{V}_s^0\right\}\times\hat{L}_f^0\times H^1\left(\hat{\Omega}_s\cup\hat{C}\right)$，$\hat{X}=\hat{V}_{f,\hat{0}}^0\times\hat{L}_s\times\hat{V}_{f,\hat{u},\hat{\Gamma}_i}^0\times\hat{V}_s^0\times\hat{L}_f^0\times H^1\left(\hat{\Omega}_s\cup\hat{C}\right)$，针对 $n=1,2,\cdots,N$，有

$$\hat{A}\left(\hat{U}^n\right)(\hat{\Psi})=0,\quad \forall\hat{\Psi}\in\hat{X} \quad (11.55)$$

其中半线性形式 $\hat{A}(\cdot)(\cdot)$ 满足：

$$\hat{A}\left(\hat{U}^n\right)(\hat{\Psi}):=\hat{A}_T\left(\hat{U}^{n,k}\right)(\hat{\Psi})+\hat{A}_I\left(\hat{U}^n\right)(\hat{\Psi})+\hat{A}_E\left(\hat{U}^n\right)(\hat{\Psi})+\hat{A}_P\left(\hat{U}^n\right)(\hat{\Psi})$$

定义 85(半线性形式分类) 半线性形式可以分为四类，分别为时间项 (包含时间导数)、隐式项 (如相场方程等)、压力项及基础项。分解如下：

$$\hat{A}_T(\hat{U})(\hat{\Psi})=\left(\hat{J}\hat{\rho}_f\partial_t\hat{v}_f,\hat{\psi}_f^v\right)_{\hat{\Omega}_f}-\left(\hat{\rho}_f\hat{J}\left(\hat{F}^{-1}\hat{w}\cdot\hat{\nabla}\hat{v}_f\right),\hat{\psi}_f^v\right)_{\hat{\Omega}_f}$$

$$+\left(\hat{\rho}_s\partial_t\hat{v}_s,\hat{\psi}_s^v\right)_{\hat{\Omega}_s}+\left(\hat{\rho}_s\partial_t\hat{u}_s,\hat{\psi}_s^u\right)_{\hat{\Omega}_s}$$

$$\hat{A}_I(\hat{U})(\hat{\Psi})=\left(\hat{\sigma}_{\text{mesh}},\hat{\nabla}\hat{\psi}_f^u\right)_{\hat{\Omega}_f}+\left(\widehat{\text{div}}\left(\hat{J}\hat{F}^{-1}\hat{v}_f\right),\hat{\psi}_f^p\right)_{\hat{\Omega}_f}$$

$$+(1-\kappa)\left(\hat{J}\hat{\varphi}_s\hat{\Sigma}:\hat{E},\hat{\psi}^\varphi\right)_{\hat{\Omega}_s}+2\left(\hat{J}\hat{\varphi}_s\hat{p}_F\hat{\nabla}\cdot\left(\hat{J}\hat{F}^{-1}\hat{u}_s\right),\hat{\psi}^\varphi\right)_{\hat{\Omega}_s}$$

$$+ G_c \left(-\frac{1}{\varepsilon} \left(\hat{J}(1-\hat{\varphi}_s), \hat{\psi}^\varphi \right) + \varepsilon \left(\hat{J} \left(\hat{\nabla}\hat{\varphi}_s \hat{F}^{-1} \right) \hat{F}^{-T}, \hat{\nabla}\hat{\psi}^\varphi \right) \right)_{\hat{\Omega}_s}$$

$$+ \left(\hat{J} \left[\hat{\Xi} + \gamma \left(\hat{\varphi}_s - \hat{\varphi}_s^{n-1} \right) \right]^+, \hat{\psi}^\varphi \right)_{\hat{\Omega}_s}$$

$$\hat{A}_P(\hat{U})(\hat{\Psi}) = \left(\hat{J}\hat{\sigma}_{f,p}\hat{F}^{-T}, \hat{\nabla}\hat{\psi}^v_f \right)_{\hat{\Omega}_f}$$

$$\hat{A}_E(\hat{U})(\hat{\Psi}) = \left(\hat{\rho}_f \hat{J} \left(\hat{F}^{-1}\hat{v}_f \cdot \hat{\nabla}\hat{v}_f \right), \hat{\psi}^v_f \right)_{\hat{\Omega}_f} + \left(\hat{J}\hat{\sigma}_{f,vu}\hat{F}^{-T}, \hat{\nabla}\hat{\psi}^v_f \right)_{\hat{\Omega}_f}$$

$$+ \left\langle \rho_f \nu \hat{J} \left(\hat{F}^{-T}\hat{\nabla}\hat{v}_f^T \right) \hat{F}^{-T}\hat{n}, \hat{\psi}^v_f \right\rangle_{\hat{\Gamma}_{\text{out}}}$$

$$+ \left(g(\hat{\varphi}_s) \hat{F}\hat{\Sigma}, \hat{\nabla}^v \hat{\psi}^v_s \right)_{\hat{\Omega}_s}$$

$$+ \left(\hat{\phi}_s^2 \hat{p}_F, \hat{\nabla} \cdot \left(\hat{J}\hat{F}^{-1}\hat{\psi}^v \right) \right)_{\hat{\Omega}_s} - \left(\hat{\rho}_s \hat{v}_s, \hat{\psi}^u_s \right)_{\hat{\Omega}_s}$$

其中流体压力项 $\hat{\sigma}_f$ 还可以进一步分解为 $\hat{\sigma}_{f,vu}, \hat{\sigma}_{f,p}$:

$$\hat{\sigma}_{f,p} = -\hat{p}_f \hat{I}, \quad \hat{\sigma}_{f,vu} = \rho_f v_f \left(\hat{\nabla}\hat{v}_f \hat{F}^{-1} + \hat{F}^{-T}\hat{\nabla}\hat{v}_f^T \right)$$

半线性形式的 (非线性) 时间项是由后向差分近似的, 即 $n = 1, 2, \cdots, N$ 时, 我们有

$$\hat{A}_T \left(\hat{U}^{n,k} \right) (\tilde{\Psi}) := \frac{1}{k} \left(\hat{\rho}_f \hat{J}^{n,\theta} \left(\hat{v}_f - \hat{v}_f^{n-1} \right), \hat{\psi}^v \right)_{\hat{\Omega}_f}$$

$$- \frac{1}{k} \left(\hat{\rho}_f \left(\hat{J}\hat{F}^{-1} \left(\hat{u}_f - \hat{u}_f^{n-1} \right) \cdot \hat{\nabla} \right) \hat{v}_f, \hat{\psi}^v \right)_{\hat{\Omega}_f}$$

$$+ \frac{1}{k} \left(\hat{\rho}_s \left(\hat{v}_s - \hat{v}_s^{n-1} \right), \hat{\psi}^v \right)_{\hat{\Omega}_s} + \left(\hat{u}_s - \hat{u}_s^{n-1}, \hat{\psi}^u \right)_{\hat{\Omega}_s} \quad (11.56)$$

这里我们令 $\theta \in [0,1]$, 参考 5.3.3 节, 我们有

$$\hat{J}^{n,\theta} = \theta \hat{J}^n + (1-\theta)\hat{J}^{n-1}$$

其中 $\hat{u}_i^n := \hat{u}_i(t_n), \ \hat{v}_i^n := \hat{v}_i(t_n), \ \hat{J} := \hat{J}^n := \hat{J}(t_n)$。

公式 28(单次 θ 分解法) 令前一时间步解为 $\hat{U}^{n-1} = \{\hat{v}_f^{n-1}, \hat{v}_s^{n-1}, \hat{u}_f^{n-1}, \hat{u}_s^{n-1}, \hat{p}_f^{n-1}, \hat{\varphi}_s^{n-1}\}$, 并给定时间步长 $k := k_n = t_n - t_{n-1}$, 则我们有

$$\hat{A}_T \left(\hat{U}^{n,k} \right)(\hat{\Psi}) + \theta\hat{A}_E \left(\hat{U}^n \right)(\hat{\Psi}) + \hat{A}_P \left(\hat{U}^n \right)(\hat{\Psi}) + \hat{A}_I \left(\hat{U}^n \right)(\hat{\Psi})$$

$$= -(1-\theta)\hat{A}_E \left(\hat{U}^{n-1} \right)(\hat{\Psi}) + \theta\hat{F}^n(\hat{\Psi}) + (1-\theta)\hat{F}^{n-1}(\hat{\Psi}) \quad (11.57)$$

公式 29(分步 θ 分解法) 令 $\theta = 1 - \frac{\sqrt{2}}{2}$, $\theta' = 1 - 2\theta$, $\alpha = \frac{1 - 2\theta}{1 - \theta}$, $\beta = 1 - \alpha$。该时间步可分解为三个连续的子时间步。令 $\hat{U}^{n-1} = \{\hat{v}_f^{n-1}, \hat{v}_s^{n-1}, \hat{u}_f^{n-1}, \hat{u}_s^{n-1}, \hat{p}_f^{n-1}, \hat{p}_s^{n-1}\}$, 时间步长为 $k_n = t_n - t_{n-1}$, 则我们有

$$\hat{A}_T\left(\hat{U}^{n-1+\theta,k}\right)(\hat{\Psi}) + \alpha\theta \hat{A}_E\left(\hat{U}^{n-1+\theta}\right)(\hat{\Psi})$$
$$+ \theta \hat{A}_P\left(\hat{U}^{n-1+\theta}\right)(\hat{\Psi}) + \hat{A}_I\left(\hat{U}^{n-1+\theta}\right)(\hat{\Psi}) = -\beta\theta \hat{A}_E\left(\hat{U}^{n-1}\right)(\hat{\Psi}) + \theta \hat{F}^{n-1}(\hat{\Psi})$$
$$\hat{A}_T\left(\hat{U}^{n-\theta,k}\right)(\hat{\Psi}) + \alpha\theta' \hat{A}_E\left(\hat{U}^{n-\theta}\right)(\hat{\Psi})$$
$$+ \theta' \hat{A}_P\left(\hat{U}^{n-\theta}\right)(\hat{\Psi}) + \hat{A}_I\left(\hat{U}^{n-\theta}\right)(\hat{\Psi}) = -\alpha\theta' \hat{A}_E\left(\hat{U}^{n-1+\theta}\right)(\hat{\Psi}) + \theta' \hat{F}^{n-\theta}(\hat{\Psi})$$
$$\hat{A}_T\left(\hat{U}^{n,k}\right)(\hat{\Psi}) + \alpha\theta \hat{A}_E\left(\hat{U}^n\right)(\hat{\Psi})$$
$$+ \theta \hat{A}_P\left(\hat{U}^n\right)(\hat{\Psi}) + \hat{A}_I\left(\hat{U}^n\right)(\hat{\Psi}) = -\beta\theta \hat{A}_E\left(\hat{U}^{n-\theta}\right)(\hat{\Psi}) + \theta \hat{F}^{n-\theta}(\hat{\Psi})$$

11.8.1.7 空间离散化

我们在 5.4 节正式介绍了空间离散化的处理方法,在这里,我们首先构造有限维子空间 $\hat{X}_h^0 \subset \hat{X}^0$ 从而寻找连续问题的近似解。在本模型中,我们使用规则网格,网格离散化参数为 $\hat{h}$。

其中速度和压力变量的空间定义如下:

$$\hat{V}_{f,s,h} := \left\{\hat{v}_h \in [C(\hat{\Omega}_h)]^d, \ \hat{v}_h|_{\hat{K}} \in [Q_2^c(\hat{K})]^d \ \forall \hat{K} \in \mathcal{T}_h, \ \hat{v}_h\big|_{\hat{\Gamma}\backslash\hat{\Gamma}_i} = 0\right\}$$

$$\hat{P}_{f,h} := \left\{\hat{p}_h \in \left[\hat{L}^2\left(\hat{\Omega}_h\right)\right], \ \hat{p}_h|_{\hat{K}} \in \left[P_1^{dc}(\hat{K})\right] \ \forall \hat{K} \in \mathcal{T}_h\right\}$$

而固体项与相场变量空间定义如下:

$$\hat{U}_{f,s,h} := \left\{\hat{u}_h \in \left[C\left(\hat{\Omega}_h\right)\right]^d, \ \hat{u}_h|_{\hat{K}} \in \left[Q_2^c(\hat{K})\right]^d \ \forall \hat{K} \in \mathcal{T}_h, \ \hat{u}_h|_{\hat{\Gamma}\backslash\hat{\Gamma}_i} = 0\right\}$$

$$\hat{\Phi}_{s,h} := \left\{\hat{\varphi}_h \in \left[C\left(\hat{\Omega}_h\right)\right]^d, \ \varphi_h|_{\hat{K}} \in \left[Q_1^c(\hat{K})\right]^d \ \forall \hat{K} \in \mathcal{T}_h, \ \varphi_h|_{\hat{\Gamma}\backslash\hat{\Gamma}_i} = 0\right\}$$

公式 30 给定时间离散解 $\hat{U}_h^{n-1} = \{\hat{v}_{f,h}^{n-1}, \hat{v}_{s,h}^{n-1}, \hat{u}_{f,h}^{n-1}, \hat{u}_{s,h}^{n-1}, \hat{p}_{f,h}^{n-1}, \hat{\varphi}_{s,h}^{n-1}\}$, 采用单次 θ 分解法处理。则我们针对 $\hat{U}_h^n = \{\hat{v}_{f,h}^n, \hat{v}_{s,h}^n, \hat{u}_{f,h}^n, \hat{u}_{s,h}^n, \hat{p}_{f,h}^n, \hat{\varphi}_{s,h}^n\} \in \hat{X}_h^0$ 有

$$\hat{A}_T\left(\hat{U}_h^{n,k}\right)(\hat{\Psi}_h) + \theta \hat{A}_E\left(\hat{U}_h^n\right)(\hat{\Psi}_h) + \hat{A}_P\left(\hat{U}_h^n\right)(\hat{\Psi}_h) + \hat{A}_I\left(\hat{U}_h^n\right)(\hat{\Psi}_h)$$
$$= -(1-\theta)\hat{A}_E\left(\hat{U}_h^{n-1}\right)(\hat{\Psi}_h), \quad \forall \hat{\Psi} \in \hat{X}_h$$

11.8.1.8 非线性求解

通过公式 30,我们获得了一个完全离散且非线性的模型。接下来我们进行非线性求解。在时间步 t_n 内,给定 $\delta \hat{U}^j \in \hat{X}_h^D$,均有如下 Jacobian 矩阵:

$$\hat{A}'\left(\hat{U}_h^j\right)\left(\delta \hat{U}_h^j, \hat{\Psi}_h\right) = -\hat{A}\left(\hat{U}_h^j\right)\left(\hat{\Psi}_h\right), \quad \forall \hat{\Psi}_h \in \hat{X}_h^0$$

$$\hat{U}_h^{j+1} = \hat{U}_h^j + \lambda_j \delta \hat{U}_h^j$$

借助单次 θ 分解法,我们可计算右手项残差:

$$\hat{A}\left(\hat{U}_h^j\right)\left(\hat{\Psi}_h\right) := \hat{A}_T\left(\hat{U}_h^{n,k}\right)\left(\hat{\Psi}_h\right) + \theta \hat{A}_E\left(\hat{U}_h^n\right)\left(\hat{\Psi}_h\right)$$

$$+\hat{A}_P\left(\hat{U}_h^n\right)\left(\hat{\Psi}_h\right) + \hat{A}_I\left(\hat{U}_h^n\right)\left(\hat{\Psi}_h\right)$$

$$+(1-\theta)\hat{A}_E\left(\hat{U}_h^{n-1}\right)\left(\hat{\Psi}_h\right)$$

对于 Jacobian 矩阵,我们需要计算方向导数:

$$\hat{A}'\left(\hat{U}_h^j\right)\left(\delta \hat{U}_h^j, \hat{\Psi}_h\right) := \hat{A}'_T\left(\hat{U}_h^{n,k}\right)\left(\delta \hat{U}_h^j, \hat{\Psi}_h\right) + \theta \hat{A}'_E\left(\hat{U}_h^n\right)\left(\delta \hat{U}_h^j, \hat{\Psi}_h\right)$$

$$+\hat{A}'_P\left(\hat{U}_h^n\right)\left(\delta \hat{U}_h^j, \hat{\Psi}_h\right) + \hat{A}'_I\left(\hat{U}_h^n\right)\left(\delta \hat{U}_h^j, \hat{\Psi}_h\right)$$

在每一个双线性形式 $\hat{A}'\left(\hat{U}_h^j\right)\left(\delta \hat{U}_h^j, \hat{\Psi}_h\right)$ 中,其方向导数均需单独计算,相关模型细节请参考文献 [428,432,203,134] 以及相关网站:https://media.archnumsoft.org/10305/; http://www.dopelib.net。

11.8.2 全 Eulerian 固定网格模型

接下来我们将介绍全 Eulerian 固定网格相场裂缝模型。首先,我们需要定义指示函数来区分固体子域与流体子域。在全 Eulerian 固定网格模型中,裂缝不与网格面对齐,因此需要借助水平集函数对其进行捕获处理。

命题 120 我们将区分域 Ω_f, Ω_s 的指示函数定义为

$$\chi_f := \begin{cases} 1, & x-u \in \hat{\Omega}_f \\ 0, & x-u \in \hat{\Omega}_s \cup \hat{\Gamma}_i \end{cases}$$

$$\chi_s := 1 - \chi_f$$

初始时刻的指示函数取值由系统给定,后续时刻 $\mathcal{A}_s(x,t)$ 的取值仅取决于固体位移。

11.8 流固耦合相场裂缝问题

备注 72 在 11.2 节中，指示函数取决于 φ，在这里还需考虑到位移解 u 的取值。

定义 86 零水平集界面定义为
$$\Gamma_i = \left\{ x \in \Omega \mid x - u \in \hat{\Gamma}_i \right\}$$

其中 $u := X_f u_f + X_s u_s$，参考图 11.9。

前文的方法均假定 u 是给定的，因此我们需要证明可以通过计算获取 u 的值。在给定 $t \in I$，$x \in \Omega$ 时，$u(x,t)$ 的值决定了该点在初始时间 $t = 0$ 的位置。因此，在 Euler 模型中，我们需要构造一个传输方程。

公式 31 令 $u \in u_0^D + V^0$，当 $t \in I$ 时：
$$(\partial_t u - w + (w \cdot \nabla)u, \psi) = 0, \quad \forall \psi \in V^0$$

其中 u_0^D 为边界上的 Dirichlet 数据。初始条件与边界条件给定如下：
$$u(x,0) = 0, \quad x \in \Omega$$
$$u(x,t) = 0, \quad x \in \partial\Omega, \quad t \in I$$

则此时 u 的值将由对流速度决定。

针对固体，采用基于命题 118 的动态相场裂缝模型。针对流体，采用公式 26 中的不可压缩流体的 Navier-Stokes 方程。前文我们处理了两种类型的界面，但现在，我们则需要在 Euler 坐标系中对其处理。

(1) 流固界面 (FSI 界面)；
(2) 固体域与裂缝域之间的界面。

在 Eulerian 坐标系中，这两种界面均须通过指示函数进行捕获，FSI 界面耦合条件如下：
$$v_f = v_s$$
$$\sigma_f n = \sigma_s n$$

其中 σ_f 与 σ_s 分别参考式 (11.48) 和 (11.13)。在裂缝界面 ∂C 处，则有
$$\sigma_s n = \sigma_F n$$

其中 $\sigma_F = -p_F I$。

在模拟过程中，界面位置需要通过高精度离散化处理来进行确定。

在整个模型中，我们共有以下几组方程：
(1) 流体 Navier-Stokes 方程；
(2) 固体位移方程；

(3) 相场参数变分不等式。

最终我们可获得以下 CVIS 耦合系统。

命题 121 基于命题 125，建立指示函数 $\chi_f(u_s), \chi_s(u_s)$。裂缝压力 $p_F \in L^\infty$，给定 $(v_f, v_s, u_f, u_s, p_f, \varphi) \in \{v_f^D + V_f^0\} \times L_s \times \{u_f^D + V_s^0\} \times \{u_s^D + V_s^0\} \times L_f^0 \times K$，且 $v_f(0) = v_f^0$，$u_f(0) = u_f^0$，$\tilde{v}_s(0) = v_s^0$，$u_s(0) = u_s^0$，$\varphi(0) = \varphi^0$，则：

$$(\chi_f \rho_f \partial_t v_f, \psi_f^v) + (\chi_f \rho_f (v_f \cdot \nabla) v_f, \psi_f^v)$$

$$+ (\chi_f \sigma_f, \nabla \psi_f^v) - \langle g, \psi_f^v \rangle - (\chi_f \rho_f f_f, \psi_f^v) = 0, \quad \forall \psi_f^v \in V_f^0$$

$$(\chi_s J \rho_s \partial_t v_s, \psi_s^v) + (\chi_s J \rho_s (v_s \cdot \nabla) v_s, \psi_s^v)$$

$$+ (\chi_s g(\varphi) \sigma_s, \nabla \psi_s^v) - (\chi_s J \rho_s f_s, \psi_s^v)$$

$$+ (\chi_s g(\varphi) p_F, \nabla \cdot \psi_s^v) = 0, \quad \forall \psi_s^v \in V_s^0$$

$$(\chi_f \nabla u_f, \nabla \psi^u) = 0, \quad \forall \psi_f^u \in V_f^0$$

$$(\chi_s (\rho_s \partial_t u_s + (v_s \cdot \nabla) u_s - v_s, \psi_s^u)) = 0, \quad \forall \psi^u \in L_s$$

$$(\chi_f \nabla \cdot v_f, \psi_f^p) = 0, \quad \forall \psi_f^p \in L_f$$

$$(\chi_s(1-\kappa)\varphi\sigma(u) : e(u), \psi - \varphi)$$

$$+ 2(1-\kappa)(\chi_s \varphi (p_F \nabla \cdot u), \psi - \varphi) - 2\langle \chi_s \varphi p_F n \cdot u, \psi - \varphi \rangle_{\Gamma_i}$$

$$+ \left(-\chi_s \frac{G_c}{\varepsilon}(1-\varphi), \psi - \varphi\right) + (\chi_s G_c \varepsilon \nabla \varphi, \nabla(\psi - \varphi)) \geqslant 0, \quad \forall \psi \in K$$

备注 73 这种全欧拉流固耦合相场裂缝模型框架是最近的研究成果，首次出现在本专著中。其离散化方案以及算法实现等内容请参考文献 [179, 174, 177, 178, 180] 以及本书其他章节。

11.8.3 Navier-Stokes 流固耦合相场裂缝模型

在这一节中，我们利用全 Eulerian 固定网格建立流固耦合相场裂缝数值模型。其中裂缝内部压力 p_F 不再给定，而要通过不可压缩流体的 Navier-Stokes 方程求解获得。同时此时的压强不再恒定，会随注入液体的变化而变化。为此我们提出以下两种模型框架。

(1) 弹性介质流固耦合相场裂缝模型。

11.8 流固耦合相场裂缝问题

基于命题 121，我们首先区分裂缝域与非裂缝域，在 FSI 裂缝边界 Γ_i 处，我们有
$$v_F = v_s$$
$$\sigma_F n = \sigma_s n$$

根据式 (11.48) $\sigma_F := \sigma_f$，指示函数为 $\chi_F(\varphi), \chi_s(\varphi)$。

(2) 多孔介质流固耦合相场裂缝模型。

与前一个框架不同，该模型中存在 3 个子域：Ω_s(固体域)，Ω_f(多孔介质流体域) 以及 Ω_F(裂缝域)。在这两种流体域中，根据材料参数以及本构规律，我们需要应用不同方程求解，耦合情况如下：

$$v_f = v_s, \quad 在 \Gamma_i 上$$
$$\sigma_f n = \sigma_s n, \quad 在 \Gamma_i 上$$
$$v_F = v_s, \quad 在 \partial\mathcal{C} 上$$
$$\sigma_F n = \sigma_s n, \quad 在 \partial\mathcal{C} 上$$

命题 122(弹性介质流固耦合相场裂缝模型) 给定 $(v_F, v_s, u_F, u_s, p_F, \varphi) \in \{v_F^D + V_F^0\} \times L_s \times \{u_F^D + V_s^0\} \times \{u_s^D + V_s^0\} \times L_F^0 \times \bar{K}$，且 $v_F(0) = v_F^0, u_F(0) = u_F^0, v_s(0) = v_s^0, u_s(0) = u_s^0, \varphi(0) = \varphi^0$，令 $p_F := p_f$。则在时间域内，有

$$(\chi_F \rho_F \partial_t v_F, \psi_F^v) + (\chi_F \rho_F (v_F \cdot \nabla) v_F, \psi_F^v) + (\chi_F \sigma_F, \nabla \psi_F^v)$$
$$- \langle g, \psi_F^v \rangle - (\chi_F \rho_F f_F, \psi_F^v) = 0, \quad \forall \psi_F^v \in V_F^0$$

$$(J\rho_s \partial_t v_s, \psi_s^v) + (J\rho_s (v_s \cdot \nabla) v_s, \psi_s^v)$$
$$+ (g(\varphi)\sigma_s, \nabla \psi_s^v) - (J\rho_s f_s, \psi_s^v)$$
$$+ (g(\varphi) p_F, \nabla \cdot \psi_s^v) + (g(\varphi) \nabla p_F, \psi_s^v) = 0, \quad \forall \psi_s^v \in V_s^0$$

$$(\chi_F \nabla u_F, \nabla \psi^u) = 0, \quad \forall \psi_F^u \in V_F^0$$

$$(\rho_s \partial_t u_s + (v_s, \nabla) u_s - v_s, \psi_s^u) = 0, \quad \forall \psi^u \in L_s$$

$$(\chi_F \nabla \cdot v_F, \psi_F^p) = 0, \quad \forall \psi_F^p \in L_F$$

$$((1-\kappa)\varphi \sigma(u) : e(u), \psi - \varphi)$$
$$+ 2(1-\kappa)(\varphi(p_F \nabla \cdot u + \nabla p_F \cdot u), \psi - \varphi)$$
$$+ \left(-\frac{G_c}{\varepsilon}(1-\varphi), \psi - \varphi\right) + (G_c \varepsilon \nabla \varphi, \nabla(\psi - \varphi)) \geqslant 0, \quad \forall \psi \in K$$

命题 123(多孔介质流固耦合相场裂缝模型) 给定 $(v_f, v_F, v_s, u_f, u_s, p_f, p_F, \varphi) \in \{v_f^D + V_f^0\} \times \{v_F^D + V_F^0\} \times L_s \times \{u_f^D + V_s^0\} \times \{u_s^D + V_s^0\} \times L_f^0 \times L_F^0 \times K$, 且 $v_f(0) = v_f^0$, $v_F(0) = v_F^0$, $u_f(0) = u_f^0$, $v_s(0) = v_s^0$, $u_s(0) = u_s^0$, $\varphi(0) = \varphi^0$, 则在时间域内, 有

$$(\chi_f \rho_f \partial_t v_f, \psi_f^v) + (\chi_f \rho_f (v_f \cdot \nabla) v_f, \psi_f^v)$$

$$+ (\chi_f \sigma_f, \nabla \psi_f^v) - \langle g, \psi_f^v \rangle - (\chi_f \rho_f f_f, \psi_f^v) = 0, \quad \forall \psi_f^v \in V_f^0$$

$$(\chi_F \rho_F \partial_t v_F, \psi_F^v) + (\chi_F \rho_F (v_F \cdot \nabla) v_F, \psi_F^v)$$

$$+ (\chi_F \sigma_F, \nabla \psi_F^v) - \langle g, \psi_F^v \rangle - (\chi_F \rho_F f_F, \psi_F^v) = 0, \quad \forall \psi_F^v \in V_F^0$$

$$(\chi_s J \rho_s \partial_t v_s, \psi_s^v) + (\chi_s J \rho_s (v_s \cdot \nabla) v_s, \psi_s^v)$$

$$+ (\chi_s g(\varphi) \sigma_s, \nabla \psi_s^v) - (\chi_s J \rho_s f_s, \psi_s^v)$$

$$+ (\chi_s g(\varphi) p_F, \nabla \cdot \psi_s^v) + (\chi_s g(\varphi) \nabla p_F, \psi_s^v) = 0, \quad \forall \psi_s^v \in V_s^0$$

$$(\chi_f \nabla u_f, \nabla \psi^u) = 0, \quad \forall \psi_f^u \in V_f^0$$

$$(\chi_s (\rho_s \partial_t u_s + (v_s \cdot \nabla) u_s - v_s), \psi_s^u) = 0, \quad \forall \psi^u \in L_s$$

$$(\chi_f \nabla \cdot v_f, \psi_f^p) = 0, \quad \forall \psi_f^p \in L_f$$

$$(\chi_F \nabla \cdot v_F, \psi_F^p) = 0, \quad \forall \psi_F^p \in L_F$$

$$(\chi_s (1 - \kappa) \varphi \sigma(u) : e(u), \psi - \varphi)$$

$$+ 2(1 - \kappa)(\chi_s \varphi (p_F \nabla \cdot u + \nabla p_F \cdot u), \psi - \varphi)$$

$$- 2 \langle \chi_s \varphi p_F n \cdot u, \psi - \varphi \rangle_{\Gamma_i}$$

$$+ \left(-\chi_s \frac{G_c}{\varepsilon}(1 - \varphi), \psi - \varphi\right) + (\chi_s G_c \varepsilon \nabla \varphi, \nabla(\psi - \varphi)) \geqslant 0, \quad \forall \psi \in K$$

11.8.4 模拟结果

如图 11.8 与图 11.9 所示, 我们分别进行了 ALE 模型以及全欧拉模型数值模拟[437], 我们注意到 ALE 方法中, 裂缝边缘网格随裂缝变形而变形, 但在欧拉模型中, 网格位置固定, 裂缝扩展情况与网格分布无关。

11.8 流固耦合相场裂缝问题

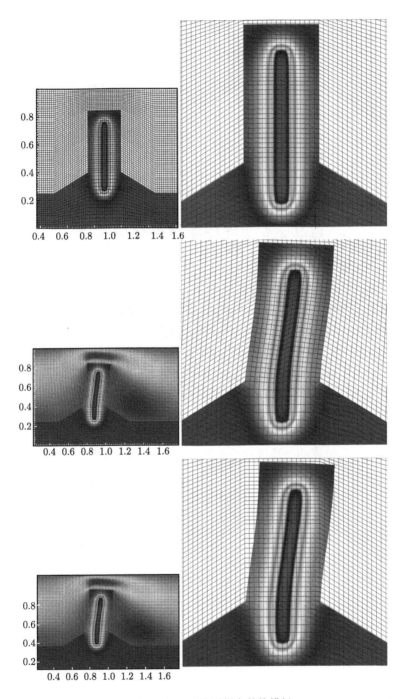

图 11.8 相场法流固耦合数值模拟
初始模型配置 (上)、ALE 模拟结果 (中) 和欧拉法模拟结果 (下)

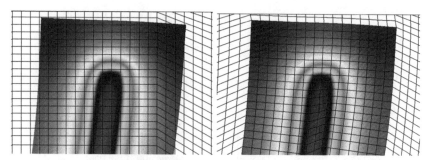

图 11.9　ALE 模拟结果 (左) 和欧拉法模拟结果 (右) 对比图

由图可知欧拉法数值模拟不涉及网格变形，处理更为简便

11.9　多物理相场裂缝模型难点

在相场裂缝模型的开发过程中，高精度求解器、并行计算方法以及优化数值模型是我们永恒的追求。但我们仍需关注一些相场裂缝模型的具体问题：

(1) 相场裂缝网格精度问题。在有限元模型中，我们需要对方程进行离散化处理。在相场裂缝模型中，$\varepsilon = 2h$ 的情况下，我们在垂直方向上仅有大概四个元素位于裂缝过渡区域，模型精度有待进一步提高。

(2) 多种类型的界面问题建模 (1.3.4 节)。

(3) 裂缝尖端区域的处理。在多孔介质中，裂缝尖端周边存在负压状态，相关研究也证实了这一点[317]。

(4) 相场裂缝模型中多物理场相互作用时的耦合处理。

(5) 裂缝边界位置的处理依赖于水平集方法，其精确度有待进一步提升。

(6) 裂缝边界上的边界条件处理[316]。

(7) 各向异性模型中的多种材料相互作用问题。

(8) 多物理场应用中应力分解定律的有效性问题。

(9) 边界条件会影响裂缝扩展方向，在处理模型边界时需要对方程以及模型做出相应修正。

第 12 章 数值模拟 IV

本章我们将提供部分多场耦合数值模型的实际算例，重点包括多物理场建模与应用以及基准测试两方面。而在多物理场多孔介质相场裂缝方面的相关研究可参考文献 [425]。

12.1 准静态裂缝模型

12.1.1 Sneddon 模型

本节我们将介绍 Sneddon 模型 [69,424,391,390]。本书 2.4 节与 3.3 节分别介绍了二维、三维模型相关内容。实际模型中，我们设定裂缝边界 ∂C 处破裂压力 $p := p_F$，随后我们参考命题 82、86 等，得到数模公式。

12.1.1.1 基准测试与结果

Sneddon 模型常被用于进行基准测试，研究人员会基于测试结果分析模型误差、求解器收敛性等特性。我们也采用了该模型对各种目标函数进行测试，主要包括：

(1) 在二维与三维模型中分别研究了网格细化程度对模拟结果的影响，同时提供了一个开源代码以供研究人员参考 [218]。

(2) 采用均匀网格细化的方法对相关目标函数进行研究 [443]。

与此同时，我们还提供以下研究结果供读者参考：

(1) 不可压缩固体中基于残差的裂缝扩展模型 [42]。
(2) Neumann 边界条件下裂缝扩展模型及边界积分的数值验证 [319]。
(3) 非线性求解器优化 [438]。
(4) 面向目标的误差估计以及局部网格自适应法 [436]。
(5) 网格尺寸对收敛性的影响 [281]。
(6) 增广 Lagrangian 补偿法的实际应用 [424]。

12.1.1.2 模型配置

该模型初始配置如图 12.1 所示。域 $B = (0,4)^2$，给定边界条件与压力如下：

$$du = 0, \quad 在 \partial B \times I 上$$
$$\varepsilon \partial_n \varphi = 0, \quad 在 \partial B \times I 上$$

初始裂缝长为 $L = 2l_0 = 0.4$，分布于

$$\Omega_F = (1.8, 2.2) \times (2-h, 2+h) \subset B$$

其中 h 为最小网格尺寸，此时我们有

$$\varphi^0 = 0, \quad 在 \Omega_F \times \{0\} 中$$

$$\varphi^0 = 1, \quad 在 B\backslash\Omega_F 中$$

由此我们可获得一个二维初始狭长缝。

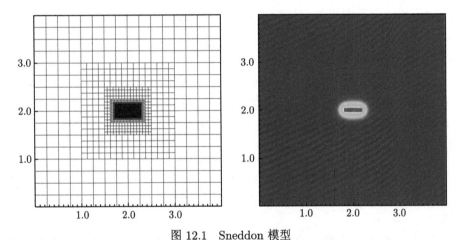

图 12.1　Sneddon 模型

左，网格配置情况；右，裂缝分布情况

接下来我们设置模型初始参数，其中杨氏模量与泊松比分别为 $E = 1.0, \nu_s = 0.2$。破裂压力为 $p = 10^{-3}$，临界能量释放率为 $G_c = 1.0$，令 $\gamma = 100 \times h^{-2}$，相场正则化参数 $\kappa = 10^{-10}$，$\varepsilon = 0.5\sqrt{h}$。

12.1.1.3　目标参数

我们主要研究空间网格细化后的裂缝张开位移 (COD) 以及裂缝体积 (TCV)，本模型中裂缝垂直于坐标轴。

裂缝体积计算公式为 $V = \pi w l_0$，裂缝开度计算公式为 $w = 4\dfrac{(1-\nu_s^2) l_0 p}{E}$，由此可得裂缝体积计算公式为

$$V = 2\pi \frac{(1-\nu_s^2) l_0^2 p}{E} \tag{12.1}$$

其中 $l_0 = 0.2$。

12.1.1.4 模拟结果

裂缝网格分布如表 12.1 所示，不同网格参数下的模拟结果如图 12.2 所示。

表 12.1 裂缝体积与网格尺寸关系

h	8.8×10^{-2}	4.4×10^{-2}	2.2×10^{-2}	1.1×10^{-2}	实际情况
V	3.02×10^{-4}	2.77×10^{-4}	2.57×10^{-4}	2.49×10^{-4}	2.41×10^{-4}

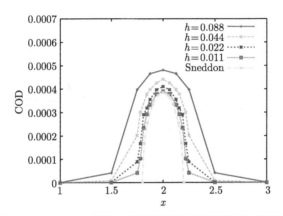

图 12.2 Sneddon 模型，不同网格尺寸下缝宽变化趋势图

12.1.2 混合边界条件

接下来我们研究混合边界条件的情况。参考 11.4.1.2 节中的式 (11.16)，非 Dirichlet 边界条件 ∂B_N 可归纳为

$$\int_{\partial B_N}(\tau+pn)\cdot u\,\mathrm{d}s$$

其中 $\tau:\partial B_N\to\mathbb{R}^d$ 为牵引力。

12.1.2.1 模型配置

模型基本参数与 12.1.1 节示例相同。我们基于扩展拉格朗日法处理裂缝不可逆约束。其中 $p=10^{-3}$，$\nabla p=0$。

12.1.2.2 边界条件

该模型中仅改变了边界条件部分。其中顶部 Γ_{top} 与底部 Γ_{bottom} 为 Neumann 边界条件 $\partial B_N=\Gamma_{\text{top}}\cup\Gamma_{\text{bottom}}$，这里我们将 Neumann 边界条件分为均质与非均质两种，并分别进行模拟。

算例 1：令 $\partial B_N=\varnothing$

算例 2：
$$u = 0, \quad 在 \partial_D B 上$$

$$\tau_{t,b} = (0,0)^{\mathrm{T}}, \quad 在 \partial B_N 上$$
$$u = 0, \quad 在 \partial_D B 上$$

算例 3：
$$\tau_t = (0, 0.001)^{\mathrm{T}}, \quad 在 \Gamma_{\text{top}} 上$$
$$\tau_b = (0, -0.001)^{\mathrm{T}}, \quad 在 \Gamma_{\text{bottom}} 上$$
$$u = 0, \quad 在 \partial_D B 上$$

算例 4：
$$\tau_t = (0, 0.1)^{\mathrm{T}}, \quad 在 \Gamma_{\text{top}} 上$$
$$\tau_b = (0, -0.1)^{\mathrm{T}}, \quad 在 \Gamma_{\text{bottom}} 上$$
$$u = 0, \quad 在 \partial_D B 上$$

以上模拟均在 7 阶细化网格上进行，其中 $h = 0.044$，初始状态与相场函数方程请参考 12.1.1 节，如图 12.3 所示。

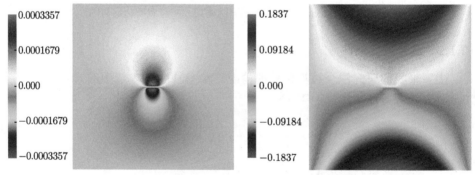

图 12.3　混合边界条件下相场裂缝模型位移参数云图
算例 1(左); 算例 4 (右)

12.1.2.3　模拟结果

在 $x = 2$ 提取最大裂缝开度 $w_{\max}$ 如下：

算例 1：$w_{\max}(x = 2; 0 \leqslant y \leqslant 4) = 5.25244 \times 10^{-4}$；

算例 2：$w_{\max}(x = 2; 0 \leqslant y \leqslant 4) = 5.52572 \times 10^{-4}$；

算例 3：$w_{\max}(x = 2; 0 \leqslant y \leqslant 4) = 1.31092 \times 10^{-3}$；

算例 4：$w_{\max}(x = 2; 0 \leqslant y \leqslant 4) = 7.67588 \times 10^{-2}$。

在算例 2 中上下边界牵引力为 0 ($\tau = 0$)，此时仅有破裂压力迫使裂缝张开，最大开度与算例 1 相似，随着后续模拟中牵引力逐渐增大，裂缝开度也随之上升。

12.2 非等温相场裂缝模型

接下来我们介绍非等温相场裂缝模型。材料参数取自文献 [410],详细结果请参考文献 [337]。

12.2.1 模型配置

模型域为 $B := (0\text{m}, 200\text{m})^2$,初始模型配置如图 12.4 所示。

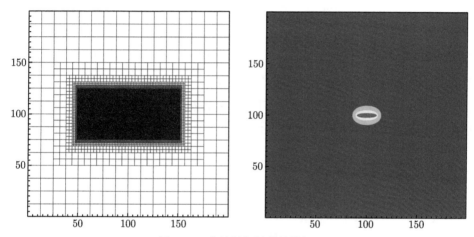

图 12.4 非等温相场裂缝模拟
网格和初始裂缝几何位置 (红色区域)

我们设定热扩散率 $\kappa_T = 10^{-6}\text{m}^2/\text{s}$,初始温度 $T_0 = 100°C$,总初始应力 $p_0 = 12130\text{kPa}$,给定一个恒定的压力 $p_F = 15834\text{kPa}$,杨氏模量与泊松比分别为 $E = 1.5 \times 10^{10}\text{Pa}$, $\nu = 0.15$,线性热膨胀系数 $\beta = 10^{-5}1/C$,临界能量释放率 $G_c = 10^{10}\text{Pa/m}$,相场参数分别为 $\kappa = 10^{-12}$, $\varepsilon = 10 \times \sqrt{h}$。

12.2.2 例 1:二维 Sneddon 模型

加载步长 $k = 1\text{s}$,采用局部自适应网格,半缝宽计算公式为

$$w(x,t) = \frac{2\left(1-\nu_s^2\right)l_0}{E}\sqrt{1-\rho^2}\left(p - p_0 - C_T\left(T_F - T_0\right)\right)$$

其中 l_0 为裂缝半缝长,$0 < \rho < 1$,$\rho = x/l_0$,模拟结果如图 12.5(a) 所示。

12.2.3 例 2:二维 Sneddon 定压恒温模型

令 $\lambda_T = 10^{-4}$,在定压恒温的模型中进行模拟,计算一个时间步后终止运算,模拟结果如图 12.5(b) 所示。

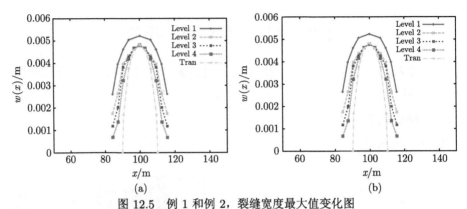

图 12.5　例 1 和例 2，裂缝宽度最大值变化图

其中 Tran 表示文献 [410] 中参考解。Level 代表网格细分程度，Level 越高，网格尺寸越小

12.2.4　例 3：二维 Sneddon 非等温模型

在例 3 中，我们引入 Hagoort 下降常数模拟温度随时间的变化，我们将时间步长设定为 $k=86400\mathrm{s}$，总时间 $T=365$ 天 (一年)，模拟结果如图 12.6 所示。

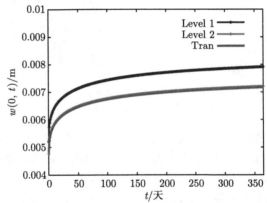

图 12.6　非等温模型中最大缝宽随时间变化的趋势图

其中 Tran 表示文献 [410] 中参考解。Level 代表网格细分程度，Level 越高，网格尺寸越小

12.3　多孔介质中的裂缝扩展模型

本节我们将讨论多孔介质中的裂缝扩展模型 [316]。

12.3.1　主要概念

本节我们研究一个带有 Biot 系数 $\alpha=1$ 的多孔介质材料，参考命题 91 与前文相关内容，对于位移与相场变量，解变量设定为 $U=(u,\varphi)$。求解时采用迭代耦合方法，得到给定时间点 t_n 的 (u,φ)。随后基于算法 23 获得流固耦合模型。为区分孔压与固体压力，我们基于相场参数的指示功能，令 $p=\chi_R p+\chi_F p$。

12.3.2 模拟结果

其裂缝形态与压力分布如图 12.7 所示，我们从相场参数以及压力分布方面定量分析模拟结果，半缝长以及最大压力如图 12.8 所示。在计算过程中令 $\varepsilon = h_{\text{coarse}} = 0.088$。

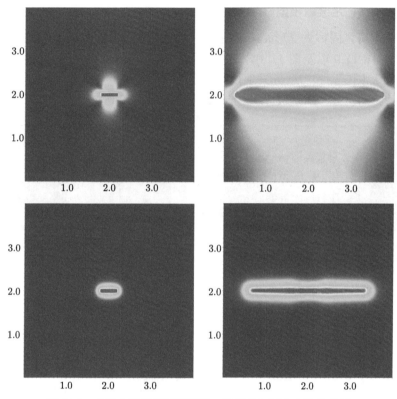

图 12.7 多孔介质裂缝扩展模拟：半缝长与压力变化趋势图

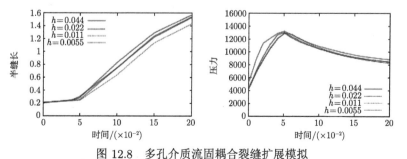

图 12.8 多孔介质流固耦合裂缝扩展模拟
$T = 0\text{s}$ 和 $T = 0.2\text{s}$ 时的裂缝形态 (上) 和压力分布 (下)[316]

12.4 Navier-Stokes 流固耦合相场裂缝模型

本节我们将介绍耦合 Navier-Stokes 流动方程的相场裂缝模型，模型配置可参考 11.8 节。

12.4.1 例 1：定流量模型

本节将研究固体大变形裂缝模型与不可压缩流体的相互作用关系 (见图 12.9、图 12.11、图 11.9 以及图 11.8)，建模过程中固定 $\varepsilon > 0$，令空间离散化参数 h 与时间步长 k 固定，模拟研究发现模型稳定性较高，鲁棒性较好 (见图 12.10)。详细信息请参考文献 [435, 437]。

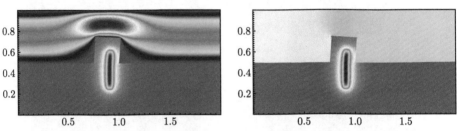

图 12.9 流固耦合相场裂缝模型 $T = 50\mathrm{s}$ 时刻流场以及裂缝分布情况。
左图：速度场。右图：压力场

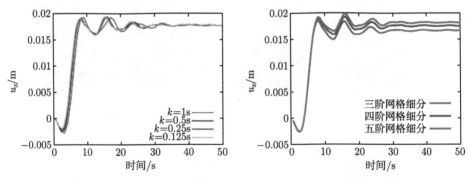

图 12.10 流固耦合模型中在点 $\hat{u}_x(1, 0.75)$ 处的收敛性分析

12.4.2 例 2：窦性流

本节我们将介绍窦性流，其特征是流场剖面是一个窦曲线，相关计算可参考文献 [437]。

我们给定一个随时间变化的抛物型流入速度剖面：

$$v_f(0, y) = \bar{U}(y - 0.25)(y - 1), \quad \bar{U} = \left[2\sin\left(\frac{2t}{\pi} - 0.15\pi\right) + 1 \right] \quad (\mathrm{m/s})$$

模拟结果如图 12.11~图 12.13 所示。

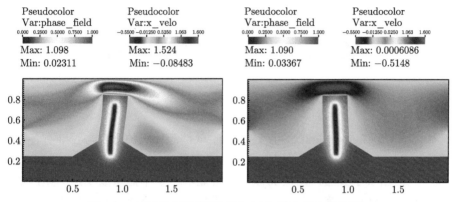

图 12.11 不同时刻下速度场分布图 (注入流与排出流)

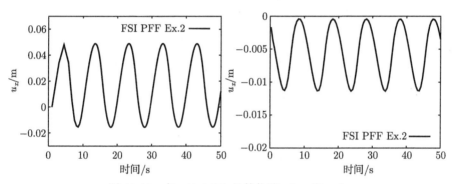

图 12.12 点 $u(1, 0.85)$ 处的位移 $u(u_x$ 和 $u_y)$

根据模拟结果可知,该点在流场的作用下做有规律的振荡运动

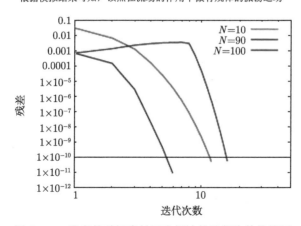

图 12.13 选定的时间步长下牛顿法的迭代次数趋势图

第 13 章　数值模拟研究软件

数模软件开发是应用数学领域的重要部分，本书所提供的数值模型主要通过 C++ 编程实现。

如 GAMM Leitartikel[440] 所述，科学软件源代码的开源与否都有各自的理由 (https://publiccode.eu/)[274,28]。在 2012 年前后，Research Software Engineer (RSE) 一词出现于英国，随后类似的协会逐渐在各个国家成立，例如，德国的 deRSE 协会。他们的首次国际会议 deRSE19 在德国举行，同时组织了一个关于可持续软件开发的研讨会。研讨会结束后，一群来自不同专业的科学家一同发布了进一步发展可持续科学软件开发的声明。

本书中的几个算例可以直接复制并使用，或者参考：

(1) T. Heister, T. Wick 基于开源有限元程序库 deal.II 开发的高性能并行计算裂缝扩展的相场模型 [218]：https://github.com/tjhei/cracks。

在最近的 deal.ii 9.2.0 版本中 (https://www.dealii.org/deal92-preprint.pdf)。Katrin Mang、Jan Philipp Thiele 和 Daniel Jodlbauer 正式参与了开发研究工作，同时 Julian Roth 与 Julius Grundmann 也在从事 deal.II 相关研究。

(2) W. Wollner, T. Wick 基于开源有限元程序库 deal.II 开发搭建 PDE/CVIS 模块化算法的工具箱 DOpElib(Differential Equations and Optimization Environment library)[203,134]：www.dopelib.net。

(3) T. Wick 基于开源有限元程序库 deal.II 建立了求解任意拉格朗日-欧拉坐标系下的流固耦合问题数值模型 [432]：https://media.archnumsoft.org/10305/。

该模型最初在 2011 年开发并用于解决流固耦合问题 [432]。在 2013 年基于该模型实现了相场模型的数值模拟，可参考文献 [217, 218]。

近年来，基于该模型开发并解决了多孔弹性力学中的问题 (线性 Biot 方程)[441] 和热孔弹性问题 [418,420]，相关信息可访问：http://www.thomaswick.org/gallery_engl.html。

(4) S. Frei, T. Richter, T. Wick 采用 deal.II 解决了基于局部修正有限元法的界面问题 [180,176]：https://zenodo.org/record/1457758#.XqsilqZCRUR。

作者同时提供了一个基于 deal.II 的详细文档，提供了立体欧拉描述的基本方法。最后，我们还提供了多孔介质中多物理场裂缝建模的内部代码。

(5) M. F. Wheeler, T. Wick, S. Lee 基于开源有限元程序库 deal.II 建立了

多孔介质中相场裂缝扩展的自适应并行计算框架：IPACS: Integrated Phase-Field Advanced Crack Propagation Simulator。

基于扩展伽辽金 (enriched Galerkin discretiations) 方法实现了多孔介质中流体流动与传输的物理离散 [395,275,279,280]，同时与裂缝扩展相耦合 [277,276]。该模型中还提供了预测-校正网格细化自适应模块和并行计算模块 [217]。

13.1 团队建设

现阶段，作者在 Hannover 以及奥地利 Linz(通过 FWF 项目 [P29181)] 合作)研究团队的成员 (博士后，博士，硕士，学士) 合作开发并使用的研究软件主要为 deal.II、DOpElib，当然也有人使用 Octave[141] 和 Python 进行开发研究。以下是团队成员各自的研究方向介绍：

(1) Bernhard Endtmayer (RICAMLinz)：多目标误差分析；

(2) Daniel Jodlbauer (RICAM Linz)：相场裂缝及流固耦合的多重网格并行求解器；

(3) Jan Philipp Thiele (LUH)：面向目标的时空自适应流动问题；

(4) Katrin Mang (LUH)：不可压缩材料中的相场裂缝问题；

(5) Sebastian Kinnewig (LUH)：麦克斯韦方程组的数值模拟；

(6) Nima Noii (LUH)：相场裂缝扩展与传热过程；

(7) 范濛 (LUH/China University of Petroleum)：相场裂缝模型中的应力分解方法；

(8) Amirreza Khodadadian (LUH)：相场裂缝问题中的贝叶斯估计应用；

(9) Denis Khimin (LUH)：相场裂缝最优化问题；

(10) Gregor Pfau (LUH)：Discontinuous Galerkin 方法；

(11) Julius Grundmann (LUH)：通量校正传输；

(12) Julian Roth (LUH)：麦克斯韦方程组的多重网格求解器。

第 14 章　结论及问题

本书中，我们讨论了多物理场相场裂缝的数模问题。其中多物理场问题属于耦合问题，是一类受变分不等式约束的非定常非线性耦合偏微分方程问题。本书对这类耦合问题进行了详细的分类，并提供了相关算法。通过数值模拟以及实际算例证明了相场模型的实用性，同时提供了大量文献供读者参考。

我们通过多种方法实现了裂缝参数正则化不可逆约束，其中基于裂缝不可逆性约束的计算处理方面的研究暂时不予讨论。原始对偶活动集法的稳定性有待进一步研究，该方法可能会在某些模型中偶尔出现死循环[210,118]。研究人员针对该问题提出了一个简单的解决方案[218]，但仍然有必要做进一步的研究。

时空严格收敛的数值模型仍有待进一步研究。读者可参考本书内容以及相关参考文献进行简单模型的数值分析[327,339,348,352,356,367,369]。但迄今为止，针对相场法的数值分析以及辅助证明的数模仍有待进一步完善[373,384,389,415,445,446,448]。

基本非线性[438,439,282,86]和线性求解[217,281,218,246]包括并行计算性能测试在内的相场裂缝问题已被广泛研究。然而在某些特殊的领域，例如对目标函数的精确评估，尚待进一步研究。事实上，在文献[436]中推导出了面向目标的误差分析，但只考虑了模型中求解器的误差，针对裂缝扩展问题中的实际误差仍然需要进一步分析[224,235,238,241]。在多物理场中，可能会同时存在多个需要研究的目标函数[248,258,269,285,290,304,324]。RICAM 与 Linz 联合提出了严格的数值分析和数值概念[151,148,150]。针对相场裂缝模型开发的自适应方法允许在计算精度与算力之间达成妥协，可在未来应用于工程领域裂缝问题的研究中。

在多物理问题的断裂与损伤方面，我们建立了应用于多孔介质和固体变形的模型设计以及数值模拟分析。对于此类问题，都可通过计算分析证明它们在空间和时间上的收敛性与稳定性[195,199,204,206]。在固体变形问题中，严格的数值分析、线性并行求解器和预处理方法仍有待进一步研究。我们提供了多物理场相场裂缝问题的一些尚待解决的问题列表：

(1) 针对"简单"相场裂缝模型的严格数值分析[41,39,40,457,161,162]。

(2) 针对相场裂缝问题开发的稳定、高效、精确的线性或者非线性求解器 (在准静态脆性相场裂缝问题上，已经完成了基于均匀细化网格的 (并行) 求解器开发)[159,218,246])。此外，我们在文献 [247, 244] 与 11.8 节提供了一个可用于参考的应用于流固耦合相场裂缝问题的模型框架。

第 14 章 结论及问题

(3) 动态相场裂缝的空间/时间自适应方法。

(4) 控制离散化误差的模型与数值求解器的自适应与并行计算；针对多目标泛函拟线性问题，研究人员基于双加权残差法进行了初步的误差分析和可靠性研究[148,150]。但适用于非定常问题的变分不等式模型尚待进一步研究。

(5) 针对局部断裂/损伤大尺度问题的多尺度建模问题。

(6) 非侵入式全局-局部相场方法的数值分析。

(7) 本构模型及其验证与分析。

(8) 多孔介质中塑性裂缝过渡区弹塑性力学分析[106,402,421]。

(9) 提高裂缝宽度计算精度的相关方法。

(10) 裂缝边界过渡区界面条件的模拟与分析。

(11) 应用于破碎带裂缝尖端的相关算法。

(12) 最优控制问题、逆反问题、参数估算、不确定性量化评价问题均有待研究。我们在文献 [331，332，255，254] 中刚刚开展了相关研究，并获得了初步的成果。由于优化增加了另一个迭代循环过程，因此开发有效的正向解决方案 (本书研究内容之一) 是非常有必要的。

第15章 结 束 语

本书完成于 2020 年春夏之交。在本书撰写期间,家庭花园的美景给予了作者无限动力与激情,在此与各位读者分享作者曾经欣赏过的诗与美景。

Johann Wolfgang von Goethe(1749—1832), German poet and naturalist
(译者：钱春绮)

Mailied

五月之歌

Wie herrlich leuchtet
自然多明媚，
Mir die Natur!
向我照耀！
Wie glänzt die Sonne!
太阳多辉煌！
Wie lacht die Flur!
原野含笑！

Es dringen Blüten
千枝复万枝，
Aus jedem Zweig
百花怒放，
Und tausend Stimmen
在灌木林中，
Aus dem Gesträuch,
万籁俱唱。

Und Freud und Wonne
万人的胸中
Aus jeder Brust.
快乐高兴，
O Erd, o Sonne!
哦，大地、太阳，
O Glück, o Lust!
幸福，欢欣！

O Lieb, o Liebe!
哦，爱啊，爱啊，

So golden schön,
灿烂如金，
Wie Morgenwolken
你仿佛朝云
Auf jenen Höhn!
漂浮山顶！

Du segnest herrlich
你欣然祝福
Das frische Feld,
青田沃野、
Im Blütendampfe
花香馥郁的
Die volle Welt.
大千世界。

O Mädchen, Mädchen,
啊，姑娘，姑娘，
Wie lieb ich dich!
我多爱你！
Wie blickt dein Auge!
你目光炯炯，
Wie liebst du mich!
你多爱我！

So liebt die Lerche
像云雀喜爱
Gesang und Luft,
太空高唱，

Und Morgenblumen
像朝花喜爱
Den Himmelsduft,
天香芬芳，

Wie ich dich liebe
我这样爱你，
Mit warmen Blut,
热血沸腾，
Die du mir Jugend
你给我勇气、

Und Freud und Mut
喜悦、青春，

Zu neuen Liedern
使我唱新歌，
Und Tänzen gibst.
翩翩起舞，
Sei ewig glücklich,
愿你永爱我，
Wie du mich liebst!
永远幸福！

参 考 文 献

[1] Abdulle A. On a priori error analysis of fully discrete heterogeneous multiscale FEM[J]. Siam Journal on Multiscale Modeling & Simulation, 2005, 4(2): 447-459.

[2] Adams R A. Sobolev Spaces[M]. New York: Academic Press, 1975.

[3] Kay A D. The approximation theory for the p-version finite element method and application to non-linear elliptic PDEs[J]. Numerische Mathematik, 1999, 82(3): 351-388.

[4] Ainsworth M, Oden J T. A posteriori error estimation in finite element analysis[J]. Computer Methods in Applied Mechanics and Engineering, 1997, 142(1-2): 1-88.

[5] Akin J E, Tezduyar T E, Ungor M. Computation of flow problems with the mixed interface-tracking/interface-capturing technique (MITICT)[J]. Computers & Fluids, 2007, 36(1): 2-11.

[6] Aldakheel F. A microscale model for concrete failure in poro-elasto-plastic media[J]. Theoretical and Applied Fracture Mechanics, 2020, 107: 102517.

[7] Aldakheel F, Hudobivnik B, Hussein A, et al. Phase-field modeling of brittle fracture using an efficient virtual element scheme[J]. Computer Methods in Applied Mechanics and Engineering, 2018, 341(NOV.1): 443-466.

[8] Aldakheel F, Hudobivnik B, Wriggers P. Virtual element formulation for phase-field modeling of ductile fracture[J]. International Journal of Multiscale Computational Engineering, 2019, 17(2): 181-200.

[9] Aldakheel F, Noii N, Wick T, et al. A global-local approach for hydraulic phase-field fracture in poroelastic media[J]. Computers & Mathematics with Applications, 2021, 91: 99-121.

[10] Alessi R, Marigo J J, Maurini C, et al. Coupling damage and plasticity for a phase-field regularisation of brittle, cohesive and ductile fracture: one-dimensional examples[J]. International Journal of Mechanical Sciences, 2018, 149: 559-576.

[11] Allaire G, Jouve F, Goethem N V. Damage and fracture evolution in brittle materials by shape optimization methods[J]. Journal of Computational Physics, 2011, 230(12): 5010-5044.

[12] Allen S M, Cahn J W. A microscopic theory for antiphase boundary motion and its application to antiphase domain coarsening[J]. Acta Metallurgica, 1979, 27(6): 1085-1095.

[13] Allgower E L, Georg K. Numerical continuation methods, an introduction[J]. Mathematics and Computers in Simulation, 1991, 33(1): 84-85.

[14] Almani T, Lee S, Wheeler M F, et al. Multirate coupling for flow and geomechanics applied to hydraulic fracturing using an adaptive phase-field technique[C]. SPE Reservoir

Simulation Conference, 2017: D031S010R001.

[15] Alvarez-Aramberri J, Pardo D, Barucq H. Inversion of magnetotelluric measurements using multigoal oriented hp-adaptivity[J]. Procedia Computer Science, 2013, 18: 1564-1573.

[16] Ambati M, Gerasimov T, de Lorenzis L. Phase-field modeling of ductile fracture[J]. Computational Mechanics, 2015, 55(5): 1017-1040.

[17] Ambati M, Gerasimov T, de Lorenzis L. A review on phase-field models of brittle fracture and a new fast hybrid formulation[J]. Computational Mechanics, 2015, 55: 383-405.

[18] Ambati M, de Lorenzis L. Phase-field modeling of brittle and ductile fracture in shells with isogeometric NURBS-based solid-shell elements[J]. Computer Methods in Applied Mechanics and Engineering, 2016, 312: 351-373.

[19] Ambrosio L, Dancer N. Calculus of Variations and Partial Differential Equations: Topics on Geometrical Evolution Problems and Degree Theory[M]. Springer Science & Business Media, 2012.

[20] Ambrosio L, Tortorelli V M. Approximation of functional depending on jumps by elliptic functional via t-convergence[J]. Communications on Pure and Applied Mathematics, 1990, 43(8): 999-1036.

[21] Braides A. Approximation of free-discontinuity problems[J]. Lecture Notes in Mathematics, 1998, 23(5): 121-132.

[22] Ambrosio L, Virga E G. A boundary-value problem for nematic liquid crystals with a variable degree of orientation[J]. Archive for Rational Mechanics & Analysis, 1991, 114(4): 335-347.

[23] Amor H, Marigo J J, Maurini C. Regularized formulation of the variational brittle fracture with unilateral contact: numerical experiments[J]. Journal of the Mechanics and Physics of Solids, 2009, 57(8): 1209-1229.

[24] Andersson J, Mikayelyan H. The asymptotics of the curvature of the free discontinuity set near the cracktip for the minimizers of the Mumford-Shah functional in the plain[J]. arXiv preprint arXiv:1204.5328, 2012.

[25] Anzt H, Bach F, Druskat S, et al. An environment for sustainable research software in Germany and beyond: current state, open challenges, and call for action[J]. F1000 Research, 2020, 9: 295.

[26] Arndt D, Bangerth W, Blais B, et al. The deal. II library, version 9.2[J]. Journal of Numerical Mathematics, 2020, 28(3): 131-146.

[27] Arndt D, Bangerth W, Davydov D, et al. The deal. II library, version 8.5[J]. Journal of Numerical Mathematics, 2017, 25(3): 137-145.

[28] Arndt D, Bangerth W, Davydov D, et al. The deal. II finite element library: design, features, and insights[J]. Computers & Mathematics with Applications, 2021, 81: 407-422.

[29] Arriaga M, Waisman H. Multidimensional stability analysis of the phase-field method

for fracture with a general degradation function and energy split[J]. Computational Mechanics, 2018, 61: 181-205.

[30] Artina M, Fornasier M, Micheletti S, et al. Anisotropic mesh adaptation for crack detection in brittle materials[J]. SIAM J. Sci. Comput., 2015, 37(4): B633-B659.

[31] Babuška I. The finite element method for elliptic equations with discontinuous coefficients[J]. Computing, 1970, 5(3): 207-213.

[32] Babuska I, Banerjee U. Stable generalized finite element method (SGFEM)[J]. Computer Methods in Applied Mechanics & Engineering, 2011, 201(1): 91-111.

[33] Babuška I, Melenk J M. The partition of unity method[J]. International journal for numerical methods in engineering, 1997, 40(4): 727-758.

[34] Babuvška I, Rheinboldt W C. Error estimates for adaptive finite element computations[J]. SIAM Journal on Numerical Analysis, 1978, 15(4): 736-754.

[35] Badnava H, Msekh M A, Etemadi E, et al. An h-adaptive thermo-mechanical phase field model for fracture[J]. Finite Elements in Analysis and Design, 2018, 138: 31-47.

[36] Bangerth W, Hartmann R, Kanschat G. deal.II – a general purpose object oriented finite element library[J]. Acm Transactions on Mathematical Software, 2007, 33(4): 24.

[37] Rannacher R. Adaptive finite element methods for partial differential equations[J]. arXiv preprint math/0305006, 2003.

[38] Barrett J W, Blowey J F, Garcke H. Finite element approximation of the Cahn-Hilliard equation with degenerate mobility[J]. SIAM Journal on Numerical Analysis, 1999, 37(1): 286-318.

[39] Bartels S. A posteriori error analysis for time-dependent Ginzburg-Landau type equations[J]. Numerische Mathematik, 2005, 99(4): 557-583.

[40] Bartels S. Error control and adaptivity for a variational model problem defined on functions of bounded variation[J]. Mathematics of Computation, 2015, 84(293): 1217-1240.

[41] Bartels S, Müller R, Ortner C. Robust a priori and a posteriori error analysis for the approximation of Allen-Cahn and Ginzburg-Landau equations past topological changes[J]. Siam Journal on Numerical Analysis, 2011, 49(1): 110-134.

[42] Basava S, Mang K, Walloth M, et al. Adaptive and Pressure-Robust Discretization of Incompressible Pressure-Driven Phase-Field Fracture[M]. Non-standard Discretisation Methods in Solid Mechanics. Cham: Springer International Publishing, 2022: 191-215.

[43] Basting S, Prignitz R. An interface-fitted subspace projection method for finite element simulations of particulate flows[J]. Computer Methods in Applied Mechanics & Engineering, 2013, 267: 133-149.

[44] Basting S, Weismann M. A hybrid level set-front tracking finite element approach for fluid–structure interaction and two-phase flow applications[J]. Journal of Computational Physics, 2013, 255(Complete): 228-244.

[45] Bause M, Radu F, Köcher U. Space-time finite element approximation of the Biot poroelasticity system with iterative coupling[J]. Computer Methods in Applied Mechanics and

Engineering, 2017, 320: 745-768.

[46] Bazilevs Y, Takizawa K, Tezduyar T E. Computational Fluid-Structure Interaction: Methods and Applications[M]. John Wiley & Sons, 2013.

[47] Becker R. An optimal-control approach to a posteriori error estimation for finite element discretizations of the Navier-Stokes equations[J]. East-West Journal of Numerical Mathematics, 2000, 8(4): 257.

[48] Braack, Malte, Becker, et al. The Finite Element Toolkit Gascoigne (v1.01)[Z]. Zenodo, 2021.

[49] Becker R, Innerberger M, Praetorius D. Optimal convergence rates for goal-oriented FEM with quadratic goal functional[J]. Computational Methods in Applied Mathematics, 2021, 21(2): 267-288.

[50] Becker R, Johnson C, Rannacher R. Adaptive error control for multigrid finite element methods[J]. Computing, 1995, 55(4): 271-288.

[51] Rannacher R, Suttmeier F T. A feed-back approach to error control in finite element methods: application to linear elasticity[J]. Computational Mechanics, 1997, 19(5): 434-446.

[52] Becker R, Rannacher R. An optimal control approach to a posteriori error estimation in finite element methods[J]. Acta Numerica, 2001, 10: 1-102.

[53] da Veiga L B, Buffa A, Sangalli G, et al. Mathematical analysis of variational isogeometric methods[J]. Acta Numerica, 2014, 23: 157-287.

[54] Bensoussan A. Asymptotic analysis for periodic structures[J]. American Mathematical Soc., 2011.

[55] Schmich M. Adaptive finite element methods for computing nonstationary incompressible flows[D]. Heidelberg: Heidelbery University, 2009.

[56] Bilgen C, Homberger S, Weinberg K. Phase-field fracture simulations of the brazil-ian splitting test[J]. International Journal of Fracture, 2019, 220: 85-98.

[57] Biot M A. Consolidation settlement under a rectangular load distribution[J]. Journal of Applied Physics, 1941, 12(5): 426-430.

[58] Biot M A. General theory of three-dimensional consolidation[J]. Journal of applied physics, 1941, 12(2): 155-164.

[59] Blum H, Suttmeier F T. An adaptive finite element discretisation for a simplified signorini problem[J]. Calcolo 1999, 37(2): 65-77.

[60] Blum H, Suttmeier F T. Weighted error estimates for finite element solutions of variational inequalities[J]. Computing, 2000, 65(2): 119-134.

[61] Handbook of Peridynamic Modeling[M]. CRC Press, 2016.

[62] Bodnár T, Galdi G P, Nečasová Š. Fluid-Structure Interaction and Biomedical Applications[M]. Springer Basel, 2014.

[63] Bonnet A, David G. Cracktip is a Global Mumford-Shah Minimizer[M]. Société mathématique de France, 2001.

[64] Borden M J, Hughes T, Landis C M, et al. A higher-order phase-field model for brittle fracture: formulation and analysis within the isogeometric analysis framework[J]. Computer Methods in Applied Mechanics & Engineering, 2014, 273(may 1): 100-118.

[65] Borden M J, Verhoosel C V, Scott M A, et al. A phase-field description of dynamic brittle fracture[J]. Computer Methods in Applied Mechanics & Engineering, 2012, 217-220(Apr.1): 77-95.

[66] Both J, Borregales M, Nordbotten J, et al. Robust fixed stress splitting for Biot's equations in heterogeneous media[J]. Applied Mathematics Letters, 2017, 68: 101-108.

[67] Bourdin B. Image segmentation with a finite element method[J]. Esaim Mathematical Modelling & Numerical Analysis, 1999, 33(2): 229-244.

[68] Bourdin B. Numerical implementation of the variational formulation for quasi-static brittle fracture[J]. Interfaces and Free Boundaries, 2007, 9: 411-430.

[69] Bourdin B, Chukwudozie C P, Yoshioka K. A variational approach to the numerical simulation of hydraulic fracturing[C]. Proceedings - SPE Annual Technical Conference and Exhibition, Society of Petroleum Engineers, 2012.

[70] Bourdin B, Francfort G A, Marigo J J. Numerical experiments in revisited brittle fracture[J]. Journal of the Mechanics & Physics of Solids, 2000, 48(4): 797-826.

[71] Bourdin B, Francfort G A, Marigo J J. The variational approach to fracture[J]. Journal of Elasticity, 2008, 91(1-3): 5-148.

[72] Bourdin B, Francfort G A. Past and present of variational fracture[J]. SIAM News, 2019, 52(9): 104-108.

[73] Bourdin B, Larsen C J, Richardson C L. A time-discrete model for dynamic fracture based on crack regularization[J]. International Journal of Fracture, 2011, 168(2): 133-143.

[74] Bourdin B, Marigo J J, Maurini C, et al. Morphogenesis and propagation of complex cracks induced by thermal shocks[J]. Physical Review Letters, 2014, 112(1): 014301.

[75] Braack M, Ern A. A posteriori control of modeling errors and discretization errors[J]. Siam Journal on Multiscale Modeling & Simulation, 2003, 1(2): 221-238.

[76] Braack M, Richter T. Solutions of 3D Navier-Stokes benchmark problems with adaptive finite elements[J]. Computers & Fluids, 2006, 35(4): 372-392.

[77] Braess D. Finite Elemente: Theorie, Schnelle Löser und Anwendungen in Der Elastizitätstheorie[M]. Springer-Verlag, 2013.

[78] Braides A. Approximation of free-discontinuity problems[J]. Lecture Notes in Mathematics, 1998, 23(5): 121-132.

[79] Bredies K, Lorenz D. Mathematische Bildverarbeitung[M]. Vieweg+Teubner, 2011. https://www.managementbuch.de/shop/magazine/44022/vieweg_teubner_verlag.html.

[80] Brenner S C. The Mathematical Theory Of Finite Element Methods[M]. Springer, 2008.

[81] Brevis I, Muga I, van der Zee K G. Data-driven finite elements methods: machine learning acceleration of goal-oriented computations[J]. arXiv preprint arXiv: 2003.04485, 2020.

[82] Brezzi F, Hager W W, Raviart P A. Error estimates for the finite element solution of variational inequalities: part i. primal theory[J]. Numerische Mathematik, 1977, 28(4): 431-443.

[83] Brezzi F, Hager W W, Raviart P A. Error estimates for the finite element solution of variational inequalities: part ii. mixed methods[J]. Numerische Mathematik, 1978, 31(1): 1-16.

[84] Bristeau M O, Glowinski R, Periaux J. Numerical methods for the navier-stokes equations. applications to the simulation of compressible and incompressible viscous flows[J]. Computer Physics Reports, 1987, 6(1-6): 73-187.

[85] Bronshtein I N, Semendyayev K A. Handbook of Mathematics[M]. Springer Science & Business Media, 2013.

[86] Mkb A, Tw B, Iba C, et al. An iterative staggered scheme for phase field brittle fracture propagation with stabilizing parameters[J]. Computer Methods in Applied Mechanics and Engineering, 2020, 361: 112752.

[87] Bryant E C, Sun W C. A mixed-mode phase field fracture model in anisotropic rocks with consistent kinematics[J]. Computer Methods in Applied Mechanics and Engineering, 2018, 342: 561-584.

[88] Bukač M, Čnić S, Glowinski R, et al. A modular, operator-splitting scheme for fluid-structure interaction problems with thick structures[J]. International Journal for Numerical Methods in Fluids, 2014, 74(8): 577-604.

[89] Bukač M, Yotov I, Zunino P. Dimensional model reduction for flow through fractures in poroelastic media[J]. ESAIM: Mathematical Modelling and Numerical Analysis, 2017, 51(4): 1429-1471.

[90] Bungartz H, Mehl M, Schäfer M. Fluid Structure Interaction II[M]. Berlin Heidelberg: Springer, 2010.

[91] Bungartz H J, Schäfer M. Fluid-Structure Interaction: Modelling, Simulation, Optimisation[M]. Springer Science & Business Media, 2006.

[92] Burke S, Ortner C, Süli E. An adaptive finite element approximation of a variational model of brittle fracture[J]. SIAM Journal on Numerical Analysis, 2010, 48(3): 980-1012.

[93] Burke S, Ortner C, Süli E. An adaptive finite element approximation of a generalised ambrosio-tortorelli functional[J]. Mathematical Models & Methods in Applied Sciences, 2013, 23(9): 1663-1697.

[94] Burman E, Claus S, Hansbo P, et al. CutFEM: discretizing geometry and partial differential equations[J]. International Journal for Numerical Methods in Engineering, 2015, 104(7): 472-501.

[95] Burstedde C, Wilcox L C, Ghattas O. P4est: scalable algorithms for parallel adaptive mesh refinement on forests of octrees[J]. Siam Journal on Scientific Computing, 2011, 33(3): 1103-1133.

[96] Cahn J W, Hilliard J E. Free energy of a nonuniform system. I. Interfacial free energy[J].

The Journal of Chemical Physics, 1958, 28(2): 258-267.
[97] Cai X C, Keyes D E. Nonlinearly preconditioned inexact Newton algorithms[J]. SIAM Journal on Scientific Computing, 2002, 24(1): 183-200.
[98] Cajuhi T, Sanavia L, De Lorenzis L. Phase-field modeling of fracture in variably saturated porous media[J]. Computational Mechanics, 2018, 61: 299-318.
[99] Carey G F, Oden J T. Finite elements. 3. Computational Aspects[M]. Prentice-Hall, 1984.
[100] Carraro T, Goll C, Marciniak-Czochra A, et al. Pressure jump interface law for the Stokes-Darcy coupling: confirmationby direct numerical simulations[J]. Journal of Fluid Mechanics, 2013, 732(732): 510-536.
[101] Carraro T, Goll C, Marciniak-Czochra A, et al. Effective interface conditions for the forced infiltration of a viscous fluid into a porous medium using homogenization[J]. Computer Methods in Applied Mechanics & Engineering, 2015, 292(aug.1): 195-220.
[102] Carstensen C, Verfürth R. Edge residuals dominate a posteriori error estimates for low order finite element methods[J]. Siam Journal on Numerical Analysis, 1999, 36(5): 1571-1587.
[103] Castelletto N, White J A, Tchelepi H A. Accuracy and convergence properties of the fixed-stress iterative solution of two-way coupled poromechanics[J]. International Journal for Numerical and Analytical Methods in Geomechanics, 2015, 39(14): 1593-1618.
[104] Ceniceros H D, Nós R L, Roma A M. Three-dimensional, fully adaptive simulations of phase-field fluid models[J]. Journal of Computational Physics, 2010, 229(17): 6135-6155.
[105] Chambolle A. Image segmentation by variational methods: mumford and shah functional and the discrete approximations[J]. Siam Journal on Applied Mathematics, 1995, 55(3): 827-863.
[106] Chambolle A, Giacomini A, Ponsiglione M. Crack initiation in brittle materials[J]. Arch. Ration. Mech. Anal., 2008, 188: 309-349.
[107] Chambolle A, Maso G D. Discrete approximation of the Mumford-Shah functional in dimension two[J]. ESAIM Mathematical Modelling and Numerical Analysis, 1999, 33(4): 651-672.
[108] Chang C, Mear M E. A boundary element method for two dimensional linear elastic fracture analysis[J]. International Journal of Fracture, 1996, 74(3): 219-251.
[109] Chen X, Elliott C M, Gardiner A, et al. Convergence of numerical solutions to the Allen-Cahn equation[J]. Applicable Analysis, 1998, 69(1): 47-56.
[110] Chukwudozie C, Bourdin B, Yoshioka K. A variational phase-field model for hydraulic fracturing in porous media[J]. Computer Methods in Applied Mechanics and Engineering, 2019, 347(APR.15): 957-982.
[111] Ciarlet P G. Lectures on Three-Dimensional Elasticity[M]. Berlin: Springer, 1983.
[112] Ciarlet P G. Mathematical elasticity: three-dimensional elasticity[J]. Society for Industrial and Applied Mathematics, 2021, 25: 5.
[113] Ciarlet P G. The Finite Element Method for Elliptic Problems[M]. Society for Industrial

and Applied Mathematics, 2002.

[114] Ciarlet P G. Linear and Nonlinear Functional Analysis with Applications[M]. Siam, 2013.

[115] Conti S, Focardi M, Iurlano F. Existence of strong minimizers for the Griffith static fracture model in dimension two[C]. Annales de l'Institut Henri Poincaré C, Analyse non linéaire, No Longer Published by Elsevier, 2019, 36(2): 455-474.

[116] Cottrell J A, Hughes T J R, Bazilevs Y. Isogeometric Analysis: Toward Integration of CAD and FEA[M]. New York: John Wiley & Sons, 2009.

[117] Coussy O. Poromechanics of freezing materials[J]. Journal of the Mechanics & Physics of Solids, 2005, 53(8): 1689-1718.

[118] Curtis F E, Han Z, Robinson D P. A globally convergent primal-dual active-set framework for large-scale convex quadratic optimization[J]. Computational Optimization & Applications, 2015, 60(2): 311-341.

[119] Maso G D, Francfort G A, Toader R. Quasistatic crack growth in nonlinear elasticity[J]. Archive for Rational Mechanics & Analysis, 2005, 176(2): 165-225.

[120] Dal Maso G, Toader R. A model for the quasistatic growth of brittle fractures: existence and approximation results[J]. Archive for Rational Mechanics and Analysis, 2002, 162: 101-135.

[121] Darcy H. Les Fontaines Publiques de La Ville De Dijon: Exposition Et Application Des Principes à Suivre Et Des Formules à Employer Dans Les Questions De Distribution d'Eau[M]. Paris: Hachette Livre, 1856.

[122] Dautray R, Lions J L. Mathematical Analysis and Numerical Methods for Science and Technology: Volume 1 Physical Origins and Classical Methods[M]. Berlin: Springer Science & Business Media, 2012.

[123] Davis T A, Duff I S. An unsymmetric-pattern multifrontal method for sparse LU factorization[J]. SIAM Journal on Matrix Analysis and Applications, 1997, 18(1): 140-158.

[124] Davydov D, Pelteret J P, Arndt D, et al. A matrix-free approach for finite-strain hyperelastic problems using geometric multigrid[J]. International Journal for Numerical Methods in Engineering, 2020, 121(13): 2874-2895.

[125] de Borst R, Verhoosel C V. Gradient damage vs phase-field approaches for fracture: similarities and differences[J]. Computer Methods in Applied Mechanics and Engineering, 2016, 312: 78-94.

[126] Dean R H, Schmidt J H. Hydraulic-fracture predictions with a fully coupled geomechanical reservoir simulator[J]. SPE Journal, 2009, 14(4): 707-714.

[127] Dembo R S, Steihaug E T. Inexact newton methods[J]. Siam Journal on Numerical Analysis, 1982, 19(2): 400-408.

[128] Deuflhard P. Newton Methods for Nonlinear Problems: Affine Invariance and Adaptive Algorithms[M]. Berlin: Springer Science & Business Media, 2005.

[129] Diening L, Rika M. Interpolation operators in Orlicz-Sobolev spaces[J]. Numerische Mathematik, 2007, 107(1): 107-129.

[130] Dittmann M, Aldakheel F, Schulte J, et al. Variational phase-field formulation of nonlinear ductile fracture[J]. Computer Methods in Applied Mechanics and Engineering, 2018, 342(DEC.1): 71-94.

[131] Donéa J, Fasoli-Stella P, Giuliani S. Lagrangian and Eulerian finite element techniques for transient fluid-structure interaction problems[J]. IASMiRT, 1977.

[132] Donea J, Giuliani S, Halleux J P. An arbitrary Lagrangian-Eulerian finite element method for transient dynamic fluid-structure interactions[J]. Computer Methods in Applied Mechanics & Engineering, 1982, 33(1-3): 689-723.

[133] Donea J, Huerta A, Ponthot J P, et al. Arbitrary Lagrangian-Eulerian methods[J]. Encyclopedia of Computational Mechanics, 2004.

[134] The Differential Equation and Optimization Environment: DOpElib[OL]. http://www.dopelib.net. [2024-1-9].

[135] Dorfler W. A convergent adaptive algorithm for poisson\"s equation[J]. Siam Journal on Numerical Analysis, 1996, 33(3): 1106-1124.

[136] Duda F P, Ciarbonetti A, Sánchez P J, et al. A phase-field/gradient damage model for brittle fracture in elastic-plastic solids[J]. International Journal of Plasticity, 2015, 65: 269-296.

[137] Dunne T. An Eulerian approach to fluid-structure interaction and goal-oriented mesh adaptation[J]. International Journal for Numerical Methods in Fluids, 2010, 51(9-10): 1017-1039.

[138] Dunne T. Adaptive finite element approximation of fluid-structure interaction based on eulerian and arbitrary Lagrangian-Eulerian variational formulations[D]. Heidelberg: Heidelberg University, 2007.

[139] Dunne T, Rannacher R, Richter T. Numerical simulation of fluid-structure interaction based on monolithic variational formulations[J]. Fundamental Trends in Fluid-Structure Interaction, 2010: 1-75.

[140] Duvaut G, Lions J L, John C W, et al. Inequalities in mechanics and physics[J]. Journal of Applied Mechanics, 1976, 44(2): 364.

[141] Eaton J W, Bateman D, Hauberg S. GNU Octave version 3.0. 1 manual: a high-level interactive language for numerical computations[M]. SoHo Books, 2007.

[142] Eck C, Garcke H, Knabner P. Mathematische Modellierung[M]. Berlin: Springer, 2008.

[143] Emmerich H. The Diffuse Interface Approach in Materials Science: Thermodynamic Concepts and Applications of Phase-Field Models[M]. Springer Science & Business Media, 2003.

[144] Endtmayer B. Multi-goal oriented a posteriori error estimates for nonlinear partial differential equations[D]. Johannes Kepler University Linz, 2021.

[145] Bea B, Ul A, In C, et al. Multigoal-oriented optimal control problems with nonlinear PDE constraints[J]. Computers & Mathematics with Applications, 2020, 79(10): 3001-3026.

[146] Endtmayer B, Langer U, Thiele J P, et al. Hierarchical DWR error estimates for

the Navier-Stokes equations: h and p enrichment[C]. Numerical Mathematics and Advanced Applications ENUMATH 2019: European Conference, Egmond aan Zee, The Netherlands, September 30-October 4. Cham: Springer International Publishing, 2020: 363-372.

[147] Endtmayer B, Langer U, Wick T. Multiple goal-oriented error estimates applied to 3d non-linear problems[J]. PAMM, 2018, 18(1): 1-2.

[148] Endtmayer B, Langer U, Wick T. Multigoal-oriented error estimates for non-linear problems[J]. Journal of Numerical Mathematics, 2019, 27(4): 215-236.

[149] Endtmayer B, Langer U, Wick T. Reliability and efficiency of DWR-type a posteriori error estimates with smart sensitivity weight recovering[J]. Computational Methods in Applied Mathematics, 2021, 21(2): 351-371.

[150] Endtmayer B, Langer U, Wick T. Two-side a posteriori error estimates for the dual-weighted residual method[J]. SIAM Journal on Scientific Computing, 2020, 42(1): A371-A394.

[151] Endtmayer B, Wick T. A partition-of-unity dual-weighted residual approach for multi-objective goal functional error estimation applied to elliptic problems[J]. Computational Methods in Applied Mathematics, 2017, 17(4): 575-599.

[152] Engwer C, Pop I S, Wick T. Dynamic and weighted stabilizations of the l-scheme applied to a phase-field model for fracture propagation[C]. Numerical Mathematics and Advanced Applications ENUMATH 2019: European Conference, Egmond aan Zee, The Netherlands, September 30-October 4. Cham: Springer International Publishing, 2020: 1177-1184.

[153] Engwer C, Schumacher L. A phase field approach to pressurized fractures using discontinuous Galerkin methods[J]. Mathematics and Computers in Simulation, 2017, 137: 266-285.

[154] Ern A, Vohralík M. Adaptive inexact Newton methods with a posteriori stopping criteria for nonlinear diffusion PDEs[J]. Siam Journal on Scientific Computing, 2013, 35(4): A1761-A1791.

[155] Ernesti F, Schneider M, Böhlke T. Fast implicit solvers for phase-field fracture problems on heterogeneous microstructures[J]. Computer Methods in Applied Mechanics and Engineering, 2020, 363: 112793.

[156] Teschl G L C. Evans: Partial differential equations[J]. Internationale Mathematische Nachrichten, 2010, 64(215): 40.

[157] Failer L, Wick T. Adaptive time-step control for nonlinear fluid-structure interaction[J]. Journal of Computational Physics, 2018, 366: 448-477.

[158] Fan M, Wick T, Jin Y. A phase-field model for mixed-mode fracture[C]. Proceedings of the 8th GACM Colloquium on Computational Mechanics for Young Scientists from Academia and Industry, 2019.

[159] Farrell P, Maurini C. Linear and nonlinear solvers for variational phase-field models of brittle fracture[J]. International Journal for Numerical Methods in Engineering, 2017,

109(5): 648-667.

[160] Feischl M, Praetorius D, Van D. An abstract analysis of optimal goal-oriented adaptivity[J]. SIAM Journal on Numerical Analysis, 2016, 54(3): 1423-1448.

[161] Feng X, Prohl A. Numerical analysis of the Allen-Cahn equation and approximation for mean curvature flows[J]. Numerische Mathematik, 2003, 94(1): 33-65.

[162] Prohl F A. Analysis of a fully discrete finite element method for the phase field model and approximation of its sharp interface limits[J]. Mathematics of Computation, 2004, 73(246): 541-567.

[163] Fernández M A, Gerbeau J F. Algorithms for Fluid-Structure Interaction Problems[M]// Cardiovascular Mathematics: Modeling and simulation of the circulatory system. Milano: Springer Milan, 2009: 307-346.

[164] Fernández M A, Mullaert J. Displacement-velocity correction schemes for incompressible fluid-structure interaction[J]. Comptes rendus - Mathématique, 2011, 349(17-18): 1011-1015.

[165] Jeffrey R. The pleasure of finding things out[J]. Primary Science Review, 1999, 10(3): 587-590.

[166] Formaggia L, Nobile F. A stability analysis for the arbitrary Lagrangian Eulerian formulation with finite elements[J]. East-West Journal Of Mathematics, 1999, 7: 105-132.

[167] Cardiovascular Mathematics: Modeling and Simulation of the Circulatory System[M]. Milano: Springer Science & Business Media, 2010.

[168] Fortin M, Glowinski R. Augmented Lagrangian Methods: Applications to the Numerical Solution Of Boundary-Value problems[M]. Amsterdam: Elsevier, 2000.

[169] Francfort G A. Un résumé de la théorie variationnelle de la rupture[J]. Séminaire Laurent Schwartz—EDP et applications, 2011: 1-11.

[170] Francfort G A, Le N Q, Serfaty S. Critical points of Ambrosio-Tortorelli converge to critical points of Mumford-Shah in the one-dimensional Dirichlet case[J]. ESAIM Control Optimisation and Calculus of Variations, 2009, 15(3): 576-598.

[171] Francfort G A, Marigo J J. Revisiting brittle fracture as an energy minimization problem[J]. J. Mech. Phys. Solids, 1998, 46(8): 1319-1342.

[172] Francfort G A, Larsen C J. Existence and convergence for quasi-static evolution in brittle fracture[J]. Communications on Pure and Applied Mathematics, 2003, 56(10): 1465-1500.

[173] Freddi F, Royer-Carfagni G. Regularized variational theories of fracture: a unified approach[J]. Journal of the Mechanics & Physics of Solids, 2010, 58(8): 1154-1174.

[174] Frei S. Eulerian finite element methods for interface problems and fluid-structure interactions[D]. Heidelberg: Heidelberg University, 2016.

[175] Basting S, Birken P, Canic S, et al. Fluid-Structure Interaction: Modeling, Adaptive Discretisations and Solvers[M]. Germany: Walter de Gruyter GmbH & Co KG, 2017.

[176] Frei S, Richter T. A locally modified parametric finite element method for interface problems[J]. SIAM Journal on Numerical Analysis, 2014, 52(5): 2315-2334.

[177] Basting S, Birken P, Canic S, et al. Fluid-Structure Interaction: Modeling, Adaptive Discretisations and Solvers[M]. Walter de Gruyter GmbH & Co KG, 2017.

[178] Frei S, Richter T. A second order time-stepping scheme for parabolic interface problems with moving interfaces[J]. Esaim Mathematical Modelling & Numerical Analysis, 2017, 51(4): 1539-1560.

[179] Frei S, Richter T, Wick T. Long-term simulation of large deformation, mechano-chemical fluid-structure interactions in ALE and fully Eulerian coordinates[J]. Journal of Computational Physics, 2016: 874-891.

[180] Frei S, Richter T, Wick T. An implementation of a locally modified finite element method for interface problems in deal.II[J]. Zenodo, 2018.

[181] Fries T P, Belytschko T. The extended/generalized finite element method: an overview of the method and its applications[J]. International Journal for Numerical Methods in Engineering, 2010, 84(3): 253-304.

[182] Frohne J, Heister T, Bangerth W. Efficient numerical methods for the large-scale, parallel solution of elastoplastic contact problems[J]. International Journal for Numerical Methods in Engineering, 2016, 105(6): 416-439.

[183] Gai X. A Coupled Geomechanics And Reservoir Flow Model On Parallel Computers[M]. The University of Texas at Austin, 2004.

[184] Galdi G P, Rannacher R. Fundamental Trends in Fluid-Structure Interaction[M]. World Scientific, 2010.

[185] Gee M W, Küttler U, Wall W A. Truly monolithic algebraic multigrid for fluid-structure interaction[J]. International Journal for Numerical Methods in Engineering, 2011, 85(8): 987-1016.

[186] Geelen R, Plews J, Tupek M, et al. An extended/generalized phase-field finite element method for crack growth with global-local enrichment[J]. International Journal for Numerical Methods in Engineering, 2020, 121(11): 2534-2557.

[187] Geelen R J M, Liu Y, Hu T, et al. A phase-field formulation for dynamic cohesive fracture[J]. Computer Methods in Applied Mechanics and Engineering, 2018, 348(MAY 1): 680-711.

[188] Gendre L, Allix O, Gosselet P, et al. Non-intrusive and exact global/local techniques for structural problems with local plasticity[J]. Computational Mechanics, 2009, 44(2): 233-245.

[189] Gerasimov T, Lorenzis L D. A line search assisted monolithic approach for phase-field computing of brittle fracture[J]. Computer Methods in Applied Mechanics & Engineering, 2015, 312(dec.1): 276-303.

[190] Gerasimov T, Lorenzis L D. On penalization in variational phase-field models of brittle fracture[J]. Computer Methods in Applied Mechanics and Engineering, 2019, 354(SEP.1): 990-1026.

[191] Tymofiy G, Nima N, Olivier A, et al. A non-intrusive global/local approach applied to phase-field modeling of brittle fracture[J]. Advanced Modeling & Simulation in Engi-

neering Sciences, 2018, 5(1): 14.

[192] Geuzaine C, Remacle J F. Gmsh: a 3-D finite element mesh generator with built-in pre-and post-processing facilities[J]. International Journal for Numerical Methods in Engineering, 2009, 79(11): 1309-1331.

[193] Ghassemi A, Kumar G S. Changes in fracture aperture and fluid pressure due to thermal stress and silica dissolution/precipitation induced by heat extraction from subsurface rocks[J]. Geothermics, 2007, 36(2): 115-140.

[194] Giacomini A. Ambrosio-Tortorelli approximation of quasi-static evolution of brittle fractures[J]. Calculus of Variations & Partial Differential Equations, 2005, 22(2): 129-172.

[195] de Giorgi E, Ambrosio L. Un nuovo tipo di funzionale del calcolo delle variazioni[J]. Atti della Accademia Nazionale dei Lincei. Classe di Scienze Fisiche, Matematiche e Naturali. Rendiconti Lincei. Matematica e Applicazioni, 1988, 82(2): 199-210.

[196] Giovanardi B, Scotti A, Formaggia L. A hybrid XFEM –Phase field (Xfield) method for crack propagation in brittle elastic materials[J]. Computer Methods in Applied Mechanics & Engineering, 2017, 320(JUN.15): 396-420.

[197] Girault V, Pencheva G, Wheeler M F, et al. Domain decomposition for poroelasticity and elasticity with dg jumps and mortars[J]. Mathematical Models & Methods in Applied Sciences, 2011, 21(1): 169-213.

[198] Girault V, Raviart P A. Finite Element Methods for Navier-Stokes Equations: Theory and Algorithms[M]. Springer Science & Business Media, 2012.

[199] Glowinski R, Leung S, Qian J. Operator-splitting based fast sweeping methods for isotropic wave propagation in a moving fluid[J]. SIAM Journal on Scientific Computing, 2016, 38(2): A1195-A1223.

[200] Glowinski R, Pan T W, Periaux J. A fictitious domain method for Dirichlet problem and applications[J]. Comput. methods Appl. Mech. Engrg, 1994, 111(3-4): 283-303.

[201] Glowinski R, Periaux J. Numerical methods for nonlinear problem in fluid dynamics[C]. INRIA Conference on Supercomputing: State-of-the-Art, 1987: 381-479.

[202] Glowinski R, Le Tallec P. Augmented Lagrangian and Operator-Splitting Methods in Nonlinear Mechanics[M]. Philadelphia: Society for Industrial and Applied Mathematics, 1989.

[203] Goll C, Wick T, Wollner W. DOpElib: differential equations and optimization environment; A goal oriented software library for solving pdes and optimization problems with pdes[J]. Archive of Numerical Software, 2017, 5(2): 1-14.

[204] Griffith A A. VI. The phenomena of rupture and flow in solids[J]. Philosophical Transactions of the Royal Society of London. Series A, Containing Papers of a Mathematical or Physical Character, 1921, 221(582-593): 163-198.

[205] Grossmann C, Roos H G, Stynes M. Numerical Treatment of Partial Differential Equations[M]. Berlin Heidelberg: Springer, 2007.

[206] Gustafsson K, Lundh M, Sderlind G. API stepsize control for the numerical solution of ordinary differential equations[J]. Bit Numerical Mathematics, 1988, 28(2): 270-287.

[207] Florez Guzman H A. Domain decomposition methods in geomechanics[D]. Austin: The University of Texas at Austin, 2012.

[208] Hagoort J. Waterflood-induced hydraulic fracturing[D]. Delft: Technische Universiteit Delft, 1981.

[209] Han W. A Posteriori Error Analysis Via Duality Theory: With Applications In Modeling And Numerical Approximations[M]. Springer Science & Business Media, 2004.

[210] Han Z. Primal-Dual Active-Set Methods for Convex Quadratic Optimization with Applications[D]. Bethlehem: Lehigh University, 2015.

[211] Hartmann R. Multitarget error estimation and adaptivity in aerodynamic flow simulations[J]. Siam Journal on Scientific Computing, 2008, 31(1): 708-731.

[212] Hartmann R, Houston P. Goal-oriented a posteriori error estimation for multiple target functionals[C]. Hyperbolic Problems: Theory, Numerics, Applications: Proceedings of the Ninth International Conference on Hyperbolic Problems held in CalTech, Pasadena, March 25-29, 2002. Berlin, Heidelberg: Springer Berlin Heidelberg, 2003: 579-588.

[213] Hausdorff F. Dimension und äußeres maß[J]. Mathematische Annalen, 1918, 79(1-2): 157-179.

[214] Heider Y, Markert B. A phase-field modeling approach of hydraulic fracture in saturated porous media[J]. Mechanics Research Communications, 2017, 80: 38-46.

[215] Heider Y, Reiche S, Siebert P, et al. Modeling of hydraulic fracturing using a porous-media phase-field approach with reference to experimental data[J]. Engineering Fracture Mechanics, 2018, 202: 116-134.

[216] Heil M. An efficient solver for the fully coupled solution of large-displacement fluid-structure interaction problems[J]. Computer Methods in Applied Mechanics and Engineering, 2004, 193(1-2): 1-23.

[217] Heister T, Wheeler M F, Wick T. A primal-dual active set method and predictor-corrector mesh adaptivity for computing fracture propagation using a phase-field approach[J]. Computer Methods in Applied Mechanics and Engineering, 2015, 290: 466-495.

[218] Heister T, Wick T. Parallel solution, adaptivity, computational convergence, and open-source code of 2D and 3D pressurized phase-field fracture problems[J]. Pamm, 2018, 18(1): e201800353.

[219] Helenbrook B T. Mesh deformation using the biharmonic operator[J]. International Journal for Numerical Methods in Engineering, 2010, 56(7): 1007-1021.

[220] Heroux M A, Bartlett R A, Howle V E, et al. An overview of the trilinos project[J]. ACM Transactions on Mathematical Software, 2005, 31(3): 397-423.

[221] Hesch C, Gil A J, Ortigosa R, et al. A framework for polyconvex large strain phase-field methods to fracture[J]. Computer Methods in Applied Mechanics and Engineering, 2017, 317(APR.15): 649-683.

[222] Hesch C, Schuss S, Dittmann M, et al. Isogeometric analysis and hierarchical refinement for higher-order phase-field models[J]. Computer Methods in Applied Mechanics

& Engineering, 2016, 303(May1): 185-207.

[223] Hestenes M R. Multiplier and gradient methods[J]. Journal of Optimization Theory & Applications, 1969, 4(5): 303-320.

[224] Rannacher H R. Finite-element approximation of the nonstationary navier-stokes problem part IV: error analysis for second-order time discretization[J]. Siam Journal on Numerical Analysis, 1990, 27(2): 353-384.

[225] Heywood J G, Rannacher R, Turek S. Artificial boundaries and flux and pressure conditions for the incompressible Navier-Stokes equations[J]. International Journal of Numerical Methods in Fluids, 1996, 22: 325-352.

[226] Hintermueller M, Hoppe R H W, Loebhard C. Dual-weighted goal-oriented adaptive finite elements for optimal control of elliptic variational inequalities[J]. Esaim, 2014, 20(2): 524-546.

[227] Hartmann R. Multitarget error estimation and adaptivity in aerodynamic flow simulations[J]. SIAM Journal on Scientific Computing, 2008, 31(1): 708-731.

[228] Hinze M, Pinnau R, Ulbrich M, et al. Optimization with PDE Constraints[M]. Springer Science & Business Media, 2008.

[229] Hirn A. Finite element approximation of singular power-law systems[J]. Math. Comp., 2013, 82(283): 1247-1268.

[230] Hirt C W, Amsden A A, Cook J L. An arbitrary Lagrangian-Eulerian computing method for all flow speeds[J]. Journal of Computational Physics, 1974, 14(2): 227-253.

[231] Hofacker M, Miehe C. Continuum phase field modeling of dynamic fracture: variational principles and staggered FE implementation[J]. International Journal of Fracture, 2012, 178(1-2): 113-129.

[232] Holzapfel G A. Nonlinear solid mechanics: a continuum approach for engineering science[J]. Kluwer Academic Publishers Dordrecht, 2002: 489-490.

[233] Hong Q, Kraus J. Parameter-robust stability of classical three-field formulation of Biot's consolidation model[J]. arXiv preprint arXiv:1706.00724, 2017.

[234] Hron J, Ouazzi A, Turek S. A computational comparison of two FEM solvers for nonlinear incompressible flow[C]. Challenges in Scientific Computing-CISC 2002: Proceedings of the Conference Challenges in Scientific Computing Berlin, October 2-5, 2002. Springer Berlin Heidelberg, 2003: 87-109.

[235] Hron J, Turek S. A Monolithic Fem/Multigrid Solver For an Ale Formulation of Fluid-Structure Interaction With Applications In Biomechanics[M]. Fluid-Structure Interaction: Modelling, Simulation, Optimisation. Berlin, Heidelberg: Springer Berlin Heidelberg, 2006: 146-170.

[236] Turek S, Hron J. Proposal for Numerical Benchmarking of Fluid-Structure Interaction between an Elastic Object and Laminar Incompressible Flow[M]. Berlin, Heidelberg: Springer, 2006.

[237] Hughes T J R. The Finite Element Method: Linear Static and Dynamic Finite Element Analysis[M]. Courier Corporation, 2012.

[238] Hughes T, Cottrell J A, Bazilevs Y. Isogeometric analysis: CAD, finite elements, NURBS, exact geometry and mesh refinement[J]. Computer Methods in Applied Mechanics & Engineering, 2005, 194(39-41): 4135-4195.

[239] Hughes T J R, Liu W K, Zimmermann T K. Lagrangian-Eulerian finite element formulation for incompressible viscous flows[J]. Computer methods in applied mechanics and engineering, 1981, 29(3): 329-349.

[240] Ito K, Kunisch K. Augmented Lagrangian methods for nonsmooth, convex optimization in Hilbert spaces[J]. Nonlinear Analysis, 2000, 41(5): 591-616.

[241] Ito K, Kunisch K. Lagrange Multiplier Approach to Variational Problems and Applications[M]. Philadelphia: Society for Industrial and Applied Mathematics, 2008.

[242] Mikelic A, Jäger W. On the interface boundary condition of Beavers, Joseph, and Saffman[J]. SIAM Journal on Applied Mathematics, 2000, 60(4): 1111-1127.

[243] Jammoul M, Wheeler M F, Wick T. A phase-field multirate scheme with stabilized iterative coupling for pressure driven fracture propagation in porous media[J]. Computers & Mathematics with Applications, 2021, 91: 176-191.

[244] Jodlbauer D, Langer U, Wick T. Parallel block-preconditioned monolithic solvers for fluid-structure interaction problems[J]. International Journal for Numerical Methods in Engineering, 2019, 117(6): 623-643.

[245] Jodlbauer D, Langer U, Wick T. Parallel matrix-free higher-order finite element solvers for phase-field fracture problems[J]. Mathematical and Computational Applications, 2020, 25(3): 40-61.

[246] Jodlbauer D, Langer U, Wick T. Matrix-free multigrid solvers for phase-field fracture problems[J]. Computer Methods in Applied Mechanics and Engineering, 2020, 372: 113431.

[247] Jodlbauer D, Langer U, Wick T. Efficient Monolithic Solvers for Fluid-Structure Interaction Applied to Flapping Membranes[M]. Domain Decomposition Methods in Science and Engineering XXVI. Cham: Springer International Publishing, 2023: 327-335.

[248] John V, Rang J. Adaptive time step control for the incompressible Navier-Stokes equations[J]. Computer Methods in Applied Mechanics & Engineering, 2010, 199(9-12): 514-524.

[249] Wrobel L C. Numerical solution of partial differential equations by the finite element method[J]. Engineering Analysis with Boundary Elements, 1987, 9(1): 106.

[250] Johnson C. Adaptive finite element methods for the obstacle problem[J]. Mathematical Models & Methods in Applied Sciences, 1992, 2(4): 483-487.

[251] Kantorovich L V. On Newton's method for functional equations[J]. Doklady Akademii Nauk SSSR, 1948, 59(7): 1237-1240.

[252] Kantorovich L V, Akilov G P. Functional Analysis in Normed Spaces [in Russian], chapter 8[M]. Pergamon Press Inc., 1959.

[253] Kergrene K, Prudhomme S, Chamoin L, et al. A new goal-oriented formulation of the finite element method[J]. Computer Methods in Applied Mechanics and Engineering,

2017, 327: 256-276.

[254] Khimin D. Numerische Modellierung von optimalsteuerungsproblemen für phasenfeld-riss-ausbreitung[D]. Master's thesis, Leibniz University Hannover, 2020.

[255] Khodadadian A, Noii N, Parvizi M, et al. A Bayesian estimation method for variational phase-field fracture problems[J]. Computational Mechanics, 2020, 66: 827-849.

[256] Kikuchi N, Oden J T. Contact Problems in Elasticity: a Study of Variational Inequalities and Finite Element Methods[M]. Philadelphia: Society for Industrial and Applied Mathematics, 1988.

[257] Kinderlehrer D, Stampacchia G. An Introduction to Variational Inequalities and Their Applications[M]. Society for Industrial and Applied Mathematics, 2000.

[258] Köcher U, Bruchhäuser M P, Bause M. Efficient and scalable data structures and algorithms for goal-oriented adaptivity of space-time FEM codes[J]. SoftwareX, 2019, 10: 100239.

[259] Königsberger K. Analysis 1[M]. New York: Springer-Verlag, 2013.[eBook ISBN:978-3-642-18490-1 Published: 07 March 2013].

[260] Königsberger K. Analysis 2[M]. New York: Springer-Verlag, 2013. [eBook ISBN: 978-3-540-35077-4 Published: 16 July 2006].

[261] Kopaničáková A, Krause R. A recursive multilevel trust region method with application to fully monolithic phase-field models of brittle fracture[J]. Computer Methods in Applied Mechanics and Engineering, 2020, 360: 112720.

[262] Kronbichler M. Computational Techniques for Coupled Flow-Transport Problems[D]. Acta Universitatis Upsaliensis, 2011.

[263] Kuhn C, Müller R. Exponential finite element shape functions for a phase field model of brittle fracture[C]. COMPLAS XI: Proceedings of the XI International Conference on Computational Plasticity: Fundamentals and Applications. CIMNE, 2011: 478-489.

[264] Kuhn C, Müller R. A continuum phase field model for fracture[J]. Engineering Fracture Mechanics, 2010, 77(18): 3625-3634.

[265] Ambati M, Gerasimov T, Lorenzis L D. Phase-field modeling of ductile fracture[J]. Computational Mechanics, 2015, 55(5): 1017-1040.

[266] Kumar A, Bourdin B, Francfort G A, et al. Revisiting nucleation in the phase-field approach to brittle fracture[J]. Journal of the Mechanics and Physics of Solids, 2020, 142: 104027.

[267] Kumar G S, Ghassemi A. Numerical modeling of non-isothermal quartz dissolution/precipitation in a coupled fracture-matrix system[J]. Geothermics, 2005, 34(4): 411-439.

[268] Ladyzhenskaia O A, Solonnikov V A, Ural'tseva N N. Linear and Quasi-Linear Equations of Parabolic Type[M]. American Mathematical Soc, 1988.

[269] Space-Time Methods: Applications to Partial Differential Equations[M]. Walter de Gruyter GmbH & Co KG, 2019.

[270] Langer U, Yang H. Robust and efficient monolithic fluid-structure-interaction solvers[J].

International Journal for Numerical Methods in Engineering, 2016, 108(4): 303-325.

[271] Langer U, Yang H. Recent development of robust monolithic fluid-structure interaction solvers[J]. Fluid-Structure Interaction. Modeling, Adaptive Discretization and Solvers. Radon Series on Computational and Applied Mathematics, 2017, 20: 169-192.

[272] Hackl K. IUTAM Symposium on Variational Concepts with Applications to the Mechanics of Materials: Proceedings of the IUTAM Symposium on Variational Concepts with Applications to the Mechanics of Materials, Bochum, Germany, September 22-26, 2008[M]. Dordrecht: Springer Science & Business Media, 2010.

[273] Imbert C, Mellet A. Existence of solutions for a higher order non-local equation appearing in crack dynamics[J]. Nonlinearity, 2011, 24(12): 3487.

[274] LeVeque R J. Top ten reasons to not share your code (and why you should anyway)[J]. Siam News, 2013, 46(3): 15.

[275] Lee S, Lee Y J, Wheeler M F. A locally conservative enriched Galerkin approximation and efficient solver for elliptic and parabolic problems[J]. SIAM Journal on Scientific Computing, 2016, 38(3): A1404-A1429.

[276] Lee S, Mikelić A, Wheeler M F, et al. Phase-field modeling of two phase fluid filled fractures in a poroelastic medium[J]. SIAM Journal on Multiscale Modeling and Simulation, 2018, 16(4): 1542-1580.

[277] Lee S, Mikeli'C A, Wheeler M F, et al. Phase-field modeling of proppant-filled fractures in a poroelastic medium[J]. Computer Methods in Applied Mechanics and Engineering, 2016, 312: 509-541.

[278] Lee S, Min B, Wheeler M F. Optimal design of hydraulic fracturing in porous media using the phase field fracture model coupled with genetic algorithm[J]. Computational Geosciences, 2018, 22(3): 1-17.

[279] Lee S, Wheeler M F. Adaptive enriched Galerkin methods for miscible displacement problems with entropy residual stabilization[J]. Journal of Computational Physics, 2017, 331: 19-37.

[280] Lee S, Wheeler M F. Enriched Galerkin methods for two-phase flow in porous media with capillary pressure[J]. Journal of Computational Physics, 2018, 367: 65-86.

[281] Lee S, Wheeler M F, Wick T. Pressure and fluid-driven fracture propagation in porous media using an adaptive finite element phase field model[J]. Computer Methods in Applied Mechanics and Engineering, 2016, 305: 111-132.

[282] Lee S, Wheeler M F, Wick T. Iterative coupling of flow, geomechanics and adaptive phase-field fracture including level-set crack width approaches[J]. Journal of Computational and Applied Mathematics, 2017, 314: 40-60.

[283] Lee S, Wheeler M F, Wick T, et al. Initialization of phase-field fracture propagation in porous media using probability maps of fracture networks[J]. Mechanics Research Communications, 2017, 80: 16-23.

[284] Lions J L, Magenes E, Magenes E. Problemes Aux Limites Non Homogenes Et Applications[M]. Paris: Dunod, 1968.

[285] List F, Radu F A. A study on iterative methods for solving Richards' equation[J]. Computational Geosciences, 2016, 20: 341-353.

[286] Liu W, Barrett J W. A further remark on the regularity of the solutions of the p-Laplacian and its applications to their finite element approximation[J]. Nonlinear Analysis: Theory, Methods & Applications, 1993, 21(5): 379-387.

[287] Luskin M, Rannacher R, Wendland W. On the smoothing properties of crank-nicholson scheme[J]. Applicable Analysis, 1982, 14(2): 117-135.

[288] Maes C M. A Regularized Active-set Method for Sparse Convex Quadratic Programming[M]. Stanford: Stanford University, 2010.

[289] Maier M, Rannacher R. A duality-based optimization approach for model adaptivity in heterogeneous multiscale problems[J]. Multiscale Modeling & Simulation, 2018, 16(1): 412-428.

[290] Mallik G, Vohralík M, Yousef S. Goal-oriented a posteriori error estimation for conforming and nonconforming approximations with inexact solvers[J]. Journal of Computational and Applied Mathematics, 2020, 366: 112367.

[291] Mandal S, Ouazzi A, Turek S. Modified Newton solver for yield stress fluids[C]. Numerical Mathematics and Advanced Applications ENUMATH 2015, Springer International Publishing, 2016: 481-490.

[292] Mandel J. Consolidation des sols (étude mathématique)[J]. Geotechnique, 1953, 3(7): 287-299.

[293] Mang K, Walloth M, Wick T, et al. Mesh adaptivity for quasi-static phase-field fractures based on a residual-type a posteriori error estimator[J]. GAMM-Mitteilungen, 2020, 43(1): e202000003.

[294] Mang K, Walloth M, Wick T, et al. Adaptive numerical simulation of a phase-field fracture model in mixed form tested on an L-shaped specimen with high poisson ratios[C]. Numerical Mathematics and Advanced Applications ENUMATH 2019: European Conference, Egmond aan Zee, The Netherlands, September 30-October 4. Cham: Springer International Publishing, 2020: 1185-1193.

[295] Mang K, Wick T. Numerical methods for variational phase-field fracture problems[J]. Hannover:Institutionelles Repositorium der Leibniz Universität Hannover, 2019.

[296] Mang K, Wick T, Wollner W. A phase-field model for fractures in nearly incom-pressible solids[J]. Computational Mechanics, 2020, 65: 61-78.

[297] Marigo J J, Maurini C, Pham K. An overview of the modelling of fracture by gradient damage models[J]. Meccanica, 2016, 51: 3107-3128.

[298] Markert B, Heider Y. Coupled multi-field continuum methods for porous media fracture[J]. Recent Trends in Computational Engineering-CE2014: Optimization, Uncertainty, Parallel Algorithms, Coupled and Complex Problems, 2015: 167-180.

[299] Maschke H G, Kuna M. A review of boundary and finite element methods in fracture mechanics[J]. Theoretical and Applied Fracture Mechanics, 1985, 4(3): 181-189.

[300] Mattis S A, Wohlmuth B. Goal-oriented adaptive surrogate construction for stochastic

inversion[J]. Computer Methods in Applied Mechanics & Engineering, 2018, 339(SEP.1): 36-60.

[301] Mehlmann C, Richter T. A modified global Newton solver for viscous-plastic sea ice models[J]. Ocean Modelling, 2017, 116: 96-107.

[302] Meidner D, Rannacher R, Vihharev J. Goal-oriented error control of the iterative solution of finite element equations[J]. Journal of Numerical Mathematics, 2009, 17(2): 143-172.

[303] Meidner D, Richter T. Goal-oriented error estimation for the fractional step theta scheme[J]. Comput. Methods Appl. Math., 2014, 14(2): 203-230.

[304] Meidner D, Richter T. A posteriori error estimation for the fractional step theta discretization of the incompressible Navier-Stokes equations[J]. Computer Methods in Applied Mechanics & Engineering, 2015, 288(may 1): 45-59.

[305] Melenk J M, Babuška I. The partition of unity finite element method: basic theory and applications[J]. Computer Methods in Applied Mechanics and Engineering, 1996, 139(1-4): 289-314.

[306] Mesgarnejad A, Bourdin B, Khonsari M M. Validation simulations for the variational approach to fracture[J]. Computer Methods in Applied Mechanics & Engineering, 2015, 290(jun.15): 420-437.

[307] Miehe C, Hofacker M, Schaenzel L M, et al. Phase field modeling of fracture in multi-physics problems. Part II. Coupled brittle-to-ductile failure criteria and crack propagation in thermo-elastic–plastic solids[J]. Computer Methods in Applied Mechanics & Engineering, 2015, 294(SEP.1): 486-522.

[308] Miehe C, Hofacker M, Welschinger F. A phase field model for rate-independent crack propagation: robust algorithmic implementation based on operator splits[J]. Computer Methods in Applied Mechanics and Engineering, 2010, 199(45-48): 2765-2778.

[309] Miehe C, Mauthe S. Phase field modeling of fracture in multi-physics problems. Part III. Crack driving forces in hydro-poro-elasticity and hydraulic fracturing of fluid-saturated porous media[J]. Computer Methods in Applied Mechanics and Engineering, 2016, 304(1): 619-655.

[310] Miehe C, Mauthe S, Teichtmeister S. Minimization principles for the coupled problem of Darcy-Biot-type fluid transport in porous media linked to phase field modeling of fracture[J]. Journal of the Mechanics & Physics of Solids, 2015, 82(SEP.): 186-217.

[311] Miehe C, Schaenzel L M, Ulmer H. Phase field modeling of fracture in multi-physics problems. Part I. Balance of crack surface and failure criteria for brittle crack propagation in thermo-elastic solids[J]. Computer Methods in Applied Mechanics and Engineering, 2015, 294: 449-485.

[312] Miehe C, Welschinger F, Hofacker M. Thermodynamically consistent phase-field models of fracture: variational principles and multi-field FE implementations[J]. International Journal for Numerical Methods in Engineering, 2010, 83(10): 1273-1311.

[313] Mikelic A, Wheeler M F, Wick T. A phase field approach to the fluid filled fracture

surrounded by a poroelastic medium[J]. ICES Report, 2013: 1315.

[314] Mikeli A, Wheeler M F. Convergence of iterative coupling for coupled flow and geomechanics[J]. Computational Geosciences, 2013, 17(3): 455-461.

[315] Mikelić A, Wheeler M F, Sanchez-Palencia E. On the interface law between a deformable porous medium containing a viscous fluid and an elastic body[J]. Mathematical Models & Methods in Applied Sciences, 2012, 22(11): 1250031.1-1250031.32.

[316] Mikelic A, Wheeler M F, Wick T. A phase-field method for propagating fluid-filled fractures coupled to a surrounding porous medium[J]. Multiscale Modeling & Simulation, 2015, 13(1): 367-398.

[317] Mikelic A, Wheeler M F, Wick T. Phase-field modeling of a fluid-driven fracture in a poroelastic medium[J]. Computational Geosciences, 2015, 19(6): 1171-1195.

[318] Mikeli A, Wheeler M F, Wick T. A quasistatic phase field approach to pressurized fractures[J]. Nonlinearity, 2015, 28(5): 1371-1399.

[319] Mikelić A, Wheeler M F, Wick T. Phase-field modeling through iterative splitting of hydraulic fractures in a poroelastic medium[J]. GEM-International Journal on Geomathematics, 2019, 10(1): 2.

[320] Mital P. The enriched Galerkin method for linear elasticity and phase field fracture propagation[D]. Austin: The University of Texas at Austin, 2015.

[321] Mital P, Wick T, Wheeler M F, et al. Discontinuous and enriched Galerkin methods for phase-field fracture propagation in elasticity[C]. Numerical Mathematics and Advanced Applications ENUMATH 2015. Springer International Publishing, 2016: 195-203.

[322] Modica L, Mortola S. Il limite nella Γ-convergenza di una famiglia di funzionali ellittici[J]. Bollettino Della Unione Matematica Italiana A, 1977, 14(3): 526-529.

[323] Moës N, Dolbow J, Belytschko T. A finite element method for crack growth without remeshing[J]. International Journal for Numerical Methods in Engineering, 1999, 46(1): 131-150.

[324] Mulder W, Osher S, Sethian J A. Computing interface motion in compressible gas dynamics[J]. Journal of Computational Physics, 1992, 100(2): 209-228.

[325] Mumford D, Shah J. Optimal approximations by piecewise smooth functions and associated variational problems[J]. Communications on Pure & Applied Mathematics, 1989, 42(5): 577-685.

[326] Nabh G. On high order methods for the stationary incompressible navier-stokes equations[D]. Heidelberg: Universität Heidelberg, 1998.

[327] Natterer F. Optimale L2-konvergenz finiter elemente bei variationsungleichungen[J]. Bonn. Math. Schrift, 1976, 89: 1-12.

[328] Negri M. The anisotropy introduced by the mesh in the finite element approximation of the Mumford-Shah functional[J]. Numerical Functional Analysis and Optimization, 1999, 20(9-10): 957-982.

[329] Negri M. Γ-convergence for high order phase field fracture: continuum and isogeometric formulations[J]. Computer Methods in Applied Mechanics and Engineering, 2020, 362:

112858.

[330] Neittaanmäki P, Repin S R. Reliable Methods for Computer Simulation: Error Control and Posteriori Estimates[M]. Amsterdam: Elsevier, 2004.

[331] Neitzel I, Wick T, Wollner W. An optimal control problem governed by a regular-ized phase-field fracture propagation model[J]. SIAM Journal on Control and Optimization, 2017, 55(4): 2271-2288.

[332] Neitzel I, Wick T, Wollner W. An optimal control problem governed by a regular-ized phase-field fracture propagation model. part ii: The regularization limit[J]. SIAM Journal on Control and Optimization, 2019, 57(3): 1672-1690.

[333] Nguyen T T, Yvonnet J, Zhu Q Z, et al. A phase-field method for computational modeling of interfacial damage interacting with crack propagation in realistic microstructures obtained by microtomography[J]. Computer Methods in Applied Mechanics & Engineering, 2015, 312(dec.1): 567-595.

[334] Ni L, Zhang X, Zou L, et al. Phase-field modeling of hydraulic fracture network propagation in poroelastic rocks[J]. Computational Geosciences, 2020, 24: 1767-1782.

[335] Nocedal J, Wright S J. Theory of Constrained Optimization[M]. New York: Springer, 2006.

[336] Noii N, Aldakheel F, Wick T, et al. An adaptive global-local approach for phase-field modeling of anisotropic brittle fracture[J]. Computer Methods in Applied Mechanics and Engineering, 2019, 361: 112744.

[337] Noii N, Wick T. A phase-field description for pressurized and non-isothermal propagating fractures[J]. Computer Methods in Applied Mechanics and Engineering, 2019, 351: 860-890.

[338] Oden J T, Prudhomme S. Estimation of modeling error in computational mechanics[J]. Journal of Computational Physics, 2002, 182(2): 496-515.

[339] Oden J T. Adaptive multiscale predictive modelling[J]. Acta Numerica, 2018, 27: 353-450.

[340] Osher S, Sethian J A. Fronts propagating with curvature-dependent speed: algorithms based on Hamilton-Jacobi formulations[J]. Journal of Computational Physics, 1988, 79(1): 12-49.

[341] Pan E. A general boundary element analysis of 2-D linear elastic fracture mechanics[J]. International Journal of Fracture, 1997, 88(1): 41-59.

[342] Pardo D. Multigoal-oriented adaptivity for -finite element methods[J]. Procedia Computer Science, 2010, 1(1): 1953-1961.

[343] Parvizian J, Düster A, Rank E. Finite cell method: h-and p-extension for embedded domain problems in solid mechanics[J]. Computational Mechanics, 2007, 41(1): 121-133.

[344] Patil R U, Mishra B K, Singh I V. An adaptive multiscale phase field method for brittle fracture[J]. Computer Methods in Applied Mechanics and Engineering, 2018, 329: 254-288.

[345] Peskin C S. The immersed boundary method[J]. Acta Numerica, 2002, 11: 479-517.

[346] Pham K, Amor H, Maurini C, et al. Gradient damage models and their use in brittle fracture, nternational J[J]. Damage Mech., 2011, 20: 618-652.

[347] Pillo G D, Liuzzi G, Lucidi S, et al. A truncated Newton method in an augmented Lagrangian framework for nonlinear programming[J]. Computational Optimization & Applications, 2010, 45(2): 311-352.

[348] Pop I S, Radu F, Knabner P. Mixed finite elements for the Richards' equation: linearization procedure[J]. Journal of Computational and Applied Mathematics, 2004, 168(1): 365-373.

[349] Powell M J D. A method for nonlinear constraints in minimization problems[J]. Optimization, 1969: 283-298.

[350] Provatas N, Elder K. Phase-Field Methods in Materials Science and Engineering[M]. John Wiley & Sons, 2011.

[351] Glowinski R, Marroco A. Sur l'approximation, par éléments finis d'ordre un, et la résolution, par pénalisation-dualité d'une classe de problèmes de Dirichlet non linéaires[J]. Revue Française d'automatique, Informatique, Recherche Opérationnelle. Analyse Numérique, 1975, 9(R2): 41-76.

[352] Rademacher A. Adaptive finite element methods for nonlinear hyperbolic problems of second order[D]. Technische Universität Dortmund, 2010.

[353] Rank E, Ruess M, Kollmannsberger S, et al. Geometric modeling, isogeometric analysis and the finite cell method[J]. Computer Methods in Applied Mechanics & Engineering, 2012, 249-252(DEC.1): 104-115.

[354] Rannacher R. Finite element solution of diffusion problems with irregular data[J]. Numerische Mathematik, 1984, 43(2): 309-327.

[355] Rannacher R. On the stabilization of the Crank-Nicolson scheme for long time calculations[J]. Preprint, August, 1986.

[356] Rannacher R. Finite Element Methods for the Incompressible Navier-Stokes Equations[M]. Birkhäuser Basel, 2000.

[357] Rannacher R. Numerische methoden der kontinuumsmechanik[J]. Numerische Mathematik, 2005, 3.

[358] Rannacher R. Numerik Gewöhnlicher Differentialgleichungen. Vorlesungsskriptum, Universität Heidelberg[J]. Heidelberg University, 2011.

[359] Rannacher R. Numerik Partieller Differentialgleichungen[M]. Heidelberg: Heidelberg University Publishing, 2017.

[360] Rannacher R. Probleme Der Kontinuumsmechanik Und Ihre Numerische Behandlung[M]. Heidelberg: Heidelberg University Publishing, 2017.

[361] Rannacher R. Special Topics in Numerics: I. FEM for Nonlinear Problems[M]. Heidelberg: Heidelberg University Publishing, 2017.

[362] Rannacher R, Suttmeier F T. A posteriori error control in finite element methods via duality techniques: application to perfect plasticity[J]. Computational Mechanics, 1998, 21(2): 123-133.

[363] Rannacher R, Suttmeier F T. A posteriori error estimation and mesh adaptation for finite element models in elasto-plasticity[J]. Computer Methods in Applied Mechanics & Engineering, 1999, 176(1/4): 333-361.

[364] Rannacher R, Suttmeier F T. Error estimation and adaptive mesh design for FE models in elasto-plasticity[J]. Error-Controlled Adaptive FEMs in Solid Mechanics, John Wiley, 2000.

[365] Rannacher R, Vihharev J. Adaptive finite element analysis of nonlinear problems: balancing of discretization and iteration errors[J]. Journal of Numerical Mathematics, 2013, 21(1): 23-62.

[366] Repin S. A Posteriori Estimates for Partial Differential Equations[M]. Berlin: Walter de Gruyter, 2008.

[367] Rice J R. Mathematical analysis in the mechanics of fracture[J]. Fracture: an Advanced Treatise, 1968, 2: 191-311.

[368] Richter T. A Fully Eulerian formulation for fluid-structure-interaction problems[J]. Journal of Computational Physics, 2013, 233(none): 227-240.

[369] Richter T. Goal-oriented error estimation for fluid-structure interaction problems[J]. Computer Methods in Applied Mechanics and Engineering, 2011, 223–224(2): 28-42.

[370] Richter T. A monolithic geometric multigrid solver for fluid-structure interactions in ALE formulation[J]. International Journal for Numerical Methods in Engineering, 2015, 104(5): 372-390.

[371] Richter T. Fluid-Structure Interactions: Models, Analysis and Finite Elements[M]. Springer, 2017.

[372] Richter T, Wick T. Finite elements for fluid-structure interaction in ALE and fully Eulerian coordinates[J]. Computer Methods in Applied Mechanics & Engineering, 2010, 199(41-44): 2633-2642.

[373] Richter T, Wick T. Optimal control and parameter estimation for stationary fluid-structure interaction problems[J]. Siam Journal on Scientific Computing, 2013, 35(5): B1085-B1104.

[374] Richter T, Wick T. On time discretizations of fluid-structure interactions[J]. 2015.

[375] Richter T, Wick T. Variational localizations of the dual weighted residual estimator[J]. Journal of Computational & Applied Mathematics, 2015, 279: 192-208.

[376] Richter T, Wick T. Einführung in Die Numerische Mathematik: Begriffe, Konzepte Und Zahlreiche Anwendungsbeispiele[M]. Springer-Verlag, 2017.

[377] Rungamornrat J, Mear M E. A weakly-singular SGBEM for analysis of cracks in 3D anisotropic media[J]. Computer Methods in Applied Mechanics and Engineering, 2008, 197(49-50): 4319-4332.

[378] Russ J B, Waisman H. Topology optimization for brittle fracture resistance[J]. Computer Methods in Applied Mechanics and Engineering, 2018, 347(APR.15): 238-263.

[379] P. Saffman. On the boundary condition at the interface of a porous medium[J]. Stud. Appl. Math., 1971, 1: 93-101.

[380] Santillán D, Juanes R, Cueto-Felgueroso L. Phase field model of fluid-driven fracture in elastic media: immersed-fracture formulation and validation with analytical solutions[J]. Journal of Geophysical Research: Solid Earth, 2017, 122(4): 2565-2589.

[381] Sargado J M, Keilegavlen E, Berre I, et al. High-accuracy phase-field models for brittle fracture based on a new family of degradation functions[J]. Journal of the Mechanics and Physics of Solids, 2018, 111: 458-489.

[382] Schäfer M, Turek S. Flow simulation with high-performance computer ii, volume 52 of notes on numerical fluid mechanics, chapter benchmark computations of laminar flow around a cylinder[J]. Vieweg, Braunschweig Wiesbaden, 1996.

[383] Schlüter A, Willenbücher A, Kuhn C, et al. Phase field approximation of dynamic brittle fracture[J]. Computational Mechanics, 2014, 14(5): 143-144.

[384] Schmich M, Vexler B. Adaptivity with dynamic meshes for space-time finite element discretizations of parabolic equations[J]. SIAM Journal on Scientific Computing, 2008, 30(1): 369-393.

[385] Schroeder A, Rademacher A. Goal-oriented error control in adaptive mixed FEM for Signorini's Problem[J]. Computer Methods in Applied Mechanics & Engineering, 2011, 200(1-4): 345-355.

[386] Schwarz H R. Methode Der Finiten elemente: Eine Einführung Unter Besonderer Berücksichtigung Der Rechenpraxis[M]. Stuttgart: Springer-Verlag, 2013.

[387] Shen J, Yang X. Energy stable schemes for Cahn-Hilliard phase-field model of two-phase incompressible flows[J]. Chinese Annals of Mathematics, Series B, 2010, 31(5): 743-758.

[388] Shen R, Waisman H , Guo L . Fracture of viscoelastic solids modeled with a modified phase field method[J]. Computer Methods in Applied Mechanics & Engineering, 2019, 346(APR.1): 862-890.

[389] Shin S, Juric D. Modeling three-dimensional multiphase flow using a level contour reconstruction method for front tracking without connectivity[J]. Journal of Computational Physics, 2002, 180(2): 427-470.

[390] Sneddon I N. The distribution of stress in the neighbourhood of a crack in an elastic solid[J]. Proceedings of the Royal Society of London. Series A, Mathematical and Physical Sciences, 1946, 187(1009): 229-260.

[391] Sneddon I N, Lowengrub M. Crack Problems in the Classical Theory of Elasticity[M]. SIAM Review, 1969: 241-242.

[392] Stein K, Tezduyar T, Benney R. Mesh moving techniques for fluid-structure interactions with large displacements[J]. J. Appl. Mech, 2003, 70(1): 58-63.

[393] Steinke C, Kaliske M. A phase-field crack model based on directional stress decomposition[J]. Computational Mechanics, 2019, 63: 1019-1046.

[394] Strobl M, Seelig T. A novel treatment of crack boundary conditions in phase field models of fracture[J]. Pamm, 2015, 15(1): 155-156.

[395] Sun S, Liu J. A locally conservative finite element method based on piecewise constant enrichment of the continuous galerkin method[J]. SIAM Journal on Entific Computing,

2010, 31(4): 2528-2548.
[396] Sussman M, Smereka P, Osher S. A level set approach for computing solutions to incompressible two-phase flow[J]. J. Comput. Phys, 1994, 114(1): 146-159.
[397] Suttmeier F T. Adaptive finite element approximation of problems in elasto-plasticity theory[D]. Universität Heidelberg. Interdisziplinäres Zentrum für Wissenschaftliches Rechnen, 1997.
[398] Suttmeier F T. Numerical Solution of Variational Inequalities by Adaptive Finite Elements[M]. Berlin: Vieweg+ Teubner Research, 2008.
[399] Suttmeier F T. General approach for a posteriori error estimates for finite element solutions of variational inequalities[J]. Computational Mechanics, 2001, 27(4): 317-323.
[400] Svolos L, Bronkhorst C A, Waisman H. Thermal-conductivity degradation across cracks in coupled thermo-mechanical systems modeled by the phase-field fracture method[J]. Journal of the Mechanics and Physics of Solids, 2020, 137: 103861.
[401] Szeri A Z. Fluid Film Lubrication[M]. Cambridge: Cambridge University Press, 2010.
[402] Tanne E, Li T, Bourdin B, et al. Crack nucleation in variational phase-field models of brittle fracture[J]. Journal of the Mechanics & Physics of Solids, 2017, 110: 80-99.
[403] Teichtmeister S, Kienle D, Aldakheel F, et al. Phase field modeling of fracture in anisotropic brittle solids[J]. International Journal of Non-Linear Mechanics, 2017, 97: 1-21.
[404] Temam R, Chorin A. Navier stokes equations: theory and numerical analysis[J]. Journal of Applied Mechanics, 1984, 2(2): 456.
[405] Tezduyar T E. Finite element methods for flow problems with moving boundaries and interfaces[J]. Archives of Computational Methods in Engineering, 2001, 8(2): 83.
[406] Tezduyar T E. Interface-tracking and interface-capturing techniques for finite element computation of moving boundaries and interfaces[J]. Computer Methods in Applied Mechanics & Engineering, 2006, 195(23/24): 2983-3000.
[407] Tezduyar T E, Takizawa K, Moorman C, et al. Space-time finite element computation of complex fluid-structure interactions[J]. International Journal for Numerical Methods in Fluids, 2010, 64(10-12): 1201-1218.
[408] Biot M A, Tolstoy I. Acoustics, elasticity, and thermodynamics of porous media: twenty-one papers[J]. Published by the Acoustical Society of America through the American Institute of Physics, 1992.
[409] Toulopoulos I, Wick T. Numerical methods for power-law diffusion problems[J]. SIAM Journal on Scientific Computing, 2017, 39(3): A681-A710.
[410] Tran D, Settari A T, Nghiem L. Predicting growth and decay of hydraulic-fracture width in porous media subjected to isothermal and nonisothermal flow[J]. SPE Journal, 2013, 18(4): 781-794.
[411] Trémolières R. Analyse Numérique des Inéquations Variationnelles[M]. Dunod, 1976.
[412] Trémolieres R, Lions J L, Glowinski R. Numerical Analysis of Variational Inequalities[M]. Elsevier, 2011.

[413] Tröltzsch F. Optimale Steuerung Partieller Differentialgleichungen[M]. Wiesbaden: Vieweg+ Teubner Verlag, 2005.

[414] Turek S. Efficient Solvers for Incompressible Flow Problems: an Algorithmic and Computational Approache[M]. Springer Science & Business Media, 1999.

[415] Turek S, Rivkind L, Hron J, et al. Numerical analysis of a new time-stepping θ-scheme for incompressible flow simulations[J]. J. Sci. Comp., 2006, 28: 2-3.

[416] Ulmer H, Hofacker M, Miehe C. Phase field modeling of brittle and ductile fracture[J]. Pamm, 2013, 13(1): 533-536.

[417] van Brummelen E H, Zhuk S, van Zwieten G J. Worst-case multi-objective error estimation and adaptivity[J]. Computer Methods in Applied Mechanics and Engineering, 2017, 313: 723-743.

[418] Duijn C V, Mikelic A, Wheeler M F, et al. Thermoporoelasticity via homogenization: modeling and formal two-scale expansions[J]. International Journal of Engineering Science, 2019, 138(MAY): 1-25.

[419] van Duijn C J, Mikelić A, Wick T. A monolithic phase-field model of a fluid-driven fracture in a nonlinear poroelastic medium[J]. Mathematics and Mechanics of Solids, 2019, 24(5): 1530-1555.

[420] van Duijn C J, Mikelić A, Wick T. Mathematical theory and simulations of thermoporoelasticity[J]. Computer Methods in Applied Mechanics and Engineering, 2020, 366: 113048.

[421] van Goethem N, Novotny A A. Crack nucleation sensitivity analysis[J]. Mathematical Methods in the Applied Sciences, 2010, 33(16): 1978-1994.

[422] Verfürth R. A review of a posteriori error estimation and adaptive mesh-refinement techniques[J]. Wiley : Teubner, 1996.

[423] Werner D. Funktionalanalysis[M]. Springer-Verlag, 2006.

[424] Mikelic A, Wheeler M F, Wick T. A phase-field method for propagating fluid-filled fractures coupled to a surrounding porous medium[J]. Multiscale Modeling & Simulation, 2015, 13(1): 367-398.

[425] Wheeler M F, Wick T, Lee S. IPACS: integrated phase-field advanced crack propagation simulator. an adaptive, parallel, physics-based-discretization phase-field framework for fracture propagation in porous media[J]. Computer Methods in Applied Mechanics and Engineering, 2020, 367: 113124.

[426] Shutov A V, Kuprin C, Ihlemann J, et al. Experimentelle Untersuchung und numerische Simulation des inkrementellen Umformverhaltens von Stahl 42CrMo4. Experimental investigation and numerical simulation of the incremental deformation of a 42CrMo4 steel[J]. Materialwissenschaft Und Werkstofftechnik, 2010, 41(9): 765-775.

[427] Wick D, Wick T, Hellmig R J, et al. Numerical simulations of crack propagation in screws with phase-field modeling[J]. Computational Materials Science, 2015, 109: 367-379.

[428] Wick T. Adaptive finite element simulation of fluid-structure interaction with applica-

tion to heart-valve dynamics[D]. University of Heidelberg , 2011.

[429] Wick T. Fluid-structure interactions using different mesh motion techniques[J]. Computers & Structures, 2011, 89(13-14): 1456-1467.

[430] Wick T. Coupling of fully Eulerian and arbitrary Lagrangian-Eulerian methods for fluid-structure interaction computations[J]. Springer-Verlag New York, Inc., 2013, 52(5): 1113-1124.

[431] Wick T. Fully Eulerian fluid-structure interaction for time-dependent problems[J]. Computer Methods in Applied Mechanics & Engineering, 2013, 255(MAR.1): 14-26.

[432] Wick T. Solving monolithic fluid-structure interaction problems in arbitrary Lagrangian Eulerian coordinates with the deal. ii library[J]. Archive of Numerical Software, 2013, 1(1): 1-19.

[433] Wick T. Flapping and contact FSI computations with the fluid-solid interface-tracking/ interface-capturing technique and mesh adaptivity[J]. Computational Mechanics, 2014, 53(1): 29-43.

[434] Wick T. Modeling, discretization, optimization, and simulation of fluid-structure interaction[Z]. Lecture notes at Heidelberg University, TU Munich, and JKU Linz available on https://wwwm17.ma.tum.de/Lehrstuhl/LehreSoSe15NMFSIEn, 2015.

[435] Wick T. Coupling fluid-structure interaction with phase-field fracture[J]. Journal of Computational Physics, 2016, 327: 67-96.

[436] Wick T. Goal functional evaluations for phase-field fracture using PU-based DWR mesh adaptivity[J]. Computational Mechanics: Solids, Fluids, Fracture Transport Phenomena and Variational Methods, 2016, 57(6): 1017-1035.

[437] Wick T. Fluid-structure interactions using different mesh motion techniques[J]. Computers & Structures, 2011, 89(13-14): 1456-1467.

[438] Wick T. Modified Newton methods for solving fully monolithic phase-field quasi-static brittle fracture propagation[J]. Computer Methods in Applied Mechanics & Engineering, 2017, 325(oct.1): 577-611.

[439] Wick T. Modified Newton methods for solving fully monolithic phase-field quasi-static brittle fracture propagation[J]. Computer Methods in Applied Mechanics & Engineering, 2017, 325(oct.1): 577-611.

[440] Wick T. Zielorientierte numerik für multiphysiksimulationen[J]. GAMM-Rundbrief, 2018, 2: 4-10.

[441] Wick T. Numerical Methods for Partial Differential Equations[M]. Hannover: Institutionelles Repositorium der Leibniz Universität Hannover, 2022.

[442] Wick T, Lee S, Wheeler M F. 3D phase-field for pressurized fracture propagation in heterogeneous media[C]. COUPLED VI: Proceedings of the VI International Conference on Computational Methods for Coupled Problems in Science and Engineering. CIMNE, 2015: 605-613.

[443] Schröder J, Wick T, Reese S, et al. A selection of benchmark problems in solid mechanics and applied mathematics[J]. Archives of Computational Methods in Engineering,

2021, 28: 713-751.
[444] Wick T, Singh G, Wheeler M F. Fluid-filled fracture propagation with a phase-field approach and coupling to a reservoir simulator[J]. SPE Journal, 2016, 21(3): 0981-0999.
[445] Wick T, Wollner W. On the differentiability of fluid-structure interaction problems with respect to the problem data[J]. Journal of Mathematical Fluid Mechanics, 2019, 21: 1-21.
[446] Wick T, Wollner W. Optimization with nonstationary, nonlinear monolithic fluid-structure interaction[J]. International Journal for Numerical Methods in Engineering, 2021, 122(19): 5430-5449.
[447] Wilson Z A, Landis C M. Phase-field modeling of hydraulic fracture[J]. Journal of the Mechanics and Physics of Solids, 2016, 96: 264-290.
[448] Winkler B J. Traglastuntersuchungen Von Unbewehrten Und Bewehrten Betonstrukturen Auf Der Grundlage Eines Objektiven Werkstoffgesetzes für Beton[M]. Innsbruck: Innsbruck University Press, 2001.
[449] Wloka J. Partial Differential Equations[M]. Cambridge: Cambridge University Press, 1987.
[450] Wohlmuth B. Variationally consistent discretization schemes and numerical algorithms for contact problems[J]. Acta Numerica, 2011, 20: 569-734.
[451] Wu C, Fang J, Zhou S, et al. Level-set topology optimization for maximizing fracture resistance of brittle materials using phase-field fracture model[J]. International Journal for Numerical Methods in Engineering, 2020, 121(13): 2929-2945.
[452] Wu J Y. A unified phase-field theory for the mechanics of damage and quasi-brittle failure[J]. Journal of the Mechanics and Physics of Solids, 2017, 103(Jun.): 72-99.
[453] Jyw A, Vpn B, Chi T, et al. Phase-field modeling of fracture[J]. Advances in Applied Mechanics, 2020, 53: 1-183.
[454] Yoshioka K, Naumov D, Kolditz O. On crack opening computation in variational phase-field models for fracture[J]. Computer Methods in Applied Mechanics and Engineering, 2020, 369: 113210.
[455] Zehnder A T. Fracture Mechanics[M]. Springer Science & Business Media, 2012.
[456] Zhang F, Huang W, Li X, et al. Moving mesh finite element simulation for phase-field modeling of brittle fracture and convergence of Newton's iteration[J]. Journal of Computational Physics, 2018, 356: 127-149.
[457] Zhang J, Du Q. Numerical studies of discrete approximations to the allen-cahn equation in the sharp interface limit[J]. Siam Journal on Scientific Computing, 2009, 31(4): 3042-3063.
[458] Zhang X, Sloan S W, Vignes C, et al. A modification of the phase-field model for mixed mode crack propagation in rock-like materials[J]. Computer Methods in Applied Mechanics and Engineering, 2017, 322(aug.1): 123-136.
[459] Zhou S, Zhuang X, Rabczuk T. A phase-field modeling approach of fracture propagation in poroelastic media[J]. Engineering Geology, 2018, 240: 189-203.

[460] Zhou S, Zhuang X, Rabczuk T. Phase-field modeling of fluid-driven dynamic cracking in porous media[J]. Computer Methods in Applied Mechanics and Engineering, 2019, 350(JUN.15): 169-198.

[461] Zhou S, Zhuang X, Rabczuk T . Phase-field modeling of fluid-driven dynamic cracking in porous media[J]. Computer Methods in Applied Mechanics and Engineering, 2019, 350(JUN.15):169-198.

[462] Ziaei-Rad V, Shen Y. Massive parallelization of the phase field formulation for crack propagation with time adaptivity[J]. Computer Methods in Applied Mechanics and Engineering, 2016, 312: 224-253.

索 引

A

Active set 活动集 26
Adaptive approach 自适应方法 153
Adaptive FEM 自适应有限元 135
Adjoint problem 伴随问题 144
AL, Augmented Lagrangian 增广拉格朗日 69
ALE, arbitrary Lagrangian-Eulerian 任意拉格朗日欧拉方法 185
All variables, symbols, etc. 所有符号及参数 18
Ambrosio-Tortorelli Ambrosio-Tortorelli 椭圆泛函 24
Augmented Lagrangian penalization 增广拉格朗日补偿 69,
Augmented Mandel's problem 广义 Mandel 问题 35

B

Balance of contact forces 接触力平衡条件 36
Benchmark 基准 74，209
Bilinear form 双线性型 17
Biot's equations Biot 方程 41
Block-diagonal preconditioner 块-对角预处理 122
Brittle fracture 脆性裂缝 49
Bulk energy 体积势能 50

C

Cauchy stress tensor 柯西应力张量 143
Céa lemma Céa 定理 84
CG, conjugate gradients 共轭梯度 38
Chain rule 链式法则 54
Characteristic function 指示函数 167
Classifications of PDEs 偏微分方程分类 29
COD 裂缝开度 169
Conservation of momentum 动量守恒 190
Constitutive laws 本构关系 192
Control parameter 控制参数 116
Convex set 凸集 17
Coupled PDE 耦合偏微分方程 1

Coupled problems 耦合问题 1
Coupling strategies 耦合策略(方法) 8
Crack energy 裂缝能量 148
Crack irreversibility 裂缝不可逆 66
Critical energy release rate 临界能量释放率 12
CVIS Coupled variational inequality system 耦合变分不等式系统 1

D

Deformation field 变形场 185
Deformation gradient 变形梯度 185
Differentiation in Banach spaces Banach 空间的微分 20
Diffraction equation 压力传播方程 182
Directional derivative 方向导数 22
Discretization 离散化 66
Displacement field 位移场 16
Divergence theorem 散度定理 19
Divergence 散度 13
Domain 域 15
DWR, dual-weighted residual 双重加权残差 136
Dynamic fracture 动态裂缝 192

E

Effectivity index 有效性指数 135
Elliptic functional 椭圆泛函 24
Energy formulations 能量公式 64
Energy release rate 能量释放率 50
Engineering applications 工程应用 7
Entropy 熵 52
Error estimator 误差估计 134
Error-oriented Newton 面向误差的牛顿迭代法 114
Eulerian coordinates 欧拉坐标系 47
Eulerian expansion formula Euler 展开式 186
Eulerian framework 欧拉框架 194
Euler-Lagrange equations Euler-Lagrange 方程 54

F

Finite element method 有限元方法 8
Fixed interfaces 固定界面 4
Fixed mesh 固定网格 194
Fixed-point iterations 固定点迭代 98
Fluid-structure interaction 流体-结构相互作用(流固耦合) 74

Fréchet derivative　　Fréchet 导数　149
Free boundary problem　　自由边界问题　26
Frobenius scalar product　　Frobenius 标量积　48
Function spaces　　函数空间　56

G

Gâteaux derivative　　Gâteaux 导数　21
Geometric multigrid,　　几何多重网格　123
Goal functional　　目标函数　133
Goal-oriented error estimation　　面向目标的误差分析　134
Gradient　　梯度　13
Green-Lagrange tensor　　Green-Lagrange 张量　187

H

Hausdorff measure　　Hausdorff 测度　51
Hilbert spaces　　Hilbert 空间　34

I

Indicator function　　指示函数　142
Inequality constraint　　不等式约束　141
Initial condition　　初始条件　158
Inner product　　内积　17
Integration by parts　　分部积分　19
Integration by substitution　　41
Interface conditions　　界面条件　167
Interface coupling　　界面耦合　35
Interface　　界面　3
Interface-capturing　　界面捕获　5
Interface-tracking　　界面跟踪　5
Interfacial thickness　　界面厚度　13
iteration error　　迭代误差　141
Iterative coupling　　迭代耦合　153

J

Jacobian matrix　　雅可比矩阵　100

K

Kinematic condition　　运动条件　177

L

Lagrangian coordinates　　拉格朗日坐标系　47

Lamé parameters　拉美常量　12
Line search algorithm　线性搜索算法　111
Linear-in-time extrapolation　线性外推　99
Linearizations　线性化　99
Linearized elasticity　线弹性　54
Loading interval　加载时间　12

M

Mandel's problem　Mandel's 问题　35
Matrix-free　无矩阵方法　123
Mesh adaptivity　自适应网格　133
Mesh refinement strategies　网格细分方案　141
Modified finite elements　修正有限元　6
Monolithic schemes　全耦合方案　37
Monotonocity tests Newton　单调性测试　108
Moving interfaces　移动界面　4
Multigoal-oriented error estimation　多目标误差分析　158
Multigrid　多重网格方法　123
Multiphysics　多物理场　1
Multiple goal functionals　多目标函数　139
Multiple goal functionals　多目标函数　140

N

Nabla operator　倒三角算子　13
Navier-Stokes　N-S 方程　142
Non-active set　非活动集　26
Nonlinear PDE　非线性偏微分方程　29
Numerical analysis　数值分析　33
Numerical optimization　数值优化　68

O

Obstacle problem　障碍问题　26
ODE　常微分方程　29

P

Optimal control　最优控制　221
Parallel computing　并行计算　2
Parameter estimation　参数估计　8
Partitioned coupling　分部耦合　37
Partition-of-unity　单位分解　138
Penalization　补偿　110

Piola transformation　　Piola 变换　　189
Poisson's ratio　　泊松比　　12
Poroelasticity　　孔弹性　　4
Porous media　　多孔介质　　34
Preconditioner　　预处理器 (预处理矩阵)　　121
Predictor-corrector global-local approach　　预测-矫正-自适应方法　　153
Pressurized fractures　　水力裂缝, 252
Primal dual active set　　原始对偶活动集　　66
Propagating interfaces　　扩展界面　　5
Prototype phase-field fracture　　原始相场模型　　23

Q

Quantity of interest　　兴趣量　　133
Quasi-monolithic scheme　　准全耦合方案　　105
Quasi-static fracture　　准静态裂缝　　73

R

Rannacher time-stepping　　Rannacher 方案　　74
Regularization parameter　　正则化参数　　12
Research software　　研究软件　　218
Residual-based Newton　　基于残差的牛顿迭代法　　111
Reynolds number　　雷诺数　　148

S

Scalar product　　标量积　　48
Schur complement　　舒尔补矩阵　　122
Screw simulations　　螺钉模拟　　7
Semi-linear form　　半线性形式　　17
Semi-smooth Newton　　半光滑牛顿法　　125
Shifted Crank-Nicolson　　CN 格式　　74
Shortcomings of phase-field fracture　　相场裂缝缺点　　3
Simple penalization　　简单补偿法　　66
Simplified phase-field fracture　　简单相场裂缝模型　　25
Single edge notched shear test　　单边剪切实验　　126
Single goal functionals　　单目标函数　　161
Slit domain　　裂缝域　　90
Sneddon's test　　Sneddon 模型 (测试)　　213
Sobolev spaces　　Sobolev 函数空间　　14
Solution variables　　解变量　　12
Spatial discretization　　空间离散　　201
Stokes problem　　Stokes 问题　　173

Strain tensor　　应变张量　　47
Stress splitting　　应力分解　　61
Strong formulation　　强形式　　40
Surface energy　　表面能　　50

T

Tensile stress　　拉应力　　59
Thermodynamic interpretations　　热力学解释　　8
Time derivative　　时间导数　　27
Time interval　　加载时间　　12
Time step control　　时间步长控制　　155
Time-lagging　　时滞　　99
Total crack volume　　总裂缝体积　　90
Toughness　　韧性　　49
Trace　　迹　　13
Types of interfaces　　界面类型　　3

V

Variational inequality　　变分不等式　　31
Variational monolithic coupling　　变分全耦合　　37
Volume coupling　　域耦合　　35

W

Weak formulation　　弱形式　　35

Y

Young's modulus　　杨氏模量　　12